Systems Science and Modeling for Ecological Economics

Systems Science and Modeling for Ecological Economics

Alexey Voinov

AMSTERDAM • BOSTON • HEIDELBERG • LONDON • NEW YORK • OXFORD
PARIS • SAN DIEGO • SAN FRANCISCO • SINGAPORE • SYDNEY • TOKYO
Academic Press is an imprint of Elsevier

Academic Press is an imprint of Elsevier
84 Theobald's Road, London WC1X 8RR, UK
30 Corporate Drive, Suite 400, Burlington, MA 01803, USA
525 B Street, Suite 1900, San Diego, CA 92101-4495, USA

First edition 2008

Notice
No responsibility is assumed by the publisher for any injury and/or damage to persons or property
as a matter of products liability, negligence or otherwise, or from any use or operation of any methods,
products, instructions or ideas contained in the material herein. Because of rapid advances in the
medical sciences, in particular, independent verification of diagnoses and drug dosages should be made

Library of Congress Cataloging in Publication Data
A catalog record for this book is available from the Library of Congress

British Library Cataloguing in Publication Data
A catalogue record for this book is available from the British Library

ISBN: 978-0-12-372583-7

For information on all Academic Press publications visit our
web site at http://elsevierdirect.com

Typeset by Charon Tec Ltd., A Macmillan Company.
(www.macmillansolutions.com)

Printed and bound by CPI Group (UK) Ltd, Croydon, CR0 4YY
Transferred to Digital Printing, 2013

To those who led — my parents, Zoe and Arkady;
and to those who follow — my sons, Anton and Ivan

Contents

Preface

Why?

As I am finishing this book, *Science* magazine is running a special issue about the sequencing of the macaque genome. It turns out that macaques share about 93 percent of their genes with us, humans. Previously it has been already reported that chimpanzees share about 96 percent of their genes with us. Yes, the macaque is our common ancestor, and it might be expected that, together with the chimps, we continued with our natural selection some 23 million years ago until, some 6 million years ago, we departed from the chimps to continue our further search for better adaptation. Actually it was not quite like this. Apparently it was the chimps that departed from us; now that we have the macaques as the starting point, we can see that the chimp's genome has way more mutations than ours. So the chimps are further ahead than we are in their adaptation to the environment.

How did that happen, and how is it then that we, and not the chimps, have spread around all the Earth? Apparently at some point a mutation put us on a different track. This was a mutation that served an entirely different purpose: instead of adapting to the environment in the process of natural selection, we started adapting the environment to us. Instead of acquiring new features that would make us better suited to the environment, we found that we could start changing the environment to better suit us – and that turned out to be even more efficient. And so it went on. It appears that not that many mutations were needed for us to start using our brainpower, skills and hands to build tools and to design microenvironments in support of the life in our fragile bodies – certainly not as many as the chimps had to develop on their road to survival. Building shelters, sewing clothing or using fire, we created small cocoons of environments around us that were suitable for life. Suddenly the rate of change, the rate of adaptation, increased; there was no longer a need for millions of years of trial and error. We could pass the information on to our children, and they would already know what to do. We no longer needed the chance to govern the selection of the right mutations and the best adaptive traits, and we found a better way to register these traits using spoken and written language instead of the genome.

The human species really took off. Our locally created comfortable microenvironments started to grow. From small caves where dozens of people were packed in with no particular comfort, we have moved to single-family houses with hundreds of square meters of space. Our cocoons have expanded. We have learned to survive in all climatic zones on this planet, and even beyond, in space. As long as we can bring our cocoons with us, the environment is good enough for us to live. And so more and more humans have been born, with more and more space occupied, and more and more resources used to create our microcosms. When microcosms are joined together and expand, they are no longer "micro." Earth is no longer a big planet with infinite resources, and us, the humans. Now it is the humans' planet, where we dominate and regulate. As Vernadskii predicted, we have become a geological force that shapes this planet. He wasn't even talking about climate change at that time. Now we can do even that, and are doing so.

Unfortunately, we do not seem to be prepared to understand that. Was there a glitch in that mutation, which gave us the mechanism and the power but forgot about the self-control? Are we driving a car that has the gas pedal, but no brake? Or we just have not found it yet? For all these years, human progress has been and still is equated to growth and expansion. We have been pressing the gas to the floor, only accelerating. But any driver knows that at high speed it becomes harder to steer, especially when the road is unmarked and the destination is unknown. At higher speeds, the price of error becomes fatal.

But let us take a look at the other end of the spectrum. A colony of yeast planted on a sugar substrate starts to grow. It expands exponentially, consuming sugar, and then it crashes, exhausting the feed and suffocating in its own products of metabolism. Keep in mind that there is a lot of similarity between our genome and that of yeast. The yeast keeps consuming and growing; it cannot predict or understand the consequences of its actions. Humans can, but can we act accordingly based on our understanding? Which part of our genome will take over? Is it the part that we share with the yeast and which can only push us forward into finding more resources, consuming them and multiplying? Or is it going to be the acquired part that is responsible for our intellect and supposedly the capacity to understand the more distant consequences of our desires and the actions of today?

So far there is not much evidence in favor of the latter. We know quite a few examples of collapsed civilizations, but there are not many good case studies of sustainable and long-lasting human societies. To know, to understand, we need to model. Models can be different. Economics is probably one of the most mathematized branches of science after physics. There are many models in economics, but those models may not be the best ones to take into account the other systems that are driving the economy. There is the natural world, which provides resources and takes care of waste and pollution. There is the social system, which describes human relationships, life quality and happiness. These do not easily fit into the linear programming and game theory that are most widely used in conventional economics. We need other models if we want to add " ecological " to " economics. "

So far our major concern was how to keep growing. Just like the yeast population. The Ancient Greeks came up with theories of oikonomika – the skills of household management. This is what later became economics – the science of production, consumption and distribution, all for the sake of growth. And that was perfectly fine, while we were indeed small and vulnerable, facing the huge hostile world out there.

Ironically, ecology, oikology – the knowledge and understanding of the household – came much later. For a long time we managed our household without knowing it, without really understanding what we were doing. And that was also OK, as long as we were small and weak. After all, what kind of damage could we do to the whole big powerful planet? However, at some point we looked around and realized that actually we were not that weak any more. We could already wipe out entire species, change landscapes and turn rivers. We could even change the climate on the planet.

It looks as though we can no longer afford " economics " – management without knowledge. We really need to know, to understand, what we are doing. And that is what ecological economics is all about. We need to add knowledge about our household to our management of it.

Understanding how complex systems work is crucial. We are part of a complex system, the biosphere, and we further add complexity to it by adapting this biosphere to our needs and adding the human component with its own complexities and uncertainties. Modeling is a fascinating tool that can provide a method to explore complex systems, to experiment with them without destroying them at the same time. The purpose of this book is to introduce some of the modeling approaches that can help us to understand how this world works. I am mostly focusing on tools and methods, rather than case studies and applications. I am trying to show how models can be developed and used – how they can become a communication tool that can take us beyond our personal understanding to joint community learning and decision-making.

Actually, modeling is pretty mundane for all of us. We model as we think, as we speak, as we read, as we communicate – and our thoughts are mental models of the reality. Some people can speak well, clearly explaining what they think. It is easy to communicate with them, and there is less chance for misunderstanding. In contrast, some people mumble incoherent sentences that it is difficult to make any sense of. These people cannot build good models of their thoughts – the thoughts might be great, but they still have a problem.

Some models are good while others are not so good. The good models help us to understand. Especially when we deal with complex systems, it is crucial that we learn to look at processes in their interaction. There are all sorts of links, connections and feedbacks in the systems that surround us. If we want to understand how these systems work, we need to learn to sort these connections out, to find the most important ones and then study them in more detail. As systems become more complex, these connections become more distant and indirect. We find feedbacks that have a delayed response, which makes it only harder to figure out their role and guess their importance.

Suppose you start spinning a big flywheel. It keeps rotating while you add more steam to make it spin faster. There is no indication of danger – no cracks, no squeaks – it keeps spinning smoothly. An engineer might stop by, see what you're doing and get very worried. He will tell you that a flywheel cannot keep accelerating, that sooner or later it will burst, the internal tension will be too high, the material will not hold. "Oh, it doesn't look that way," you respond, after taking another look at your device. There is no evidence of any danger there. But the problem is that there is a delayed response and a threshold effect. Everything is hunky-dory one minute, and then "boom!" – the flywheel bursts into pieces, metal is flying around and people are injured. How can that happen? How can we know that it will happen?

Oh, we know, but we don't want to know. Is something similar happening now, as part of the global climate change story and its denial by many politicians and ordinary people? We don't want to know the bad news; we hate changing our lifestyle. The yeast colony keeps growing till the very last few hours.

Models can help. They can provide understanding, visualization, and important communication tools. The modeling process by itself is a great opportunity to bring together knowledge and data, and to present them in a coherent, integrated way. So modeling is really important, especially if we are dealing with complex systems that span beyond the physical world and include humans, economies, and societies.

What?

This book originated from an on-line course that I started some 10 years ago. The goal was to build a stand-alone Internet course that would provide both access to the knowledge base and interaction between the instructor and the students. The web would also allow several instructors at different locations to participate in a collaborative teaching process. Through their joint efforts the many teachers could evolve and keep the course in the public domain, promoting truly equal opportunity in education anywhere in the world. By constantly keeping the course available for asynchronous teaching, we could have overlapping generations of students involved at the same time, and expect the more advanced students to help the beginners. The expectation was that, in a way that mimics how the open source paradigm works for software development, we would start an open education effort. Clearly, the ultimate test of this idea is whether it catches on in the virtual domain. So far it is still a work in progress, and there are some clear harbingers that it may grow to be a success.

While there are always several students from different countries around the world (including the USA, China, Ireland, South Africa, Russia, etc.) taking the course independently, I also use the web resource in several courses I teach in class. In these cases I noticed that students usually started with printing out the pages from the web. This made me think that maybe after all a book would be a good idea.

The book has gone beyond the scope of the web course, with some entirely new chapters added and the remaining ones revised. Still, I consider the book to be a companion to the web course, which I intend to keep working and updated. One major advantage of web tutorials is that new facts and findings can be incorporated almost as soon as they are announced or published. It takes years to publish or update a book, but only minutes to insert a new finding or a URL into an existing web structure. By the time a reader examines the course things will be different from what I originally wrote, because there are always new ideas and results to implement and present. The virtual class discussions provide additional material for the course. All this can easily become part of the course modules. The book allows you to work off-line when you don't have your computer at hand. The on-line part offers interaction with the instructor, and downloads of the working models.

Another opportunity opened by web-based education can be described as distributed open-source teaching, which mimics the open-source concept that stems from the hacker culture. A crucial aspect of open-source licenses is that they allow modifications and derived works, but they must also be distributed under the same terms as the license of the original software. Therefore, unlike simply free code that could be borrowed and then used in copyrighted commercial distributions, the open-source definition and licensing effectively ensures that the derivatives stay in the open-source domain, extending and enhancing it. Largely because of this feature, the open-source community has grown very quickly.

The open-source paradigm may also be used to advance education. Web-based courses could serve as a core for joint efforts of many researchers, programmers, educators and students. Researchers could describe the findings that are appropriate for the course theme. Educators could organize the modules in subsets and sequences that would best match the requirements of particular programs and curricula, and develop ways to use the tools more effectively. Programmers could contribute software tools for visualization, interpretation and communication. Students would test the materials and contribute their feedback and questions, which is essential for improvements of both content and form.

Some of this is still in the future. Perhaps if you decide to read the book and take the course on-line, you could become part of this open-source, open-education effort.

How?

I believe that modeling cannot be really taught, only learned, and that it is a skill and requires a lot of practice – just as when babies learn to speak they need to practice saying words, making mistakes, and gradually learning to say them the right way. Similarly, with formal modeling, without going through the pitfalls and surprises of modeling, it is not possible to understand the process properly. Learning the skill must be a hands-on experience of all the major stages of modeling, from data acquisition and building conceptual models to formalizing and iteratively improving simulation models. That is why I strongly recommend that you look on the web, get yourself a trial or demo version of some of the modeling software that we are working with in this book, then download the models that we are discussing. You can then not just read the book, but also follow the story with the model. Do the tests, change the parameters, explore on your own, ask questions and try to find answers. It will be way more fun that way, and it will be much more useful.

Best of all think of a topic that is of interest to you and start working on your individual project. Figure out what exactly you wish to find out, see what data are available, and then go through the modeling steps that we will be discussing in the book.

The web course is at http://www.likbez.com/AV/Simmod.html, and will remain open to all. You may wish to register and take it. You will find where it overlaps with the book, you will be able to send your questions, get answers and interact with other students.

At the end of each chapter, you will find a bibliography. These books and articles may not necessarily be about models in a conventional sense, but they show how complex systems should be analyzed and how emergent properties appear from this analysis. Check out some of those references for more in-depth real-life examples of different kind of models, systems, challenges and solutions.

Best of all, learn to apply your systems analysis and modeling skills in your everyday life when you need to make small and big decisions, when you make your next purchase or go to vote. Learn to look at the system as a whole, to identify the elements and the links, the feedbacks, controls and forcings, and to realize how things are interconnected and how important it is to step back and see the big picture, the possible delayed effects and the critical states.

Acknowledgements

Many people have contributed to my understanding of modeling and to this effort. Professor Yuri Svirezhev, who passed away in 2007, was my teacher, and he certainly played a great role in shaping my vision of modeling – of what it should be, and what it can and what it can't do. My colleagues on many modeling projects in various parts of the world helped me to learn many important modeling skills. I am grateful to my students, especially those who took the on-line modeling course and contributed by asking questions, participating in on-line discussions, and letting me know what kind of improvements were needed. The Gund Institute for Ecological Economics and its director, Robert Costanza, provided a stimulating and helpful environment for developing various ideas and applications. I very much appreciate that. For almost a decade I have been teaching a modeling course as part of the MSc Program in Environmental and Natural Resource Economics at Chulalongkorn University in Bangkok. I am grateful to Jiragorn and Nantana Gajaseni for inviting me and helping with the course. My thanks are due to the Thai students who took the course and helped me improve it in many respects.

Several people have reviewed various chapters of the book and provided very useful comments. My thanks are due to Helena Vladich, Carl Fitz, Urmila Diwekar, Evan Forward and Nathan Hagens. I appreciate the suggestions I received from Andrey Ganopolski, Dmitrii O. Logofet, and Jasper Taylor. I am grateful to Erica Gaddis, who helped with several chapters and co-authored Chapter 9. Joshua Farley encouraged me to write the book, and has been the resource on all my questions on ecological economics. Finally, my thanks go to Tatiana Filatova, who diligently read the whole manuscript and provided valuable comments on many occasions.

1. Models and Systems

1.1   Model

1.2   System

1.3   Hierarchy

1.4   The modeling process

1.5   Model classifications

1.6   Systems thinking

SUMMARY

What's a model? Why do we model? How do we model? These questions are addressed in this chapter. It is a very basic introduction to the trade. We shall agree on definitions – what is a system, what are parameters, forcing functions, and boundaries? We will also consider some other basic questions – how do we build a conceptual model? How are elements connected? What are the flows of material, and where is it actually information? How do interactions create a positive feedback that allows the system to run out of control or, conversely, how do negative feedbacks manage to keep a system in shape? Where do we get our parameters from? We shall then briefly explore how models are built, and try to come with some dichotomies and classes for different models.

Keywords

Complexity, resolution, spatial, temporal and structural scales, physical models, mathematical models, Neptune, emergent properties, elements, holism, reductionism, Thalidomide, flows, stocks, interactions, links, feedbacks, global warming, structure, function, hierarchy, sustainability, boundaries, variables, conceptual model, modeling process.

1.1   Model

> *A model is a simplification of reality*

We model all the time, even though we don't think about it. With words that we speak or write, we build models of what we think. I used to have a poster in my office of a big gorilla scratching his head and saying: "You think you understood what I said, but I'm not sure that what I said is what I thought." One of the reasons it is sometimes hard to communicate is that we are not always good at modeling our thoughts by the words that we

1

pronounce. The words are always a simplification of the thought. There may be certain aspects of the thought or feeling that are hard to express in words, and thus the model fails. Therefore, we cannot understand each other.

The image of the world around us as we see it is also a model. It is definitely simpler than the real world; however, it represents some of its important features (at least, we think so). A blind person builds a different model, based only on sound, smell and feeling. His model may have details and aspects different from those in the model based on vision, but both models represent reality more simply than it actually is.

We tend to get very attached to our models, and think that they are the only right way to describe the real world. We easily forget that we are dealing only with simplifications that are never perfect, and that people are all creating their own simplifications in their particular unique way for a particular purpose.

People born blind have different ideas about space, distance, size and other features of the 3D world than do the rest of us. When eye surgeons learned to remove cataracts, some people who had been blind from birth suddenly had the chance to see. They woke up to a new world, which was totally foreign and even hostile to them. They did not have any idea of what form, distance and perspective were. What we take for granted was unknown in their models of reality. They could not imagine how objects could be in front or behind other objects; to them, a dog that walked behind a chair and re-emerged was walking out of the room and then coming back. They were more comfortable closing their eyes and feeling for objects with their hands to locate them, because they could not understand how objects appear smaller when they are farther away. They seemed to change size, but not location.

Annie Dillard, Pilgrim at Tinker Creek

The way we treat reality is indeed very much a function of how our senses work. For instance, our perception of time might be very different if we were more driven by scent than by vision and sound. Imagine a dog that has sensitivity to smells orders of magnitude higher than humans do. When a dog enters a room, it will know much more than we do about who was there before, or what was happening there. The dog's perception of the present moment would be quite different from ours. Based on our visual models, we clearly distinguish the past, the present and the future. The visual model, which delivers a vast majority of information to our brain, serves as a snapshot that stands between the past and the future. In the case of a dog driven by scent, this transition between the past and the future becomes blurred, and may extend over a certain period of time. The dog's model of reality would be also different. Similarly in space – the travel of scents over distances and around obstacles can considerably alter the spatial model, making it quite different from what we build based on the visible picture of our vicinity.

Another example of a model we often deal with is a map. When a friend explains how to get to his house, he draws a scheme of roads and streets, building a model for you to better understand the directions. His model will surely lack a lot of detail about the landscape that you may see on your way, but it will contain all the information you need to get to his house.

Note that the models we build are defined by the purposes that they serve. If, for example, you only want to show a friend how to get to your house, you will draw a very simple diagram, avoiding description of various places of interest on the way. However, if you want your friend to take notice of a particular location, you might also show her a photograph, which is also a model. Its purpose is very different, and so are the implementation, the scale and the details.

The best explanation is as simple as possible, but no simpler.

Albert Einstein

The best model, indeed, should strike a balance between realism and simplicity. The human senses seem to be extremely well tuned to the levels of complexity and resolution that are required to give us a model of the world that is adequate to our needs. Humans can rarely distinguish objects that are less than 1 mm in size, but then they hardly need to in their everyday life. Probably for the same reason, more distant objects are modeled with less detail than are the close ones. If we could see all the details across, say, a 5-km distance, the brain would be overwhelmed by the amount of information it would need to process. The ability of the eye to focus on individual objects, while the surrounding picture becomes somewhat blurred and loses detail, probably serves the same purpose of simplifying the image the brain is currently studying. The model is made simple, but no simpler than we need. If our vision is less than 20/20, we suddenly realize that there are certain important features that we can no longer model. We rush to the optician for advice on how to bring our modeling capabilities back to certain standards.

As in space, in time we also register events only of appropriate duration. Slow motion escapes our resolution capacity. We cannot see how a tree grows, and we cannot register the movement of the sun and the moon; we have to go back to the same observation point to see the change. On the other hand, we do not operate too well at very high process rates. We do not see how the fly moves its wings. Even driving causes problems, and quite often the human brain cannot cope with the flow of information when driving too fast.

Whenever we are interested in more detail regarding time or space, we need to extend the modeling capabilities of our senses and brain with some additional devices – microscopes, telescopes, high-speed cameras, long-term monitoring devices, etc. These are required for specific modeling goals, specific temporal and spatial scales.

The image created by our senses is static; it is a snapshot of reality. It is only changed when the reality itself changes, and as we continue observing we get a series of snapshots that gives us the idea of the change. We cannot modify this model to make it change in time, unless we use our imagination to play "what if?" games. These are the mental experiments that we can make. The models we create outside our brain, physical models, allow us to study certain features of the real-life systems even without modifying their prototypes – for example, a model of an airplane is placed in a wind tunnel to evaluate the aerodynamic properties of the real airplane. We can study the behavior of the airplane and its parts in extreme conditions; we can make them actually break without risking the plane itself – which is, of course, many times more expensive than its model. (For examples of wind tunnels and how they are used, see http://wte.larc.nasa.gov/.)

Physical models are very useful in the "what if?" analysis. They have been widely used in engineering, hydrology, architecture, etc. In Figure1.1 we see a physical model developed to study stream flow. It mimics a real channel, and has sand and gravel to

Figure 1.1   A physical model to study stream flow in the Main Channel Facility at the St Anthony Falls
Laboratory (SAFL) in Minnesota.
The model is over 80 m long, has an intake from the Mississippi River with a water
discharge capacity of 8.5 m^3 per second, and is configured with a sediment (both gravel
and sand) recirculation system and a highly accurate weigh-pan system for measuring
bedload transport rates (http://www.nced.umn.edu/streamlab06_sed_xport).

represent the bedforms and allow us to analyze how changes in the bottom profiles can
affect the flow of water in the stream. Physical models are quite expensive to create
and maintain. They are also very hard to modify, so each new device (even if it is fairly
similar to the one already studied) may require the building of an entirely new physical
model.

Mathematics offers another tool for modeling. Once we have derived an adequate mathematical relationship for a certain process, we can start analyzing it in

many different ways, predicting the behavior of the real-life object under varying conditions. Suppose we have derived a model of a body moving in space described by the equation

$$S = V \cdot T$$

where S is the distance covered, V is the velocity and T is time.

This model is obviously a simplification of real movement, which may occur with varying speed, be reciprocal, etc. However, this simplification works well for studying the basic principles of motion and may also result in additional findings, such as the relationship

$$T = \frac{S}{V}$$

An important feature of mathematical models is that some of the previously derived mathematical properties can be applied to a model in order to create new models, at no additional cost. In some cases, by studying the mathematical model we can derive properties of the real-life system which were not previously known. It was by purely mathematical analysis of a model of planetary motion that Adams and Le Verrier first predicted the position of Neptune in 1845. Neptune was later observed by Galle and d'Arrest, on 23 September 1846, very near to the location independently predicted by Adams and Le Verrier. The story was similar with Pluto, the last and the smallest planet in the Solar System (although, as of 2006, Pluto is no longer considered to be a planet; it has been decided that Pluto does not comply with the definition of a planet, and thus it has been reclassified as a "small planet"). Actually, the model that predicted its existence turned out to have errors, yet it made Clyde Tombaugh persist in his search for the planet. We can see that analysis of abstract models can result in quite concrete findings about the real modeled world.

> *All models are wrong ... Some models are useful.*
>
> **William Deming**

All models are wrong because they are always simpler than the reality, and thus some features of real-life systems get misrepresented or ignored in the model. What is the use of modeling, then? When dealing with something complex, we tend to study it step by step, looking at parts of the whole and ignoring some details to get the bigger picture. That is exactly what we do when building a model. Therefore, models are essential to *understand* the world around us.

If we understand how something works, it becomes easier to *predict* its behavior under changing conditions. If we have built a good model that takes into account the essential features of the real-life object, its behavior under stress will likely be similar to the behavior of the prototype that we were modeling. We should always use caution when extrapolating the model behavior to the performance of the prototype because of the numerous scaling issues that need be considered. Smaller, simpler models do not necessarily behave in a similar way to the real-life objects. However, by applying appropriate scaling factors and choosing the right materials and media, some very useful results may be obtained.

When the object performance is understood and its behavior predicted, we get additional information to *control* the object. Models can be used to find the most sensitive components of the real-life system, and by modifying these components we can efficiently tune the system into the desired state or set it on the required trajectory.

In all cases, we need to compare the model with the prototype and refine the model constantly, because it is only the real-life system and its behavior that can serve as a criterion for model adequacy. The model can represent only a certain part of the system that is studied. The art of building a useful model consists mainly of the choice of the right level of simplification in order to match the goals of the study.

Exercise 1.1

1. Can you think of three other examples of models? What is the spatial/temporal resolution in your models?
2. Can you use an electric lamp as a model of the sun? What goals could such a model meet? What are the restrictions for using it? When is it not a good model?

1.2   System

Any phenomenon, either structural or functional, having at least two separable components and some interaction between these components may be considered a system.

Hall and Day, 1977

When building models, you will very often start to use the word "system." Systems approach and systems thinking can help a lot in constructing good models. In a way, when you start thinking of the object that you study as a system, it disciplines your mind and arranges your studies along the guidelines that are essential for modeling. You might have noticed that the term *system* has been already used a number of times above, even though it has not really been defined. This is because a system is one of those basic concepts that are fairly hard to define in any way other than the intuitively obvious one. In fact, there may be numerous definitions with many long words, but the essence remains the same – that is, a *system is a combination of parts that interact and produce some new quality in their interaction.*

Thus there are three important features:

1. Systems are made of parts or elements
2. The parts interact
3. Something new is produced from the interaction.

All three features are essential for a system to be a system. If we consider interactions, we certainly need more than one component. There may be many matches in a matchbox, but as long as they are simply stored there and do not interact, they cannot be termed a system. Two cars colliding at a junction of two roads are two components that are clearly interacting, but do they make a system? Probably not, since there is hardly any new quality produced by their interaction. However, these same two cars may be part of a transportation system that we are considering to analyze the flow or material and people through a network of roads. Now the cars are delivering a new quality, which is the new spatial arrangement of material and people. The safe movement of cars is essential for the system to perform. There are new emergent properties (such as traffic jams) which consist of something that a single car or a simple collection of cars (say, sitting in a parking lot) will never produce.

Two atoms of hydrogen combine with one atom of oxygen to produce a molecule of water. The properties of a water molecule are entirely different from those of hydrogen or oxygen, which are the elements from which water is constructed.

> *The whole is more than the sum of parts.*
>
> von Bertalanffy, 1968

We may look at a water molecule as a system that is made of three elements: two hydrogen atoms and one oxygen atom. The elements interact. This interaction binds the elements together and results in a new quality displayed by the whole.

Exercise 1.2

1. Think of examples of three systems. How would you describe these systems?
2. Describe chicken noodle soup as a system. What are the elements? What is the function? What makes it a system?

Elements ←→ whole

A system may be viewed as a *whole* or as a combination of *elements*. An element is a building block of a system that can be also considered separately, having its own properties and features. If a cake is cut into pieces, these pieces are not called elements of a cake because they have no particular features to separate them from one another – there may be any number of pieces that cannot be distinguished from another. Besides, the pieces do not offer any other properties except those delivered by the cake as a whole. The only difference is in size. Therefore, just a piece of a whole is not an element.

If you separate the crust, the filling and the topping of the cake, we will get something quite different from the whole cake. It makes much more sense to call these elements of the whole. The taste and other properties of different elements will be different, and so there are ways to distinguish one element from another.

Parts brought together do not necessarily make a system. Think of the 32 chess pieces piled on a table. They are elements in terms of being separable, looking different and carrying some unique properties. However, they could hardly be called a system. Adding another element (the chess board), as well as rules of interaction (how figures move over the board, and how they interact with each other), makes a system – the chess game. There are some additional *emergent properties* from the whole, which none of the elements possess.

Exercise 1.3

1. List five elements for each of the following systems:
   a. A steam engine,
   b. An oak tree,
   c. A Thanksgiving turkey,
   d. A city.
2. What is the system that has the following elements: water, gravel, three fish, fish feed, aquatic plants? What if we add a scuba diver to this list? Can elements entirely describe a system?

Reductionism ← → holism

We may look at a system as a whole and focus on the behavior of elements in their interconnectivity within the system. This approach is called holism. In this case it is the behavior of the whole that is important, and this behavior is to be studied within the framework of the whole system – not the elements that make it. On the contrary, reductionism is the theory that assumes that we can understand system behavior by studying the elements and their interaction.

> *The features of the complex, compared to those of the elements, appear as "new" or "emergent."*
>
> von Bertalanffy, 1968

Like analysis and synthesis, both approaches are important and useful. The reductionist approach allows reduction of the study of a complex system to analysis of smaller and presumably simpler components. Though the number of components increases, their complexity decreases and they become more available for experiments and scrutiny. However, this analysis may not be sufficient for understanding the whole system behavior because of the emergent features that appear at the whole system level. The holistic approach is essential to understanding this full system operation. It is much simpler, though, to understand the whole system performance if the behavior of the elements is already well studied and understood.

For example, most of modern medicine is very much focused on studies of the biochemistry and processes within individual organs at a very detailed level that considers what happens to cells and molecules. We have achieved substantial progress in developing sophisticated drugs that can treat disease, attacking microbes and fixing particular biochemical processes in the organism. At the same time, we have found that by treating one problem we often create other, sometimes even more severe, conditions at the level of the whole organism. While understanding how elements perform, we may still be unaware of the whole system functioning. Listen to almost any drug commercial on the TV. After glamorous reports about successful cures and recoveries, closer to the end you may notice a rapid and barely readable account of all the horrible side-effects, which may include vomiting, headache, diarrhea, heartburn, asthma – you name it. The whole system can react in a way that it is sometimes hard or even impossible to predict when looking at the small, local scale.

One of the saddest stories came about through the use of a drug called thalidomide. It was originally synthesized in 1954, marketing started in 1957, and its use rapidly spread to many countries in Europe, Asia, Australia, America and Africa. Thalidomide was presented as a "wonder drug" that provided "safe, sound sleep." It was a sedative that was found to be effective when given to pregnant women to combat many of the symptoms associated with morning sickness. It was not realized that thalidomide molecules could cross the placental wall and affect the fetus. Eight months later, an epidemic of malformations started in babies born to mothers who had taken the drug during their pregnancies. Those babies born with thumbs with three joints, with only three fingers or with distorted ears can probably be considered lucky. Many others had hands growing directly from their shoulders. Other babies suffered from malformations of the internal organs – the heart, the bowel, the uterus and the gallbladder. About 40 percent of thalidomide victims died before their first birthday.

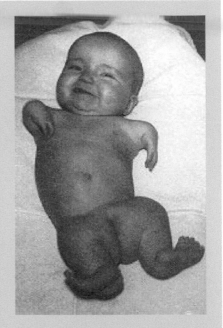

It was particularly difficult to make the connection because of the important time factor: the sensitive period is from days 35 to 49 of the pregnancy. Indeed, a holistic approach can be very important.

What were the boundaries of the system in this case? It would appear to be the whole organism of a patient. Apparently in this case the fetus also needed to be included in the thorough studies. Lawsuits were followed by some ugly denials and manipulations by the producer, but that is another very sad story …

(http://www.thalidomide.ca/en/information/index.html)

Listing all the elements (Figure 1.2A) is not enough to describe a system. Elements may be connected or related to each other in a variety of different ways, and the *relationships* between elements are essential to describe a system. The simplest is to acknowledge the existence of a relationship between certain elements, as is done in a graph (Figure 1.2B). In this case, a node presents an element, and a link between any two nodes shows that these two elements are related. An element can be also connected to itself, to show that its behavior depends on its state. However, in this diagram there is no evidence of the direction of the relationship: we do not distinguish between element x influencing element y and *vice versa*. This relationship can be further specified by an oriented graph that shows the direction of the relationship between elements (Figure 1.2C). Next, we can describe the relationships by identifying whether element x has a positive or negative effect on element y.

There may be two types of relationships between elements:

1. Material flows
2. Flows of information.

Material flows connect elements between which there is an exchange of some substance. This can be some kind of material (water, food, cement, biomass, etc.),

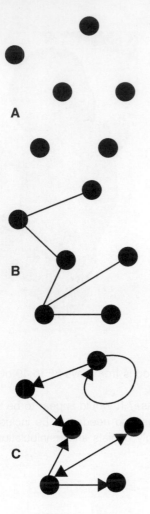

Figure 1.2 Elements and interactions.
We first identify elements in the system (A), then figure out
which ones are connected (B). Next we start describing the
types of interactions (C – which element influences which, and
how). By putting together these kinds of relationship diagrams
we can better understand and communicate how systems
work.

energy (light, heat, electricity, etc.), money, etc. It is something that can be measured and tracked. Also, if an element is a donor of this substance the amount of
substance in this element will decrease as a result of the exchange, while at the same
time the amount of this substance will increase in the receptor element. There is
always a mass or energy conservation law in place. Nothing appears from nothing,
and nothing can disappear to nowhere.

The second type of exchange is an information flow. In this case, element A gets
the information about element B. Element B at the same time may have no information about element A. Even when element A gets information about B, element
B does not lose anything. Information can be about the state of an element, about
the quantity that it contains, about its presence or absence, etc. For example, when
we sit down for breakfast, we eat food. As we eat, there is less food on the table and
more food in our stomachs. There is a flow of material. At some point we look at the
clock on the wall and realize that it is time to stop eating and go to work. There is a
flow of information from the clock to us. Nothing has been taken from the clock, yet
we learned something from the information flow that we used.

When describing flows in a system it is useful to identify when the flows play a
stimulating or a dampening effect. For example, consider a population growth process.

The larger the number of individuals in a population, the more potential births are occurring, the larger the number of individuals in a population, etc. This is an example of a *positive feedback*. There are numerous examples of systems with a positive feedback. When a student learns to read, the better she can read the more books she reads, and the better she learns to read. Another one: the heavier a man is, the less fun it is for him to walk, so the less he enjoys hiking or going somewhere, so the less he burns calories, and the more he gains weight. And so on.

On the contrary, a system with *negative feedback* tries to stabilize itself according to the rule: the larger something is, the smaller something becomes. For example, looking again at populations, if there is a limited supply of food and the population grows, there is less food for each individual. The more the population grows, the less food there is for individuals. At some point there is not enough food to support the population, and some individuals die. Eventually growth shuts down completely, and the population equilibrates at a level that can be sustained by the supply of food.

Systems with positive feedback end up in uncontrolled exponential growth or collapse.

Systems with negative feedback tend to stabilize at a certain level.

In some cases identifying the positive and negative feedbacks in a system allows us to make a quick ballpark projection of the possible future for the system.

The more money I have in my savings account, the more interest it will bring. And then I'll have more money – and then I'll have more interest. That's "positive feedback," "the more – the more" story. There is also "negative feedback." That's when, after I flush the toilet, the water starts to run into the tank. There's a little float in the tank attached to a valve. So the more water runs in, the higher the float goes, the less water flows through the valve. Then, at some point the valve shuts, the flow stops. Now we have "the more – the less" situation.

It is easy to see how these systems are different. If there is positive feedback, "the more – the more" case, we will have a problem – unless it's our money in the bank. Something gets out of control. There is no mechanism to stop. It's a runaway situation. It's cancer, it's an explosion. It does not stop by itself.

"The more – the less" or similarly, "the less – the more" keeps everything nicely under control. It's "the more – the more" that is kinda scary. These processes just don't know how to stop. They are fueling themselves until they run out of steam, or simply blow up everything.

Now this is what makes the patterns of global climate change look pretty dim. The only kind of processes that we find in the climate system are "the more – the more" ones. Here's an example. When it's hot, I prefer to dress in white. I don't like wearing black because it just feels much hotter in black. White seems to reflect more sunshine, and I don't get so hot. This is called albedo. White has high albedo; black has low albedo. So the Arctic has lost almost 30 percent of its ice because now it's warmer. But this means that there is less white over there and more dark – but the darker it gets the hotter it gets, and the more ice is melted, and the more dark it gets, and ... Well, you can see, "the more – the more." Not good, and not going to stop until all the ice is gone.

What else? Well there is permafrost that is thawing, especially in Siberia. There are all those huge bogs, which used to be frozen. Now they are melting and releasing huge amounts of some gases into the atmosphere, and apparently these gases happen to be just the ones that are causing the temperature to rise. These are called greenhouse gases. The temperature started to rise because we have put too much of these gases into the atmosphere already.

So now it is becoming warmer and the bogs are melting. And the more it warms up, the more bogs melt and the more gases they pump into the atmosphere and the more temperature rises, and the more … Gosh, there we go again. There's no way this will stop until all the frozen bogs have melted.

So where is this going to take us? How much is the climate supposed to change before we regain some equilibrium?

Structure ←→ function

The elements of a system and their interactions define the system *structure*. The more elements we distinguish in a system and the more interactions between them we present, the more complex is the system structure. Depending on the goal of our study, we can present the system in a very general way with just a few elements and relations between them, or we may need to describe many detailed elements and interactions. One and the same system can be presented in many different ways. Just as with temporal and spatial resolution, the choice of structural resolution and the amount of detail regarding the system that you include in your description depend upon the goals that you want to accomplish.

Whereas the elements are important to define the structure of the system, the analysis of a system as a whole is essential to figure out the *function* of a system. The function is the emerging property that makes a system a *system*. Putting together all the components of a birthday cake, including the candles on top, generates the new function, which is the taste and the celebration that the cake delivers. Separate elements have other functions, but in this combination they create this new function of the system.

1.3   Hierarchy

Subsystem ←→ System ←→ Supersystem

In any hierarchical structure the higher levels embrace or "comprehend" the lower, but the lower are unable to comprehend the higher. This is what may be called the "hierarchical principle."

Haught, 1984

Every system is part of a larger system, or supersystem, while every element of a system may be also viewed as a system, or subsystem, by itself (Figure 1.3). By gradually decomposing an object into smaller parts and then further decomposing those parts into smaller ones, and so on, we give rise to a *hierarchy*. A hierarchy is composed of levels. The entries that belong to one level are assumed to be of similar complexity and to perform a somewhat similar function. *New emergent functions appear when we go from one level of a hierarchy to another.*

When analyzing a system, it is useful to identify where it can be placed in a hierarchy. The system is influenced by the higher levels and by similar systems at the

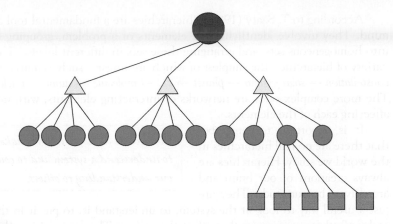

Hierarchies in systems.

Systems may be presented as interacting subsystems. Systems themselves interact as parts of supra-systems. There are various hierarchical levels that can be identified to improve the descriptions of systems in models. Elements in the same hierarchical level are usually presented in the same level of detail in the space–time–structure dimensions.

same level. However, lower levels of those similar systems are hardly important for this system. They enter the higher levels in terms of their function; the individual elements may be negligible but their emergent properties are what matter. Fiebleman describes this in his theory of integrative levels as follows: "For an organism at any given level, its mechanism lies at the level below and its purpose at the level above" (Fiebleman, 1954: 61).

For example, consider a student as a system. The student is part of a class, which is the next hierarchical level. The class has certain properties that are emergent for the set of students that enter it. At the class level, the only thing that is important about students is their learning process. It does not matter what individual students had for breakfast, or whether they are tall or short. On the other hand, their individual ability to learn is affected by their individual properties. If a student has a headache after the party on the night before, he or she probably will not be able to study as well as a neighbor who went to the gym instead. The class as a whole may be characterized by a certain degree of academic achievement that will be different from the talents and skills of individual students, yet that will be the benchmark that the teacher will consider when working with the class. Each student affects this emergent property to a certain extent, but not entirely. On the contrary, the class average affects each individual student, setting the level of instruction that is to be offered by the teacher. Different classes are assembled into a school, which is the next level in the hierarchy. Schools may be elements in a Regional Division, and so on.

At the other end of this hierarchy, we can start by "decomposing" each individual student, looking at his or her body organs and considering their functions – and so on, until we get to molecules and atoms. There are many ways we can carry out the decomposition. Instead of considering a student as an element of a class, we may look at that student as an element of a family and build the hierarchy in a different way. As with modeling in general, the type of hierarchy that we create is very much driven by the goals of our study. The hierarchical approach is essential in order to place the study object within the context of the macro- and micro-worlds – that is, the super- and subsystems – relative to it.

According to T. Saaty (1982), "hierarchies are a fundamental tool of the human mind. They involve identifying the elements of a problem, grouping the elements into homogeneous sets, and arranging these sets in different levels." There may be a variety of hierarchies, the simplest of which are linear – such as *universe* → *galaxy* → *constellation* → *solar system* → *planet* → … → *molecule* → *atom* → *nucleus* → *proton*. The more complex ones are networks of interacting elements, with multiple levels affecting each of the elements.

It is important to remember that there are no real hierarchies in the world we study. Hierarchies are always creations of our brain and are driven by our study. They are

> *Hierarchies do not exist. We make them up to understand a system and to communicate our understanding to others.*

just a useful way to look at the system, to understand it, to put it in the context of scale, of other components that affect the system. There is nothing objective about the hierarchies that we develop.

For example, consider the hierarchy that can be assumed when looking at the Earth system. Clearly, there are ecological, economic and social subsystems. Neo-classical economists may forget about the ecological subsystem and put together their theories with only the economic and social subsystems in mind. That is how you would end up with the Cobb–Douglas production function that calculates output as a function of labor (social system) and capital (economic system).

Environmental economists would certainly recognize the importance of the eco-logical system. They would want to take into account all three subsystems, but would think about them as if they were acting side-by-side, as equal components of the whole (Figure 1.4A). For them, the production function is a product of population (labor), resources (land) and capital. All three are equally important, representing the social, natural (ecological) and economic subsystems, respectively. They are also substitutable: you can either work more or invest more to get the same result. You can also come up

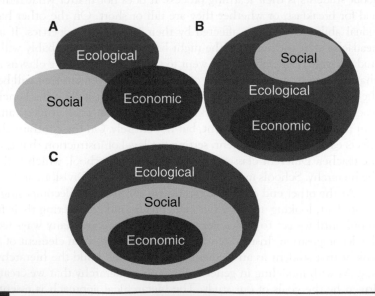

Figure 1.4   Different ways to present the ecological–economic–social hierarchy in the Earth system. Hierarchies are subjective and serve particular purposes of the analysis.

with a price for ecological goods or services, and calculate how much money you need to pay to compensate for a ruined wetland or an extinct species.

Ecological economists would argue that this is not the right way to look at this hierarchy, since there is no way labor or capital can exist in the absence of natural resources. They would argue that you cannot compensate for a natural resource – say, land – by working more or investing more money. Therefore, the economic and social systems are actually subsystems of the ecological system. A different hierarchy would therefore emerge (Figure 1.4B).

A social or behavioral scientist would argue that all the economic relationships are produced within the social system. There is no reason to talk about the economy where the social system is gone. What are the economics in a buffalo herd? We need humans to develop the economic subsystem. Therefore we may come up with yet another hierarchy (Figure 1.4C). Which hierarchy is the "real" one? It depends upon the focus of the research.

We will be considering sustainability and sustainable development in more detail in Chapter 7. Here, let us use this notion to demonstrate how systems and hierarchies may be a useful tool for some far-reaching conclusions. The World Commission on Environment and Development (WCED, 1987) introduced the idea of sustainability several decades ago, but there is still no single agreed definition for it. Most would agree that it implies that a system is to be maintained at a certain level, held within certain limits. Sustainability denies run-away growth, but also precludes any substantial set backs or cuts. While most – probably all – natural systems go through a renewal cycle, where growth is followed by decline and eventual disintegration, sustainability in a way has the goal of preventing the system from declining and collapsing. Originally the Brundland Commission came up with the concept of sustainability at the global level, as a way to protect our biosphere from becoming uninhabitable by humans, and human lives becoming full of suffering and turmoil because of the lack of natural resources and assimilative capacity of the planet.

However, somehow in the environmental movement the goal of sustainability was translated into the regional and local levels. Indeed, the famous Schumacher idea of "Think globally – act locally" apparently means that the obvious path to global sustainability is through making sure that our local systems are sustainable. Is that really the case? Let us apply some of the ideas about hierarchies and systems.

Keep in mind that renewal allows for readjustment and adaptation. However, it is the next hierarchical level that benefits from this adaptation. Renewal in components helps a system to persist; therefore, for a hierarchical system to extend its existence, to be sustainable, its subsystems need to go through renewal cycles. In this way, the death of subsystems contributes to the sustainability of the supra-system, providing material and space for reorganization and adaptation. Costanza and Patten (1995: 196), looking at sustainability in terms of component longevity or existence time, recognized that "evolution cannot occur unless there is limited longevity of the component parts so that new alternatives can be selected."

Sustainability of a system borrows from sustainability of a supra-system and rests on lack of sustainability in subsystems. This might be hard to perceive, because at first glance it seems that a system made of sustainable, lasting components should be sustainable as well. However, in systems theory it has been long recognized that "the whole is more than the sum of parts" (von Bertalanffy, 1968: 55), that a system function is not provided only by the functions of its components, and therefore, in fact, system sustainability is not a product of sustainable parts and *vice versa*. This is especially true for living, dynamically evolving systems. "You cannot sum up the behavior of the whole from the isolated parts, and you have to take into account the relations between the various subordinated systems and the systems which are super-ordinated to them in order to understand the behavior of parts" (von Bertalanffy, 1950: 148).

One way to resolve this contradiction between sustainability of a socioeconomic ecological system and its components is to agree that there is only one system for which sustainability will be sought, and that is the top level system – which in this case is the biosphere as a whole. The global scale in this context seems to be the maximal that humans·can influence at the present level of their development. It is also the scale that affects the humanity as a whole, the system that is shared by all people, and should therefore be of major concern to all.

Probably the famous Schumacher slogan ("Think globally – act locally") should also include: "When acting locally – keep thinking globally." We do not want locally sustainable systems (cities, counties, regions, farms, industries). We want to let them renew, so that at the global level we can have material for adaptation and evolution, which is essential for sustainability.

Exercise 1.4

1. Consider a tree in a forest and describe the relevant hierarchy. What would be the levels "above" and "below"? How do you decide what to include in each hierarchical level?
2. Think of an example when a system is affected by a system 3 levels above in the hierarchy, but is not affected by the system 2 levels above in the hierarchy. Is this possible?
3. If a system collapses (dies off) can subsystems survive?

1.4   The modeling process

At this point we will give a general overview of the modeling process. This will be illustrated in numerous applications further on. Try not to get frustrated by the overall complexity of the process and by some of the terminology that will be introduced with no due explanation. It is not as difficult as it may seem.

Building a model is an *iterative* process, which means that a number of steps need be taken over and over again. As seen in Figure 1.5, you start from setting the goal for the whole effort. What is it that we want to find out? Why do we want to do it? How will we use the results? Who will be involved in the modeling process? To whom do we communicate the results? What are the properties of the system that need to be considered to reach the goal?

Next, we start looking at the information that is available regarding the system. This can consist of data gathered either for the particular system in mind, or from similar systems studies elsewhere and at other times. Note that immediately we are entering the iterative mode, since once we start looking at the information available we may very quickly realize that the goals we have set are unrealistic with the available data about the system. We need either to redefine the goal, or to branch out into more data collection, monitoring and observation – undertakings that may shadow the modeling effort, being much more time- and resource-consuming. After studying

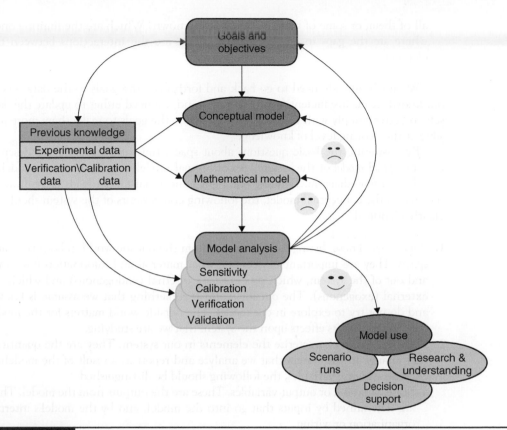

Figure 1.5   The modeling process.

Note the iterative nature of the process; from almost any step we should be prepared to go back and start from the beginning, as we improve our understanding of the system while we model it.

the available information, and with the goal in mind, we start identifying our system in its three main dimensions: space, time and structure.

1. *Space.* What is the specific size of the object that we need to analyze, in which level of the hierarchy is our system embedded? How far spatially does that system extend? What will be the spatial resolution of the processes that we need to consider? How does the system evolve in space? Is it static, like a map, or dynamic, like the "game of life" (see http://www.bitstorm.org/gameoflife/ or http://www.mindspring.com/~alanh/life/index.html)?

2. *Time.* What is the specific time span of the system? Are we looking at it over years, days, or seconds? How fast are the processes? Which processes are so slow that they may be considered constant, and which other processes are so fast that they may be considered at equilibrium? Do we need to see how the system evolves in time, like in a movie, or do we just need a snapshot of the reality, like a photo? If the system is evolving, how does it change from one state to another? Is it a continuous process or a discrete, instantaneous one? Is the next state of the system totally defined by its current one, or is it a stochastic process, where future states occur spontaneously with certain probability?

3. *Structure.* What are the elements and processes in our system? How much detail about them do we need and can we afford? Do we have enough information about

all of them, or some of them are entirely unknown? Which are the limiting ones, where are the gaps in our knowledge? What are the interactions between the elements?

We might already need to go back and forth from the goals to the data sets. If our knowledge is insufficient for the goal in mind, we need either to update the data sets to better comply with the goals, or to redefine the goals to make them more feasible at the existing level of knowledge.

By answering the basic questions about space, time and structure, we describe the *conceptual model* of the system. A conceptual model may be a mental model, a sketch or a flow diagram. Building the right conceptual model leads us halfway to success. In the conceptual model, the following components of the system should be clearly identified.

1. *Boundaries.* These distinguish the system from the outside world in both time and space. They are important in deciding what material and information flows into and out of the system, which processes are internal (endogenous) and which are external (exogenous). The outside world is something that we assume is known and do not try to explore in our model. The outside world matters for the model only in terms of its effects upon the system that we are studying.
2. *Variables.* These characterize the elements in our system. They are the quantities that change in the system that we analyze and report as a result of the modeling exercise. Among variables, the following should be distinguished:
   - State variables, or output variables. These are the outputs from the model. They are determined by inputs that go into the model, and by the model's internal organization or wiring.
   - Intermediate or auxiliary variables. These are any quantities defined and computed in the model. They usually serve only for intermediate calculations; however, in some cases looking at them can help us to understand what happens "under the hood" in the model.
3. *Parameters.* These are generally all quantities that are used to describe and run a model. They do not need to be constant, but all their values need to be decided before the model runs. These quantities may be further classified into the following categories:
   - Boundary conditions. These describe the values along the spatial and temporal boundaries of a system. For a spatially homogeneous system we have only initial conditions, which describe the state of the variables at time $t = 0$ when we start the model, and the length of the model run. For spatially distributed systems, in addition we may need to define the conditions along the boundary, as well as the geometry of the boundary itself.
   - Constants or parameters in a narrow sense. These are the various coefficients and constants measured, guessed or found. We may want to distinguish between real constants, such as gravity, g, and, say, the half-saturation coefficient, K, in the Michaelis–Menten function that we will consider in the next chapter. While both of them take on constant values in a particular model run, g will be always the same from one run to another, but K may change quite substantially as we improve the model. Even if K comes from observations, it will normally be measured with certain error, so the exact value will not be really known.
   - Forcing functions. These are parameters that describe the effect of the outside world upon the system. They may change in time or space, but they do

not respond to changes within the system. They are external to it, driven by processes in the higher hierarchical levels. Climatic conditions (rainfall, temperature, etc.) certainly affect the growth of tomatoes in my garden, but the tomatoes hardly affect the temperature or the rainfall patterns. If we build a model of tomato growth, the temperature will be a forcing function.
- Control functions. These are also parameters, except that they are allowed to change to see how their change affects systems dynamics. It is like tuning the knob on a radio set. Every time the knob is dialed to a certain position, but we know that it may vary and will result in a different performance by the system.

Note that in some texts parameters will be assumed only in the narrow sense of constants that may sometimes change, like the growth rate or half-saturation coefficients. However, this may be somewhat confusing, since forcing functions are also such parameters if they are fixed. Suppose we want to run a model with the temperature held constant and equal to the mean over a certain period of time – say, the 6 months of the growth season for a crop. Then suppose later on we want to feed into the model the actual data that we have measured for temperature. Temperature is now no longer a constant, but changes every day according to the recorded time series. Does this mean that temperature will no longer be a parameter? For any given moment it will still be a constant. It will only change from time to time according to the data available. Probably, it would make sense still to treat it as a parameter, except now it will be no longer constant but will change accordingly.

Suppose now that we approximate the course of temperatures by a function with some constants that control the form of this function. Suppose we use the sine function and have parameters for the amplitude and the period. Now temperature will no longer be a parameter. Note that we no longer need to define all the values for temperature *before* we hit the "Run" button. Instead, temperature will become an intermediate variable, while we will have two new parameters in the sine function that now specifies temperature – one parameter ($B = 4$) will make the period equal to 6 months, the other parameter (A) will define the amplitude and make the temperature change from a minimal value (0) to the maximal value (40, if $A = 20$) and back over this period of time, as in the function:

$$\text{Temperature} = A * \text{SIN}\left(t * \frac{B * \pi}{365} + \frac{3 * \pi}{2}\right) + A$$

where t is time, π is a constant $\pi = 3.14$, and A and B are parameters. If $B = 2$, then the period will change from 6 to 12 months. Both A and B are set before we start running the model.

There may be a number of ways to determine model parameters, including the following.

1. Measurements in situ. This is probably the best method, since the measurements define the value of exactly what is assumed in the model. However, such measurements are the most labor- and cost-intensive, and they also come with large margins of error. Besides, in many cases such measurements may not be possible at all, if a parameter represents some aggregated value or an extreme condition that may not occur in reality (for example, the maximal temperature for a population to tolerate – this may differ from one organism to another, and such conditions may be hard to find in reality).

2. Experiments in the lab (in vitro). These are usually performed when in situ experiments are impossible. Say we take an organism and expose it to high temperatures to find out the limits of its tolerance. We can create such conditions artificially in a lab, but we cannot change the temperature for the whole ecosystem.

3. Values from previous studies found from literature, web searches or personal communications. If data are available for similar systems, it certainly makes sense to use them. However, always keep in mind that there are no two identical ecosystems, so it is likely that there will be some error in the parameters borrowed from another case study.

4. Calibration (see Chapter 4). When we know what the model output should look like, we can always tweak some of the parameters to make the model perform at its best.

5. Basic laws, such as conservation principles and therefore mass and energy balances.

6. Allometric principles, stoichiometry, and other chemical, physical, etc., properties. Basic and derived laws may help to establish relationships between parameters, and therefore identify at least some of them based on the other ones already measured or estimated.

7. Common sense. This always helps. For example, we know that population numbers cannot be negative. Setting this kind of boundary on certain parameters may help with the model.

Note that in all cases there is a considerable level of uncertainty present in the values assigned to various model parameters. Further testing and tedious analysis of the model is the only way to decrease the error margin and deal with this uncertainty.

Creating a conceptual model is very much an artistic process, because there can hardly be any exact guidelines for that. This process very much resembles that of perception, which is individual to every person. There may be some recommendations and suggestions, but eventually everybody will be doing it in his or her own personal way. The same applies to the rest of the modeling process.

When a conceptual model is created, it may be useful to analyze it with some tools borrowed from mathematics. In order to do this we need to *formalize* the model – that is, find adequate mathematical terms to describe our concepts. Instead of concepts, words and images, we need to come up with equations and formulas. This is not always possible, and once again there is no one-to-one correspondence between a conceptual model and its mathematical formalization. One formalism can turn out to be better for a particular system or goal than another. There are certain rules and recommendations, but no ultimate procedure is known.

Once the model is formalized, its further analysis becomes pretty much technical. We can first compare the behavior of our mathematical object with the behavior of the real system. We start solving the equations and generate trajectories for the variables. These are to be compared with the data available. There are always some parameters that we do not know exactly and that can be changed a little to achieve a better fit of the model dynamics to the one observed. This is the so-called *calibration* process.

Usually it makes sense to first identify those parameters that have the largest effect on system dynamics. This is done by performing *sensitivity analysis* of the model. By incrementing all the parameters and checking out the model input, we can identify to which ones the model is most sensitive. We should then focus our attention on these parameters when calibrating the model. Besides, if the model has already been tested and found to be adequate, then model sensitivity may be translated into system sensitivity: we may conclude that the system is most sensitive to certain parameters and therefore the processes that these parameters describe. If the calibration does not look good enough, we need to go back to some of the previous steps of our modeling process (reiterate). We may have got the wrong conceptual model, or we did not formalize it properly, or there is something wrong in the data, or the goals do not match the resources. Unfortunately, once again we are plunged into the imprecise "artistic" domain of model reevaluation and reformulation.

> *Once you gain new understanding with your model, you may realize that something is missing. It's OK: go back and improve the model. You don't build a model going down a straight path. You build a model going in circles.*

If the fit looks good enough, we might want to do another test and check if the model behaves as well on a part of the data that was not used in the calibration process. We want to make sure that the model indeed represents the system and not the particular case that was described by the data used to tweak the parameters in our formalization. This is called the *validation* process. Once again, if the fit does not match our expectations we need to go back to the conceptualization phase.

However, if we are happy with the model performance we can actually start using it. Already, while building the model, we have increased our knowledge about the system and our understanding of how the system operates. That is probably the major value of the whole modeling process. In addition to that we can start exploring some of the conditions that have not yet occurred in the real system, and make estimates of its behavior in these conditions. This is the "what if?" kind of analysis, or the scenario analysis. These results may become important for making the right decisions.

1.5   Model classifications

There may be several criteria used to classify models. We will consider examples of many of the models below in much more detail in the following chapters. Here we give a brief overview of the kinds of models that are out there, and try to figure ways to put some order in their descriptions. Among many ways of classifying the models we may consider the following:

1. *Form*: in which form is the model presented?
   - Conceptual (verbal, descriptive) – only verbal descriptions are made. Examples include the following.
     - A description of directions to my home: *Take Road 5 for 5 miles East, then take a left to Main Street and follow through two lights. Take a right to Cedar Lane. My house is 3333 on the left.* This is a spatial model of my house location relative to a certain starting point. I describe the mental model of the route to my house in verbal terms.
     - A verbal portrait of a person: *He is tall with red hair and green eyes, his cheeks are pale and his nose is pimpled. His left ear is larger than the right one and one of his front teeth is missing.* This is a static verbal model of a person's face.
     - Verbal description of somebody's behavior: *When she wakes up in the morning, she is slow and sleepy until she has her first cup of coffee. After that she starts to move somewhat faster and has her bowl of cereal with the second cup of coffee. Only that brings her back to her normal pace of life.* This is a dynamic conditional verbal model.

- A verbal description of a rainfall event: *Rainfall occurs every now and then. If temperature is below 0°(C) (32 F) the rain is called snow and it is accumulated as snow or ice on the terrain. Otherwise it comes in liquid form and part of it infiltrates into the subsurface layer and adds to the unsaturated storage underground. The rest stays on the surface as surface water.*

- Conceptual (diagrammatic) – in some cases a good drawing may be worth a thousand words. Examples include the following.

  - A diagram that may explain your model even better than words.
  - A drawing or an image is also a model. In some cases it can offer much more information than the verbal description, and may be also easier to understand and communicate among people. Also note that in some cases a diagram can exclude some of the uncertainties that may come from the verbal description. For example, the verbal model cited above mentioned the left ear, but did not specify whether it is the person's left ear or the person's left ear as seen by the observer. This ambiguity disappears when the image is offered.

  - Dynamic features can be included in an animation or a cartoon.
  - A conceptual model of the hydrologic cycle.

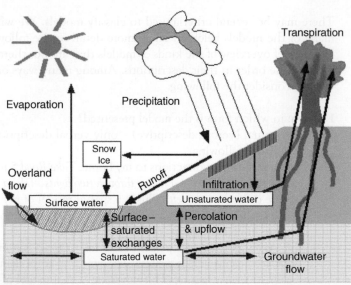

- Physical – a reconstruction of the real object at a smaller scale. Examples include the following.
  - Matchbox toy cars.
  - Remember those mannequins they put in cars to crash them against a brick wall and see what happens to the passengers? Well, those are models of

humans. They are no good for studying IQ, but they reproduce certain features of a human body that are important to design car safety devices.

- An airplane model in a wind tunnel.
- A fairly large (about 50-m long) model was created in the 1970s to analyze currents in Lake Balaton (Hungary). Large fans blew air over the model and currents were measured and documented.
- A physical model to study stream flow (see Figure 1.1).
- Formal (mathematical) – that is when equations and formulas reproduce the behavior of physical objects. Examples include the following.
  - $Q = m\,C\,(t_1 - t_2)$ – a model of heat emitted by a body of mass m, when cooling from temperature t_1 to temperature t_2. C is the heat capacity parameter.
  - $Y = Y_0* \, 2^{t/d}$ – a model of an exponentially growing population, where Y_0 is the initial population and d is doubling time.

2. *Time*: how is time treated in the model?
   - Dynamic vs static. A static model gives a snapshot of the reality. In dynamic models, time changes and so do the variables in the model. Examples include the following.
     - A map is a static model; so is a photo.
     - A cartoon is a dynamic model.
     - Differential or difference equations are dynamic models.
   - Continuous vs discrete. Is time incremented step-wise in a dynamic model, or is it assumed to change constantly, in infinitesimally small increments? Examples include the following:
     - You may watch a toy car roll down a wedge. It will be a physical model with continuous time.
     - Generally speaking, systems of differential equations represent continuous time models.
     - A difference equation is a discrete model. Time can change, but it is incremented in steps (1 minute, 1 day, 1 year, etc.)
     - A movie is a discrete model. Motion is achieved by viewing separate images, taken at certain intervals.
   - Stochastic vs deterministic. In a deterministic model, the state of the system at the next time step is entirely defined by the state of the system at the current time step and the transfer functions used. In a stochastic model, there may be several future states corresponding to the same current state. Each of these future states may occur with a certain probability.

3. *Space*: how is space treated in the model?
   - Spatial vs local (box-models). A point model assumes that everything is homogeneous in space. Either it looks at a specific locality or it considers averages over a certain area. A spatial model looks at spatial variability and considers spatially heterogeneous processes and variables. Examples include the following.
     - A demographic model of population growth in a city. All the population may be considered as a point variable, the spatial distribution is not of interest, and only the total population over the area of the city is modeled.
     - A box model of a small lake. The lake is considered to be a well-mixed container, where spatial gradients are ignored and only the average concentrations of nutrients and biota are considered.
     - A spatial hydrologic model. The watershed is presented as an array of cells with water moving from one cell to another downhill, along the elevation gradient.

- Continuous vs discrete. Like time, space may be represented either as continuous or as a mosaic of uniform objects. Examples include the following.
  - A painting vs a mosaic. Both represent a spatial picture and both look quite similar from a distance. However, at close observation it is clear that smooth lines and color changes in a painting are substituted by discrete uniform elements in the mosaic, which change their color and shape in a stepwise manner.
  - Differential equations or equations in partial derivatives are used for continuous formalizations.
  - Finite elements or difference schemes are used to formalize discrete models.

4. *Structure*: how is the model structure defined?
   - Empirical (black-box) vs process-based (simulation) models. In empirical models, the output is linked to the input by some sort of a mathematical formula or physical device. The structure of the model is not important as long as the input signals are translated into the output ones properly – that is, as they are observed. These models are also called black-box models, because they operate as some closed devices on the way of the information flows. In process-based models, individual processes are analyzed and reproduced in the model. In any case, it is not possible to go into all the details or to describe all the processes in all their complexity (it would not be a model then). Therefore, a process-based model may be considered as being built from numerous black boxes. The individual processes are still presented as closed devices or empirical formulas; however, their interplay and feedbacks between them are taken into account and analyzed.
   - Simple vs complex. Though qualitatively clear, this distinction might turn out to be somewhat hard to quantify. It is usually defined by the goals of the model. Simple models are built to understand the system in general over long time intervals and large areas. Complex models are created for detailed studies of particular system functions. The increased structural complexity usually has to be compensated by coarser temporal and spatial resolutions.

5. *Method*: how is the model formulated and studied?
   - Analytic vs computer models. Analytical models are solved by finding an analytical mathematical solution to the equations. Mathematical models easily become too complex to be studied analytically. Instead, numerical methods are derived that allow solving equations on a computer.
   - Modeling paradigm.
     - Stock-and-flows or systems dynamics models assume that the system can be represented as a collection of reservoirs (that accumulate biomass, energy, material, etc.) connected by pipes (that move the material between reservoirs).
     - Individual- (or agent-) based models. These describe individual organisms as separate entities that operate in time and space. There are rules that define the behavior of these agents, their growth, movement, etc.
     - Network-based models.
     - Input/output models.
     - Artificial neural networks.

6. *Field-related classification*: what field is the model in (e.g. ecology)?
   - Population models. These are built to study the dynamics and structure of populations. A population is easily characterized by its size, which may be why population ecology is probably the most formalized branch of ecology.

- Community models take several populations and explore what happens when they interact. The classic predator–prey or host–parasite systems and models of trophic interactions are the most prominent examples.
- Ecosystem models attempt to represent the whole ecosystem, not just some components of it. For example, a model has been developed for the wetland ecosystem in the Florida Everglades (http://my.sfwmd.gov/pls/portal/url/page/PG_SFWMD_HESM/PG_SFWMD_HESM_ELM?navpage=elm). It includes the dynamics of water, nutrients, plants, phytoplankton, zooplankton and fish. The goal is to understand how changes in the hydro-period affect the biota in that area, and how the biota (plants) affects hydrology.

7. *Purpose*: what is the model built for?
   - Models for understanding would normally be simple and qualitative, focusing on particular parts or processes of a system – for example, the predator–prey model that we consider in Chapter 5.
   - Models for education or demonstration. These are built to demonstrate particular features of a system, to educate students or stakeholders. For example, the well-known Daisy World model is used to demonstrate how the planet can self-regulate its temperature, using black and white daisies (See http://www.informatics.sussex.ac.uk/research/projects/daisyworld/daisyworld.html for more about the model or http://library.thinkquest.org/C003763/flash/gaia1.htm for a nice Flash animation).
   - Predictive models are detailed and scrupulously tested simulations that are designed to make real decisions. A perfect example is a weather model that would be used for weather forecasts.
   - Knowledge bases. Models can serve as universal repositories of information and knowledge. In this case, the model structure puts various data in a context providing conceptual links between different qualitative and quantitative bits of information. For example, the Multi-scale Integrated Models of Ecosystem Services (MIMES – http://www.uvm.edu/giee/mimes/) organizes an extensive body of information relevant to ecosystem services valuation in five spheres: anthroposphere, atmosphere, biosphere, hydrosphere, and lithosphere.

1.6   Systems thinking

In more recent years, people have really started to appreciate the importance of the systems approach and systems analysis. We are now talking about a whole new mindset and worldview based on this understanding of systems and the interconnectedness between components and processes. With systems we can look at connections between elements, at new properties that emerge from these connections and feedbacks, and at the relationships between the whole and the part. This worldview is referred to as "systems thinking."

The roots of systems thinking go back to studies on systems dynamics at MIT led by Jay Forrester, who was also the inventor of magnetic-core memory, which evolved into the random access memory used in all computers today. Even though back in 1956 he never mentioned systems thinking as a concept, the models he was building clearly chiseled out the niche that would be then filled by this type of holistic, integrative, cross-disciplinary analysis. With his background in electrical and computer engineering, Forrester has successfully applied some of the same engineering principles to social, economic and environmental problems. You can find a certain resemblance between electric circuits and systems diagrams that Forrester has introduced. The titles of his most famous books, *Industrial Dynamics* (1962), *Urban Dynamics* (1969) and *World Dynamics* (1973), clearly show the types of applications that have been studied using this approach. The main idea is to focus on the system as a whole. Instead of traditional analytical methods, when in order to study we disintegrate, dig inside and study how parts work, now the focus is on stud=ying how the whole works, how the parts work together, what the functions are, and what the drivers and feedbacks are.

Forrester's works led to even more sophisticated world models by Donella and Dennis Meadows. Their book, *Limits to Growth* (1972), was published in paperback and became a national bestseller. Systems dynamics got a major boost when Barry Richmond at High Performance Systems introduced Stella, the first user-friendly icon-based modeling software.

Despite all the power and success of the systems dynamics approach, it still has its limits. As we will see later on, Stella should not be considered to be the ultimate modeling tool, and there are other modeling systems and modeling paradigms that are equally important and useful. It would be wrong to think that systems approach and the ideas of systems thinking are usurped by the systems dynamics methods. Systems can be described in a variety of different ways, not necessarily using the stock-and-flow formalism of Stella and the like.

Systems thinking is more than just systems dynamics. For example, the so-called Life Cycle Assessment (LCA) is clearly a spin-off of systems thinking. The idea of LCA is that any economic production draws all sorts of resources from a wide variety of areas. If we want to assess the true cost of a certain product, we need to take into account all the various stages of its production, and estimate the costs and processes that are associated with the different other products that went into the production of this one. The resulting diagrams become very complex, and there are elaborate databases and econometric models now available to make these calculations. For example, to resolve the ongoing debate about the efficiency of corn-based ethanol as a substitute for oil, we need to consider a web of interactions (Figure 1.6) that determine the so-called Energy Return on Energy Invested (EROEI). The idea is that you always need to invest energy to derive new energy. If you need to invest more than you get, it becomes meaningless to run the operations. That is exactly why we are not going to run out of oil. What will happen is it will become more expensive in terms of energy to extract it than we can gain from the product. That is when we will stop pumping oil to burn it for energy, but perhaps will still extract it for other purposes, such as the chemical industry or material production.

So if e_{out} is the amount of energy produced and e_{in} is the amount of energy used in production, then EROEI, $e = e_{out}/e_{in}$. In some cases the net EROEI index is used, which is the amount of energy we need to produce to deliver a unit of net energy to the user: $e' = e_{out}/(e_{out} - e_{in})$. Or $e' = e/(e-1)$. As we unwind the various chains of products and processes that go into the production of energy from corn, the EROEI dramatically falls. The current estimate stands at about 1.3, and there are still some processes that have not been included in this estimate. A true systems thinking approach would require that we go beyond the processes in Figure 1.6 and also look at social impacts as well as further ecological impacts, such as the eminent deforestation that is required for expanded corn production, and the loss of wildlife that will follow. Taking all that into account, the question arises: with an EROEI of 1.3 or less, is it worth it? To compare, the EROEI for crude oil used to be about 100; nowadays it is falling to about 10.

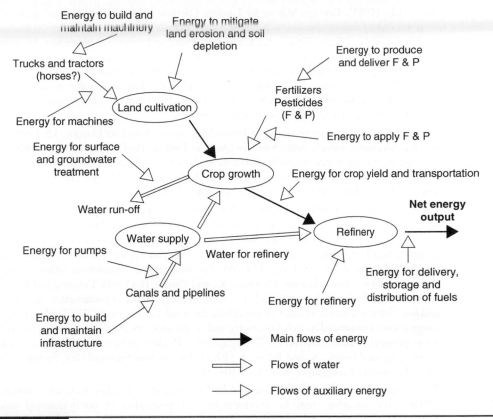

Figure 1.6 The lifecycle of energy from biomass production.

Further reading

von Bertalanffy, L. (1968). *General System Theory.* George Braziller. – *This is an introduction to systems theory, and one of the classics of this approach.*

Hall, C. and Day, J. (1977). *Ecosystem Modeling in Theory and Practice. An Introduction with Case Histories.* Wiley. – *An excellent collection of papers on the theory of modeling illustrated by a variety of models from very different tiers of life.*

Ford, A. (1999). *Modeling the Environment.* Island Press. – *An entirely Stella based textbook. Gives a neat introduction to modeling the way it can be performed without really knowing what is going on beneath the Stella interface. The book is perfect for a mathematically deprived modeler.*

Berlinski, D. (1978). *On Systems Analysis. An Essay Concerning the Limitations of Some Mathematical Methods in the Social, Political, and Biological Sciences.* MIT Press. – *A curious collection of critiques of some very famous models. May be recommended to better understand that models are not the ultimate solution, that there are always certain limitations to their use and that these limitations should be explicitly made part of the model.*

There is some controversy about who actually said, "All models are wrong … Some models are useful". According to some texts it was William Deming; at least that is what McCoy claims in his collection of quotes McCoy, R. (1994). *The Best of Deming. Statistical Process Control Press. However, others attribute it to George E.P. Box in "Robustness in the strategy of scientific model building", page 202 of Robustness in Statistics (1979), Launer, R.L. and Wilkinson G.N., Editors. Academic Press. After all it doesn't really matter who said it first; it is certainly very true.*

Saaty, T.L. (1982). *Decision Making for Leaders*. Lifetime Learning Publications. p.28. – *This is not exactly related to modeling but gives a good analysis of hierarchies and their applications in decision making. Saaty distinguishes between structural and functional hierarchies. In structural hierarchies systems are decomposed into their constituent parts in descending order according to structural properties such as size, shape, color, age, etc. This is the type of hierarchies most useful in building process-based models. Saaty is analyzing functional hierarchies that are created according to the essential relationships between the elements. His hierarchies are essential to analyze the decision making process and help in conflict resolution*

Some philosophical interpretations of hierarchies can be found in Haught, J.F. (1984). *The Cosmic Adventure: Science, Religion and the Quest for Purpose*. Paulist Press (also available online at http://www.religion-online.org/showchapter.asp?title=1948&C=1814). *An interesting analysis of hierarchical levels and their interaction is performed by* Fiebleman, J. (1954). Theory of Integrative Levels. *The British Journal for the Philosophy of Science*, 5 (17): 59–66.

There is more on sustainability in Chapter 7. The Bruntland Commission report gives a good introduction to the concept: WCED (World Commission on Environment and Development, 1987). *Our Common Future*. Oxford University Press. *Some of the issues related to hierarchy theory are presented by* von Bertalanfy, L. (1950). An Outline of General System Theory. *The British Journal for the Philosophy of Science*, 1 (2): 134–165. *For more on how sustainability relates to longevity and eventually – to hierarchies see:* Costanza, R., and Patten, B. (1995). Defining and predicting sustainability. *Ecological Economics:* 15, 193–196. *For an overview of sustainability, its definitions, and how different it can be in different hierarchical levels see* Voinov, A. (2007). *Understanding and communicating sustainability: global versus regional perspectives.* Environ. Dev. and Sustain. (http://www.springerlink.com/content/e77377661p8j2786/). *If you want to see how this can be related to discounting, see* Voinov, A. and Farley, J. (2007). Reconciling Sustainability, Systems Theory and Discounting. *Ecological Economics*, 63:104–113.

The modeling process is very well described by Jakeman, A.J., Letcher R.A. and Norton J.P. (2006). Ten iterative steps in development and evaluation of environmental models. *Environmental Modelling and Software:* 21(5): 602–614.

There are quite a few web sites on Systems Thinking. http://www.thesystemsthinker.com/systemsthinkinglearn.html *gives a good overview of the field. Another good introduction is available at* http://www.thinking.net/Systems_Thinking/Intro_to_ST/intro_to_st.html

The several classic books that led to many concepts of systems thinking are by J.Forrester: (1969). *Urban Dynamics*. Pegasus Communications, Inc.; (1962) *Industrial Dynamics*, The M.I.T. Press & John Wiley & Sons.; (1973) World Dynamics, Wright-Allen Press; *and a general overview in his 1968 book: Principles of Systems*, Pegasus Communications.

Another classic is the book by Meadows D. H., Randers J., and Meadows D. L. (1972). *Limits to Growth*, Signet. *Its paperback edition was a bestseller at that time. More recently the topic was revisited in the 2004 edition: Limits to Growth: The 30-Year Update*, Chelsea Green, 368 p.

2. The Art of Modeling

2.1 Conceptual model
2.2 Modeling software
2.3 Model formalization

"How to avoid false proof?

1. *allow no hasty and predetermined judgment;*
2. *decompose each difficult problem into simple ones that you can resolve;*
3. *always start with simple and clear, and gradually move on to more complex;*
4. *make complete surveys of all done before and make sure that nothing is left aside."*

Descartes

SUMMARY

There is really a lot of art in building a good model. There are no clear rules, only guidelines for good practice. These are constantly modified when required by the goals of modeling, the data available, and the particular strengths and weaknesses of the research team. In many cases it is possible to achieve the same level of success coming from very different directions, choosing different solutions. However, there are certain steps or stages that are common to most models. It is important to understand these and learn to apply them. Any system can be described in the spatial, temporal and structural context. It is important to be clear about these three dimensions in any model, to avoid inconsistencies or even errors.

We start with a conceptual model describing the system in general terms, qualitatively. If needed, we will then find the right quantitative formalizations for the processes involved. We may apply theoretical knowledge or rely on data from another similar system to do this, or we can base our search on data that are available only for the particular system we are studying and try to reproduce these in our equations. As a result, we will get either process-based or empirical "black-box" models. They both have their strengths and weaknesses.

A brief introduction to Stella, a systems dynamics modeling package, is presented, with step-by-step instructions for model building using this formalism.

It is important to have a version of this or other similar software (Madonna, Simile, Vensim, or the like) and start practicing, since modeling is like playing a piano – it is hard to learn to do it only by reading books and listening to lectures. You have to get your hands dirty and do it yourself.

Keywords

Time, space, structure, conceptual models, resolution, Superfund, biological time, grids, black box models, empirical models, process-based models, Bonnini's paradox, systems dynamics, software, Stella, stocks and flows, exponential growth, limiting factors, Michaelis–Menten function.

* * *

There is no predefined prescription for how to build a good model. It is the model-building process itself that is most valuable for a better understanding of a system, for exploring the interactions between system components, and for identifying the possible effects of various forcing functions upon the system. Once the model has been built it is a useful tool to explain the system properties, and in some cases may lead to new findings about the system, but it is clearly the process of modeling that adds most to our knowledge about and understanding of the system.

Even though we do not know the ultimate model-making algorithm, we are aware of some key rules that are always useful to keep in mind when creating a model. By adhering to them, a great deal of frustration and various crises can be avoided. The list of such rules can be quite long, and varies slightly for every modeler and every modeled system. Therefore, as in art in modeling – experience is probably the most valuable asset, and there is no way to avoid all errors. We can only try to decrease their number.

2.1   Conceptual model

In most cases, the modeling process starts with a conceptual model. A conceptual model is a qualitative description of the system, and a good conceptual model is half the modeling effort. To create a conceptual model, we need to study the system and collect as much information as possible both about the system itself and about similar systems studied elsewhere. When creating a conceptual model, we start with the goal of the study and then try to explain the system that we have in terms that would match the goal. In designing the conceptual model, we decide what temporal, spatial and structural resolutions and ranges are needed for our study to reach the goal. Reciprocally, the conceptual model eventually becomes important to refine the goal of model development. In many cases the goal of the study is quite vague, and it is only after the conceptual model has been created and the available data sets evaluated that the goals of modeling can become clear. Modeling is an essentially *iterative* process. We cannot prescribe a sequence of steps that take us to the goal; it is an adaptive process where the target is repeatedly adjusted and moved as we go along, depending both on our modeling progress and on the external conditions that may be changing the scope of the study. It is like shooting at a moving target – we cannot make the target stop to take a good aim and then start the process; we need to learn to readjust, to refine our model as we go. Building a good conceptual model is an important step on this path.

Temporal domain

In the temporal domain, we first figure out the specific rates (resolution) of the main processes that we are to model and decide for how long (range) we want to observe the system. If we are looking at bacterial processes with microorganisms developing and changing the population size within hours, it is unlikely that we would want to track such a system for over a hundred years. On the other hand, if we are modeling a forest we can probably ignore the processes that are occurring within an hour, but we would want to watch this system for decades or even centuries.

If there is little change registered over the study period, the model may not need to be dynamic. It may be static and focus on other aspects of the system. For example, a photo can be a snapshot that captures the state of the system at a particular moment, or a series of snapshots can be averaged over a certain time interval. In a way every photo is like that, since it is never really instantaneous. Some time needs to pass between the moment the shutter opens and the moment it closes. A picture on a photo can be just a little bit blurred, representing the change in the system while the shutter was open, and we may not even notice it if the exposure was short enough. In some photos where the exposure was not set right this comes out quite

A photo as a snapshot may not be an instantaneous model. It may actually represent averages over time for certain system components.

clearly, and in most cases these photos end up in the trash bin, except when we actually wanted to see the trajectory of the object while it was moving and intentionally kept the shutter open for a while. That would be a static representation of a dynamic system, in a way showing its average state.

If temporal change is important, we need to identify how this change occurs. In reality, time is continuous. However, in some cases it may be useful to think of time as being discrete and to describe the system using event-based formalism. Or we can think of change in time as a sequence of snapshots. A series of snapshots creates a better representation of dynamics. That is how a movie is made, when by alternating many static images we create the feeling of moving objects.

If a dynamic approach is selected and warranted to answer the questions posed in the study, we should start thinking about the appropriate resolution of our temporal model. To have the movie smoothly rolling, we need to display at least 16 snapshots per second. In this model that we watch on the screen, the temporal resolution is then one-sixteenth of a second. This resolution is dictated by the goal, which is in this case to create a representation of reality that the human eye can see as continuous. In most TV systems we use even a finer resolution to further improve the image – 24 snapshots per second. Apparently cats do not like to watch TV programs because their eye has less inertia, and this rate is not fast enough to create a continuous image; they see the movie as a series of still images replacing each other. This is similar to our image of old movies, where we see everybody move in jerks and jumps, as we actually track the change of images. However, if the process we film is slow enough, it makes little sense to stare at a series of pictures that are virtually the same. Suppose, for example, we want to visualize the growth of a plant. At 24 snapshots per second there is little to see, so for this model we may want to look at one snapshot per hour. By displaying these snapshots one after another, we can "see" how the plant grows. On the other hand, this resolution would be quite inappropriate if we are looking at the flight of a bullet, or trying to see how a fly moves its wings. Here, we would probably need a resolution of hundreds of snapshots per second.

The model's goal tells us how often we need to update the model to match the expected temporal detail of the study: is once a year enough, as when we model forests, or do we need to track the dynamics every one-hundredth of a second, as when we model the movement of a wing of a fly? At some point we may also consider the data that we have. The goal may be quite ambitious – say, to represent the dynamics on an hourly timescale – but if we have only one annual value describing the state of the system, we may need to reconsider the goal and save the hassle of developing models of fine temporal resolution for this system.

An example of a conceptual diagram depicting the sequence of events in a landscape model is presented in Figure 2.1.

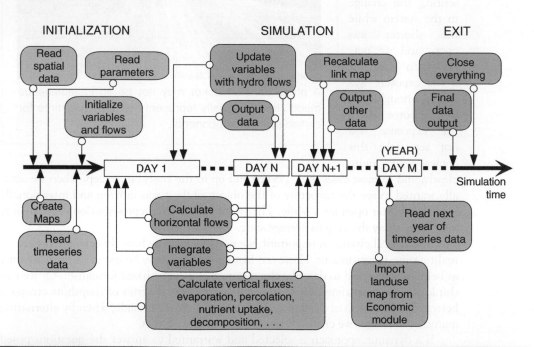

Figure 2.1   A conceptual diagram for time in a landscape model.

Note that while many processes actually occur at the same time, we need to sequence them when putting them in a model. Sequencing of processes is important and may produce quite different results depending upon which process is put in the model first and which next, etc.

Time can be different in different systems. We are very used to the time that we live in, and tend to think that this is the only time that really matters. Actually, each system evolves according to its own time counter, and humans represent only one such system with its own counter. Everything that happens within the timeframe of an average human lifespan seems to matter much more than what happens over other periods of time – whether longer or shorter.

The lifespan of most elementary particles is less than 1^{-6} s. This value is totally meaningless to us; it does not relate to any processes that we know and care about. At the other extreme, there are stars with lifespans of 10^9 or more years. Now, this number is so huge that we still cannot associate much with it and think about it in totally abstract terms. We seem to care less about systems that evolve in considerably different timescales. Large mammals live for 10 and more years, and we care a lot more about them, than, say, about insects, who live for months or less. Squashing a mosquito is no big deal to us, nor is stepping on an ant.

While some religions, such as Buddhism, are quite concerned about life in general and consider it a sin to kill even an insect, they do not and cannot care about the myriads of bacteria that live around us, the life cycles of which we purposefully or inadvertently interfere with. Just imagine how many bacteria you kill when you take antibiotics! Similarly, our concerns seem to fade when we start looking at longer timescales. Most people certainly care for their children, perhaps to a large extent because the children's lifespan overlaps considerably with our own. The more distant the generation, the less concerned we seem to be about them. How else would we explain our obsession with the idea of economic growth, which is almost always associated with further resource consumption at one end and pollution at the other? There was certainly a period in human history when economic growth created more possibilities for future generations: it fed the growing population of people, it helped to fight disease, and it has significantly decreased mortality. There used to be clear correlation between the size of the economy, measured by, say, the Gross National Product (GNP), and life expectancy or quality of life in general. But is that still the case, or are we now exceeding the carrying capacity and mostly borrowing from the future, from our children, grandchildren and future generations? Clearly, we are taking more resources than we are returning back to the pool. The global footprint goes way beyond the size of this planet, but since we are not used to thinking in multiple timescales, we do not seem to care about that. We are already reaping the consequences of similar thoughtlessness by previous generations, and investing millions and billions of dollars to clean up the mess that was left for us by our predecessors.

One such example is the Superfund – an environmental program established to take care of the abandoned hazardous waste sites. It is also the name of the fund established by the Comprehensive Environmental Response, Compensation and Liability Act of 1980 (CERCLA). This law was enacted in the wake of the discovery of toxic waste dumps such as Love Canal and Times Beach in the 1970s. The EPA uses this Act and the money to clean up such sites, and to compel responsible parties to perform clean-ups or reimburse the government for EPA-led clean-ups. At this time there are some 1,623 Superfund sites, and the number is steadily increasing. While the amount of money that goes into the fund varies from year to year (in 1995 the program received $1.43 billion in appropriations; 12 years later it received $1.25 billion), we are still talking about billions and billions of dollars. That is the legacy of previous generations. We are already a generation that suffers from the unwise environmental decisions of our predecessors.

Yet at the same time we continue to advocate economic growth and are tending to leave even more problems, and fewer resources for our ancestors to cope with them. Interestingly, in some other cultures, which we may want to call more "primitive" than Western civilizations, there used to be more interest in long-term effects. The Great Law of the Iroquois Confederacy states: "In our every deliberation we must consider the impact of our decisions on the next seven generations." Are there any modern societies where decisions are made based on models with such a temporal extent?

Actually, in some systems time evolves in quite a different way than we are used to. For certain processes in plants, for example, the number of elapsed minutes and days does not matter. Plants do not do something simply because it is 8 o'clock in the morning; what triggers their processes is the total number of fair days when temperature is higher than a certain threshold – say, 5°C. In such systems it makes more sense to think in terms of such time, also called biological time, to figure out when the plants start to grow, when seeds sprout, when leaves appear, etc. The time then is calculated as

$$T = \sum_i \max(0, t_i - 5)$$

where t_i is the temperature on day i. So a seed can stay dormant until there will be a series of warm days over which its biological time exceeds a certain value. If the spring is warm, then it may be only 5 or 6 days before it starts to sprout. During a cold spring it can take much longer for the necessary active temperature to accumulate. Temperature in this system becomes time. Time is measured as temperature. Clearly it is a totally different timescale, in which our normal time can accelerate or slow down depending on climatic conditions.

Spatial domain

In the spatial domain, we are to make similar decisions about the representation of space in the model. Is there enough spatial heterogeneity in the system to justify a spatially explicit description, or can the system be considered spatially homogeneous? Are there large segments in which the system parts are uniform, and can the whole system be described as a collection of these spatially homogeneous compartments? How big are these compartments? And how far do we need to look? Is it just this river reach that we want to model, or the whole river, or perhaps the whole watershed? Where do we draw the spatial boundaries of our system?

The picture in Figure 2.2 describes how a lake ecosystem can be modeled in space. It shows that there are large parts of the system that may be considered homogeneous and represented by some average values. The geometry of the lake is described; the central, deeper part is considered to be a separate compartment and is subdivided in depth. The shallow littoral segments are assumed to be entirely mixed to the bottom, and are therefore described by one layer. As in most water ecosystems, the spatial range is clearly defined by the lake boundaries.

In this compartmental approximation, we show how different parts of the ecosystem are connected in space, but we do not describe them spatially. Each compartment may be seen as a node, and the nodes are connected in a particular way. It is only their spatial arrangement that matters; we do not show their spatial configuration, as we would do it on a map. If we need that level of detail, immediately we encounter the next set of questions. What should be the resolution of our map? Can we describe the system spatially with a map of a 1 : 1,000,000 scale, or should it be 1 : 100? How do we turn continuous contours on the map into digitized data that the computer can handle?

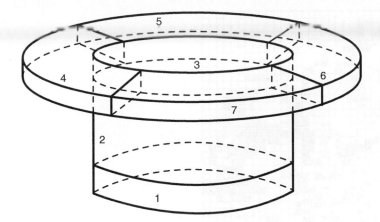

Figure 2.2   Conceptual model of a spatial representation of a lake ecosystem.
1 – hypolimnion, the deepest portion of the lake below the thermocline; 2 – metalimnion and 3 – epilimnion, the upper part where photosynthesis occurs; 4–7 – the littoral zones, the shallow waters, that also receive the waters from tributaries. Note that the shape of the segments does not really matter, since in any case they will look quite different in the real lake.

There is an interesting connection between spatial and temporal resolution that we can explore by looking at how photos used to be formed before the digital era. We have noted above that blurred images are a result of a moving object projecting over several pixels or grains on the film. Grain size refers to the size of the silver crystals in the emulsion. Therefore, with the same amount of light and length of exposure you will likely get a sharper image if the grain size is big enough that the object can travel in space while still projecting onto the same grain particle on the film. "Slow" films require much more exposure to produce a usable image than do "fast" films. Fast films have larger grains and are thus better suited to low-light situations and action shots (where the short exposure time limits the total light received). The slower films have finer grains and better color rendition than fast films. The smaller the crystals, the finer the detail in the photo and the slower the film – so the higher the spatial resolution, the longer it takes for the object to imprint on the film, the slower the object should move to avoid the blur. It looks as though there is a certain trade-off between temporal and spatial resolution in this kind of model. In this case, this is stipulated by chemical properties of the film. In other models, such as computer models, you will find similar arrangements, but then it will be the computer speed, for example, that will likely limit your ability to have high temporal and spatial resolution at the same time.

The resolution of my computer display is 1152×768 pixels. Each pixel is homogeneous, its color is the same across the whole pixel, and its shape is the same as all the others. However, with this resolution I can enjoy quite complex graphics that are almost indistinguishable from real-life pictures. The early VGA cards had 800×400 pixels. Small fonts and details of some pictures were blurred, they were almost

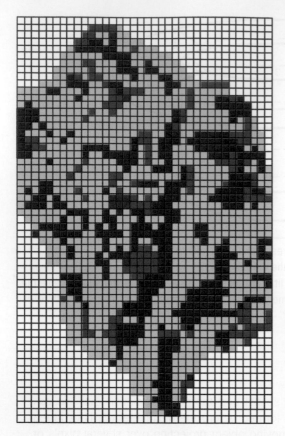

Figure 2.3   Uniform grid of equal square cells.
A watershed is represented by different landuse types. The grid simply mimics the raster information that comes from a landuse map. The cell is given the attribute of the landuse that covers the largest proportion of the area of a cell.

impossible to work with. Again, depending upon the system specifics and the goal of our model, we would want to use different spatial resolutions.

It is not just the size of the grain that is important; the form of the grain also matters. Should we use a grid of uniform, equal-sized square cells, as we would do on a rasterized map (Figure 2.3)? Or perhaps the cells should not be uniform, representing the actual configuration of the ecosystem? And where do we draw the boundary in this case – especially if there is an exchange of material across the boundary, as is the case at the outlet of a bay? And how small should the cells be (Figure 2.4)? Perhaps a triangulated grid would be better (Figure 2.5)? This certainly works better if we have non-uniform spatial complexity and need to describe certain spatial entities in more detail than others. Suppose we model a watershed. It makes sense to have finer resolution along the river to capture some of the effects of the riparian zone. On the other hand, vast stretches of forest or agricultural land may be presented as spatially homogeneous entities – there is no need to subdivide them into smaller areas. The boundaries may also need a higher resolution. A triangular grid serves these purposes really well. But then other considerations also come into play: What data do we have? How much complexity can we afford with the type of computer that we have at our disposal? What are the visualization tools that we have to make the most of the model results?

Maybe instead of triangles we prefer to use hexagons (Figure 2.6). These can do a better job describing dispersion, since they measure about the same distance from

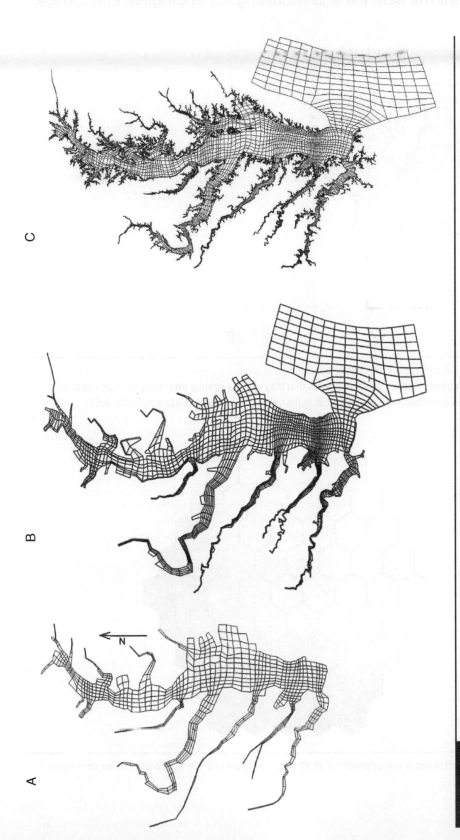

Figure 2.4 A non-uniform grid of quadrangles used to model the Chesapeake Bay. There is also the third, depth dimension (as in Figure 2.2), which is not visible in this picture.

A. The original grid of 4,000 cells. B. An advanced grid of 10,000 cells, which also expanded into the ocean to give a better description of the boundary conditions. C. The latest 14,000-cell grid, with finer resolution in tributaries. Ironically, changing from one grid to another did not make the model any more accurate, however it did help incorporate more processes.

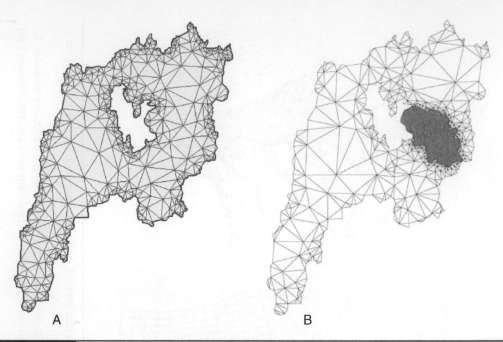

A   B

Figure 2.5   A triangular non-structured grid.
A.This grid allows higher definition along elements that may require special attention, such as rivers or boundaries. B. Subwatersheds may be defined at higher resolution if needed (Qu and Duffy, 2007).

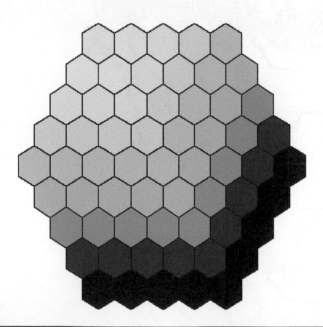

Figure 2.6   Hexagons are symmetrical in all diagonal flows and are convenient when describing dispersion.

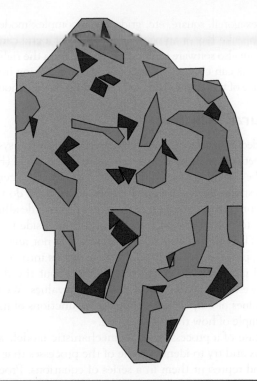

Figure 2.7   Polygons as spatial compartments.
The area is described by much smaller number of entities. Flows between compartments need special attention. In most cases they are connected with some other processes, like river flow, for example.

the center to the boundary and they are symmetrical in terms of diagonal flows. No matter in which of the six directions we go, the links with neighboring cells will be the same. This is not so in the case of square cells, if we want them to communicate with eight surrounding cells, assuming diagonal flows.

Perhaps polygons could be used, as in the case of vector-based Geographic Information Systems (GIS) such as ArcINFO (Figure 2.7). Here the space is described by polygons, which are presented in terms of vectors of coordinates for all vertices of the polygons. Converting regular continuous geographic maps into vector-based digital sets is usually performed in a tedious process of "digitizing," using special equipment that registers the coordinates of various points chosen along the boundary. The more points you choose to describe the polygon that will approximate the area digitized, the higher the precision of the digital image and the closer it is to the original.

Polygons are good for map and image processing, since they create a digital image that is more accurate with far less information to store. To achieve the same accuracy with raster maps, we would need many more cells and therefore much larger data sets. However, for purposes of modeling, polygons are quite hard to handle if you want to streamline processing. Each polygon is unique, and needs to be specially defined. If something changes – say, land is converted from one landuse type to another – then the model may need to be reinitialized. Triangular grids seem to present a good compromise, offering greater flexibility: the size of each triangle might vary, yet it is still a triangle, with three boundaries and three vertices, and each can be handled in a similar way in the model.

Choosing the right spatial representation and designing a good spatial grid is a craft in its own right. As you have seen, there are uniform and non-uniform grids, triangular, hexagonal, square, etc. grids. For a complex model of a large spatial object, say the Chesapeake Bay or an ocean, the design of a grid can take many months if not years. There are also software tools that help to design the right grid. Model performance and even results can change substantially when switching from one grid to another, so the importance of this step in the model-building process should not be underestimated.

Structural domain

Finally, we decide how to represent the structure of the system. One important distinction is between empirical and process-based models (Figure 2.8). An empirical model may be considered as a "black box," which takes certain inputs and produces outputs in response to the inputs. Perhaps because we do not know, or do not care, or cannot afford greater computing resources or laxer deadlines, we make a deliberate decision not to consider what happens and how inside the black box that presents the system. The internal structure in this case is not analyzed, and our only goal is to find the appropriate function to translate inputs into outputs. This is usually done by statistical methods. We have information about the data sets that describe the inputs, and we have the data regarding output values. We then try to represent the numerical values of outputs as mathematical functions of inputs. Below, we will consider an example of how this can be done.

In the case of a process-based or mechanistic model, we attempt to look inside the black box and try to identify some of the processes that occur in the system, analyze them and represent them in a series of equations. Process-based models employ the additional information about the system that we may have from previous studies of analogous systems, or about the individual processes that we are looking at. They may use certain theoretical knowledge coming from a variety of disciplines. In this sense, a process-based model may be even more useful than the information available about the system studied.

It should be noted, though, that all process-based models are still empirical, in a sense. We can never describe all the details of all the processes in a system. It is

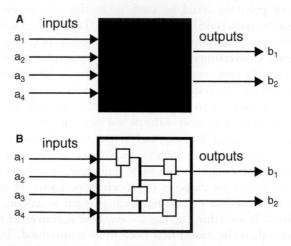

Figure 2.8   A. A black-box model, where the output is calculated as a function of the inputs: $b_i = f_i(a_1, a_2, a_3, a_4)$, without looking at what is happening inside the system. B. A white-box model, where the structure of the system is analyzed and represented in the model.

just that we go into further depth in the system, providing more detail about the processes in it. Yet we still end up with certain black boxes, which we do not wish to or cannot consider in any more detail. If that were not the case, we would hardly be accomplishing the major goal of any modeling effort, which is a simplification of the system description. We would be ending up with models as complex as the original systems, and therefore delivering little value for purposes of synthesis.

If we choose to build a process-based model, we may start describing the structure by using a diagram, representing the major components of the system: variables, forcing functions and control functions.

When deciding on the model structure, it is important to match the structural complexity with the goals of the study, the available data, and the appropriate temporal and spatial resolution. For example, if we are modeling fish populations (Figure 2.9), which grow over several years, there is little use in considering the dynamics of bacterial processes, which have a specific rate of hours. In this case we may consider the bacterial population to be in equilibrium, quickly adapting to any changes occurring in the system in "fish time", which is weeks or months. We may still want to consider the bacterial biomass for mass balance purposes, but in this case it makes perfect sense to aggregate it with the detrital biomass.

However, certain fast processes may have a detrimental effect upon the system. For example, it is well known that fish kills may occur during night-time and in the early morning hours, when there is still no photosynthesis, but only respiration from algae in the system. As a result, the oxygen content may fall below certain threshold levels. The oxygen concentrations in this case vary from hour to hour, whereas fish

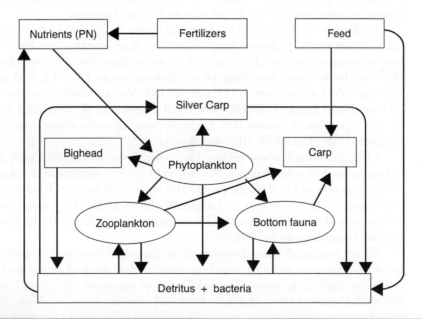

Figure 2.9   Conceptual model of a fishpond.

This model is not very detailed. It represents the chosen state variables and some of the forcing functions (fertilizers, feed). It is not clear what the other forcing functions involved are, such as environmental, climatic conditions. There is also no indication of the spatial and temporal scale. Apparently this information is contained in the narrative about the model that usually goes together with the diagram.

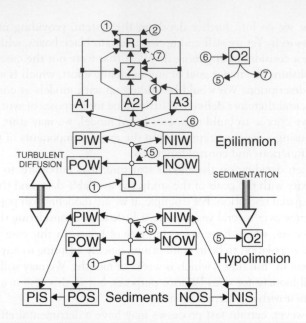

Figure 2.10   Conceptual model of a lake ecosystem structure.
The model structure is different for the different spatial segments used in the model.

biomass changes much more slowly. We might want to consider oxygen as part of the system, to make sure that we do not miss such critical regimes.

In the lake ecosystem model shown in Figure 2.10, in addition to trophic relations certain spatial properties are present. The diagram shows how the model structure is presented in the three vertical segments that describe the pelagic part of the lake. In the upper part three phytoplankton groups (A1, A2, A3) are present; they are food for zooplankton (Z) and fish (R). Various forms of nutrients (organic and inorganic nitrogen (NOW, NIW) and phosphorus (POW, PIW) are supplied by decomposition of detritus (D). In the bottom segments, there are no biota, only nutrients (PIS, POS, NOS, NIS) and detritus.

Conceptual models may present more than flows of material. Figure 2.11 shows a diagram used in a simple model developed to analyze sustainable development in a socio-economic and ecological system. The model will be considered in more detail in Chapter 7. Here, note that, in addition to the variables, the diagram also contains information about the processes and their causes. It describes both the flows of material and information in the system.

When making all these decisions about the model structure, its spatial and temporal resolution, we should always keep in mind that the goal of any modeling exercise is to simplify the system, to seek the most important drivers and processes. If the model becomes too complex to grasp and to study, its utility drops. There is little advantage in substituting one complex system that we do not understand with another complex system that we also do not understand. Even if the model is simpler than the original system, it is quite useless if it is still too complex to shed new light on the system and to add to the understanding of it. Even if you can perform experiments on this model that you might not be able to do in the real world, is there much value in that if you cannot explain your results, figure out the causes, and have any trust in what you are producing?

Bonnini's paradox

I thought that this was an original way to phrase this: the danger and low utility of substituting one complex system with another. But then I learned that something similar has been already described by Dutton and Starbuck in 1971 as the "Bonnini's paradox."

> *A model is built in order to achieve understanding of an observed causal process, and the model is stated as being a simulation program in order that the assumptions and functional relations may be as complex and realistic as possible. The resulting program produces outputs resembling those observed in the real world, and inspires confidence that the real causal process has been accurately represented. However, because the assumptions incorporated in the model are complex and their mutual interdependencies are obscure, the simulation program is no easier to understand than the real process was.*
>
> Starbuck and Dutton, 1971

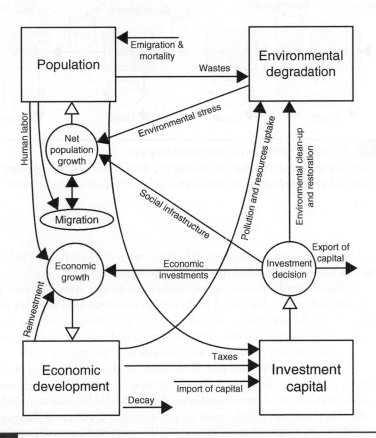

Figure 2.11   Conceptual model of a socio-economic and ecological system designed to analyze sustainable development. See Chapter 7 for more details about this model.

Forrester put some formalism into the diagrams, using a series of simple symbols or icons associated with various processes. The two main ones were *level* and *rate* (Figure 2.12). Levels were used to identify stores of material or energy, while rates were controlling the flows between them. Using this formalism, Forrester created complex

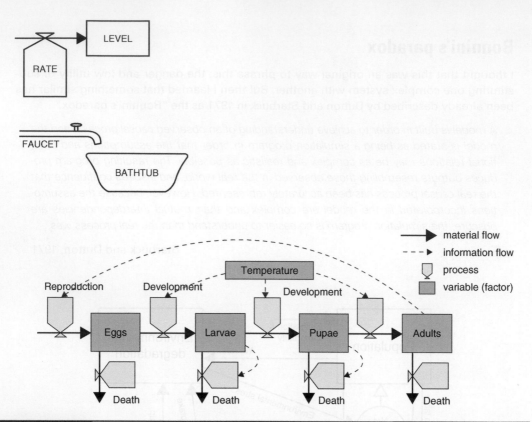

Figure 2.12   Forrester's formalism for conceptual diagrams.
The rate and the level are two main icons that can be used to put together more complex diagrams such as the one for the insect population (from: http://www.ento.vt.edu/~sharov/PopEcol/lec1/struct.html).

models for such systems as cities, industries and even the whole world. Similar formalism was later used in several modeling software packages.

Odum created another set of symbols to model systems based on the energy flows through them. He called them energy diagrams, and used six main icons (see Figure 2.13). All systems are described in terms of energy, assuming that for all variables and processes we can calculate the "embodied" energy. In this case, energy works as a general currency to measure all processes and "things."

In many software packages (like some of those considered in the next paragraph), conceptual diagrams are used to input the model. For example, one of the reasons that systems dynamics software such as Stella became so popular in modeling is that they are also handy tools to put together conceptual diagrams, and, moreover, these diagrams are then automatically converted into numeric computer models. Figure 2.14 presents a sample conceptual model for a river system put together on the Stella interface. It describes the river network as a combination of subwatersheds, river reaches and reservoirs. The Stella interface can be used as a drawing board to put together various conceptual diagrams and discuss them with other people in a process known as participatory modeling. In this case, the major value of the interface is that it is possible to easily add or delete variables and processes and immediately

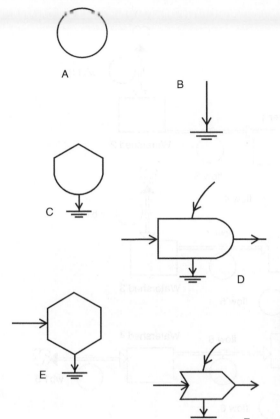

Figure 2.13   Odum's formalism for energy-based conceptual diagrams.
A. Source of energy, B. Sink (loss of energy from system), C. Storage tank, D. Production unit (takes in energy and information to create other quality of energy), E. Consumption unit, F. Energy mixer or work gate.

see the impact on model performance. The model itself becomes a tool for deliberation and consensus building.

Very similar diagrams can be put together using other systems dynamics software, such as Madonna, Vensim, Powersim or Simile. In these software packages, "stock-and-flow" formalism is used to describe the system. The diagrams are also known as flow diagrams, because they represent how material flows through the system.

As we will see below, a somewhat different formalism is used in such packages as GoldSim, Simulink and Extend. Here we have more flexibility in describing what we wish to do in the model, and the model does not present only stocks and flows. Groups of processes can be defined as submodels and encapsulated into special icons that become part of the icon set used to put together the diagrams. As usual, we get more functionality and versatility at the expense of a steeper learning curve and higher complexity of design.

Yet another option in building conceptual diagrams is provided by the Universal Modeling Language (UML), which is a standardized specification language for object modeling. It is designed as a diagrammatic tool that can be used to build models as diagrams, which can be then automatically converted into a number of object-oriented languages, such as Java, C++, Python, etc. In this case you are actually

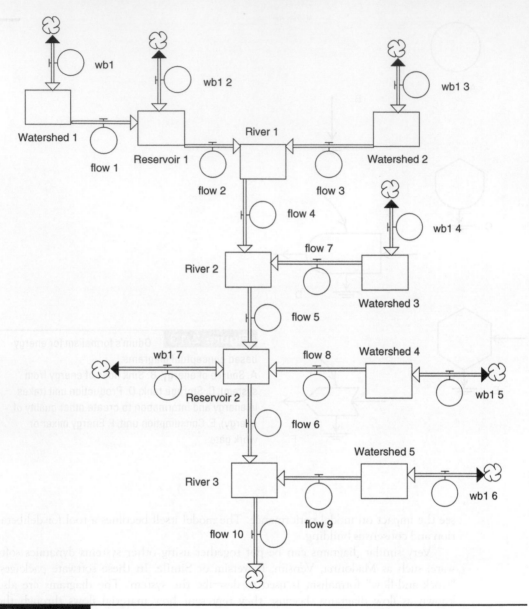

almost writing computer code when developing the conceptual model. Once again,
even more universality and almost infinite flexibility is achieved at the price of yet
greater effort spent in mastering the tool. Figure 2.15 presents a sample conceptual
diagram created in UML to formulate an agent-based model of a landscape used by
sheep farmers, foresters and National Park rangers, who are interacting on very different temporal and spatial scales with different development objectives (sheep production, timber production and nature conservation, respectively).

There are several types of diagrams that can be created using UML. One of them
is the activity diagram, which describes the temporal dimension of your model. The

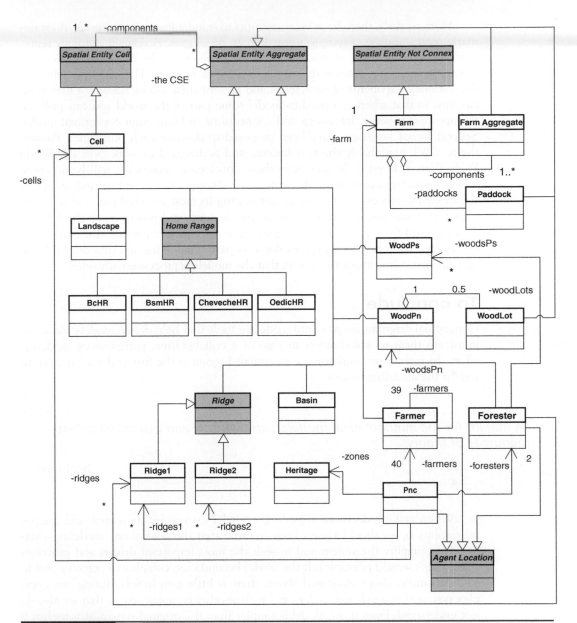

Figure 2.15   A UML class diagram of a system can be used both as a conceptual diagram and as a way to program the model (from: http://jasss.soc.surrey.ac.uk/6/2/2.html, reproduced with kind permission of the Journal of Artificial Societies and Social Simulation, Centre for Research on Social Simulation, Surrey).

class diagram presented in Figure 2.15 in a way corresponds to the structural dimension, but also has elements of the spatial representation such as that displayed in the lake model in Figure 2.10. Most software tools designed to create UML diagrams, such as Visual Paradigm (http://www.visual-paradigm.com/product/vpuml/), also provide code generators that will convert your UML diagram into computer code in a language of your choice.

More recently, there have been attempts to standardize the conceptual, diagrammatic representation of systems using domain ontologies. A domain ontology represents a certain domain, ecosystem or part of an ecosystem by defining the meaning of various terms, or names as they apply to those ecosystems. The idea is to define all the various components of ecosystems and present their interactions in a hierarchical way, so that when you need to model some part of the world you can pull out the appropriate set of definitions and connections and have your conceptual model. Several formal languages have been proposed to describe such ontologies. Among them, OWL is probably the best known, and is designed to work over the World Wide Web. It is yet to be seen how these ontological approaches will be accepted by the modeling community. As with other attempts to streamline and automate the modeling process, we may be compromising its most essential part – that is, the exploration and research of the system, its elements and processes, at the level of detail needed for a particular study goal. Any attempt to automate this part of the modeling process may forfeit the exploratory part of modeling and thus diminish the new understanding about the system that the modeling process usually offers.

To conclude...

Conceptual diagrams are powerful modeling tools that help design models and communicate them to stakeholders in case of a collaborative, participatory modeling effort. In most cases, building a conceptual diagram is the first and very important step in the modeling process.

"A maxim for the mathematical modeler: start simply and use to the fullest resources of theory."

Berlinski

When making decisions regarding a model's structure, its spatial and temporal resolution, we should always keep in mind that the goal of any modeling exercise is to simplify the system and to seek the most important drivers and processes (Descartes's second principle). If the model becomes too complex to perceive and to study, its utility drops. As stated above, there is little gain in substituting one complex system that we do not understand with another complex system that we also do not understand. Even if the model is simpler than the original system, it is useless if it is still too complex to shed new light and to add to the understanding of the system. So our first rule is:

KEEP IT SIMPLE

It is better to start with a simplified version, even if you know it is unrealistic, and then add components to it. It helps a lot when you have a model that always runs and the performance of which you understand. This is much better than putting together a model that has everything in it to satisfy the most general goals and requirements. Complex models are hard to handle, they tend to go out of control, they behave counter-intuitively and produce unreliable and uncertain results. At every step of model development you should try to have a running and tested version of the model, and you can then build more into it. You will then always know at what point the model fails and no longer produces something reasonable, and thus what kind of recent changes have caused the problem. Our second rule is:

KEEP IT RUNNING. KEEP TESTING IT

Everything you know about the system is good for the model. The more you know about the system, the better the model. However, that does not mean that all the available data and information from previous or similar studies have to end up as part of the model. Modeling and data collection are iterative processes; one drives another. You never know which data at what stage of the modeling study will be required, and how these will modify your interpretation and understanding of the system. At the same time, one of the most important values of the modeling effort is that it brings together all the available information about the system in an organized and structured format. The model then checks that these data are full and consistent. Even if the model turns out to be a failure and does not produce any reliable predictions and conclusions, by bringing the data together new understanding is created and important gaps in our knowledge may be identified. So the third rule is:

THE DATA DRIVE THE MODEL. THE MODEL DRIVES THE DATA

No matter whether the goal of the model is reached or the model fails to produce the expected results, the modeling effort is always useful. When building a model, a great deal of information is brought together, new understanding is created, and new networks and collaborations between researchers, experimenters, stakeholders, and decision-makers are emerging. This clearly brings a study to a new level. We therefore conclude that:

THE MODELING PROCESS MAY BE MORE IMPORTANT THAN THE PRODUCT

2.2 Modeling software

There is a lot of software currently available that can help to build and run models. Between the qualitative conceptual model and the computer code, we could place a variety of software tools that can help to convert conceptual ideas into a running model. Usually there is a trade-off between universality and user-friendliness. At one extreme we see computer languages that can be used to translate any concepts and any knowledge into working computer code, while at the other we find realizations of particular models that are good only for the particular systems and conditions that they were designed for. In between, there is a variety of more or less universal tools (Figure 2.16).

We can distinguish between *modeling languages*, which are computer languages designed specifically for model development, and *extendable modeling systems*, which are modeling packages that allow specific code to be added by the user if the existing methods are not sufficient for their purposes. In contrast, there are also *modeling systems*, which are completely prepackaged and do not allow any additions to the methods provided. There is a remarkable gap between closed and extendable systems in terms of their user-friendliness. The less power the user has to modify the system, the fancier the graphic user interface is and the easier it is to learn the system. From modeling systems we go to *extendable models*, which are actually individual models that can be adjusted

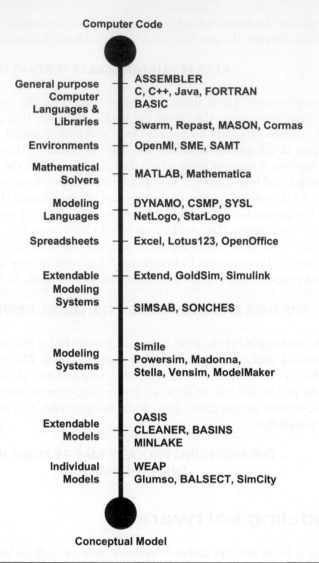

Computer Code

General purpose — ASSEMBLER
Computer   C, C++, Java, FORTRAN
Languages & BASIC
Libraries

— Swarm, Repast, MASON, Cormas

Environments — OpenMI, SME, SAMT

Mathematical — MATLAB, Mathematica
Solvers

Modeling — DYNAMO, CSMP, SYSL
Languages   NetLogo, StarLogo

Spreadsheets — Excel, Lotus123, OpenOffice

Extendable — Extend, GoldSim, Simulink
Modeling
Systems   — SIMSAB, SONCHES

Modeling — Simile
Systems   Powersim, Madonna,
           Stella, Vensim, ModelMaker

Extendable — OASIS
Models   CLEANER, BASINS
           MINLAKE

Individual — WEAP
Models   Glumso, BALSECT, SimCity

Conceptual Model

Figure 2.16   Hierarchy of modeling software.

---

for different locations and case studies. In these, the model structure is much less flexible, the user can make choices from a limited list of options, and it is usually just the parameters and some spatial and temporal characteristics that can be changed.

Models

Any model we run on a computer comes as a piece of software. Therefore, in some cases, to solve a particular modeling task we may try to find an appropriate model that has been developed previously for a similar case, and see if this software package, if available, can be adapted to the needs of your project. This can save you time and money; another benefit is that the model may have already been calibrated in a variety of locations and circumstances, and thus be more easily accepted by a group of stakeholders. Some models are distributed for a price, while others are available free of charge. The Register of Ecological Models (REM – http://eco.wiz.uni-kassel. de/ecobas.html) is a meta-database for models in ecology. It can be a good starting point if you are looking for a particular model. In some cases you will be able to download the executables from the website, in others you will have to contact the authors. For the vast majority of models the source code is unlikely to be available, and we can never be sure what actually goes on inside the processor. We can only look at the output and the documentation, run scenarios and analyze trends, but ultimately we have to trust the model developers that the model is programmed properly. We also can make no changes to this kind of model.

The fact that models come as software black boxes may be one of the reasons that model re-use is not very common. It may take a long time to learn and understand an off-the-shelf model, and it can be quite frustrating if, after this investment of time and effort, we find that the model is not quite applicable to our case. It certainly helps when models are well documented, have good user guides and tutorials, and come with nice graphic user interfaces (GUI). Most of the models that are commercially distributed have very slick GUIs that help set up these tools for particular applications. For example, the WEAP (Water Evaluation and Planning system – http://www.weap21. org/index.asp) is a user-friendly software tool that helps with an integrated approach to water resources planning. The core of the model is a water balance model that calculates the dynamics of supply and demand in a river system. To set up the model the user is guided through a series of screens, which start with a river schematic that can be arranged on top of an ArcView map, and then takes care of data input with a series of dialogue boxes for water use, loss and re-use, demand management, priorities, etc. The results are then displayed in the same GUI in charts and tables, and on the schematic of the river system. Scenarios that describe different demand and supply measures are driving the system, and are connected with the various results.

These user interfaces certainly help with using the models; however, extending the model capabilities is not a straightforward task, if it is possible at all. In particular when models are not open source, it is usually an "all or nothing" deal – you either use the model as it is, or drop it entirely if it does not have some of the features needed for your study.

Some models are deliberately designed as games, with special emphasis put on the graphic interface and ease of use. One good example is the *SimCity* computer game, which has a sophisticated socio-economic and ecological model at its core, but no one other than the model developers has ever seen this model and users do not know whether the model was calibrated or validated. The purpose here is to enhance the interactive utility of the program, to maximize its user-friendliness and simplify the learning process.

Extendable models

Some models and modeling systems are designed in such a way that they allow additions to their structure. For example, OASIS (Operational Analysis and Simulation of Integrated Systems) is a software package designed to model river, reservoir and hydropower systems to develop operating policies and optimize water use. OASIS has a graphical user interface that allows easy configuration of the system. You can describe how the river system looks, locate the inputs and withdrawals, and enter historical data sets that the system is to work with. In addition, there is an Operation Control Language (OCL) – a special language used to enter rules and constraints that are specific for your case study. OCL also acts as a bridge from OASIS to other computer programs. Users can express all operating rules as operating goals or operating constraints, and can account for both human control and physical constraints on the system. This takes care of all sorts of "if–then" operations, which can go beyond just operational rules. To model any system, the problem must simply be approached as a set of goals and constraints. The software then works out the best means of moving water through the system to meet these goals and constraints. OCL allows data to be sent and received between OASIS and other programs while the programs are running, and each program can then react to the information provided by the other. Thus you are dealing with a prefabricated "closed" system, yet have some flexibility to modify it to the particular needs of a study. There is clearly more flexibility than in case of a pre-packaged model; however, the user is still operating within the set of assumptions and formalizations embedded in the model core of the software. There are also limitations to what OCL can handle as extensions to the OASIS system.

Modeling systems

Unlike pre-fabricated models, which are after all developed for specific systems, there are also generic software tools that can help to build models of any systems. These are probably most interesting to consider when a new modeling task is in order. However, the more versatile and powerful the system gets, the harder it becomes to master it and the more inclined modelers will be to stick to what they already know how to use – the well-known "hammer-and-nail" paradox, which we will revisit in Chapter 9.

Here, we will give a brief overview of some software tools that are available for modeling, along with some recommendations about their applicability. It should be noted that there is a great deal of development in progress, and new features are being added to the software packages quite rapidly, so it is always recommended that you check out the latest developments on the respective web pages. Note that neither of these tools implies any kind of core model; they can be used to put together any models. However, each one assumes a particular modeling paradigm and therefore has certain limitations.

Systems dynamics tools

Most of these appeared as an outgrowth of the systems dynamics approach of Jay Forrester and his DYNAMO language. Stella was inspired by Forrester's formalism, and quickly gained worldwide recognition. In the following years a number of other software packages have appeared that are better than Stella in many aspects, and are certainly worth investigating and comparing prior to any purchase decisions.

STELLA – isee systems (formerly HPS), http://www.iseesystems.com/ – Free Player and 1-month trial version – Mac/Win

Most used in academia, and has much legacy code developed. Over the past decade has been heavily prioritizing the User Interface features with nice capabilities to create game-like models, where the modeling part can be hidden from the user and only the front-end, which is similar to a Flight Simulator dashboard, is provided. Recent addition of isee.NET Framework promises more integration with other tools, but is not extensively used and tested yet.

VENSIM – Ventana Systems, http://www.vensim.com/ – Free Vensim PLE (personal learning edition) – Mac/Win

Same basic features for stock-and-flow modeling as Stella, with recent addition of some important functionality, such as calibration (will automatically adjust parameters to get the best match between model behavior and the data), optimization (efficient Powell hill-climbing algorithm searches through the parameter space looking for the largest cumulative pay-off), Kalman filter, Monte Carlo analysis, Causal Tracing (a tree diagram shows a selected variable and the variables that "cause" it to change), etc. Vensim DLL is a way to communicate with other applications such as Visual Basic, C, C++, Excel, multimedia authoring tools, etc. The DLL allows access to a Vensim model from custom-built applications; it can send data to Vensim, simulate a model, make changes to model parameters, and collect the simulation data for display.

POWERSIM – Powersim, http://www.powersim.com/ – Free Player and trial version, Win

This modeling tool has mostly been catering for the business community. Communicates with MS Excel. Powersim Solver is a companion product that handles calibration, optimization, risk analysis and risk management.

MADONNA – UC Berkeley, http://www.berkeleymadonna.com/ – Free Run Time version, Win/Mac

Runs many times faster than Stella. Will do parameter calibration (curve fitting), and optimization. Has several more numeric methods to solve ordinary differential equations. Stella compatible: will take Stella equations almost as is and work with them.

MODELMAKER4 – Exeter Software (formerly Cherwell), http://www.exetersoftware.com/cat/modelmaker.html – No free versions, Win

Same as the others in this category, plus quite extensive optimization and numeric methods, including Marquardt or Simplex methods, simulated annealing and grid search methods of initial parameter estimation; ordinary, weighted, and extended least squares methods of error scaling; comprehensive statistical reporting; Monte Carlo global sensitivity with 14 distribution types, 5 different integration methods – Runge-Kutta, Mid-Point, Euler, Bulirsch-Stoer and Gear. Gear's is an appropriate solver for stiff simulations where processes happen on very different timescales.

SIMILE – Simulistics (formerly open-source AME, Agroforestry Modelling Environment), http://www.simulistics.com/ – Free Evaluation Edition, Mac/Win/Linux

Allows object-based representation that handles disaggregation and individual-based modeling, auto-generates C++ model code, plug-and-play modules. Supports modular modeling: any part of a model can be extracted and used separately. Has plug-in displays, allowing field-specific graphics. Also has options for spatial models with some basic links to GIS.

The basic mathematical formalism and the interface conventions used in all these packages are quite similar, so once you have mastered one of them it should be quite easy to switch to another if you are looking for certain special features.

Pros: The development of all this modeling software has certainly simplified the process of building models, to the extent that programming is no longer needed to put together models, and only very basic numeric and mathematical skills are required. Systems dynamics has become widely used in a variety of applications.

Cons: There is also a reverse side to it. Most of the software developers advertising their products will tell you that building a model is now as simple as clicking your mouse. Unfortunately this is still not quite so, and can hardly ever be so, since modeling is primarily a research process that requires knowledge and understanding of the system to generate more knowledge and more understanding. By simply putting together diagrams and pretending that now you can run a model of your system, you may generate false knowledge and illusions. The modeling systems are indeed very helpful if you know how to build models; otherwise, they can become deceptive distractions.

Systems diagrams

An outgrowth of the systems dynamics approach is what we called systems diagrams tools. The software discussed here has many more icons than the stocks, flows and parameters that the systems dynamics tool operates with. Whole submodels or solvers for mathematical equations, say, partial differential equations, may be embedded into specially designed icons that later on become part of the toolbox for future applications. Once again we get more functionality and flexibility, but certainly at the expense of a much steeper learning curve.

EXTEND – Imagine-That, http://www.imaginethatinc.com/index.html – Free Demo, Win/Mac

As follows from the name of the product, the system is extendable. It encourages modularity, providing the functionality to encapsulate certain processes and subsystems into blocks that can be further reused. Extend models are constructed with library-based iconic blocks. Each block describes a calculation or a step in a process. Interprocess communication allows two applications to communicate and share data with one another. This feature allows the integration of external data and applications into and out of Extend models. Information is automatically updated between Extend and Excel, can be connected with databases (Open DataBase Connectivity), has embedded ActiveX or OLE (Object Linking and Embedding), and works with DLL (Dynamic-Link Library). Block-building is based on ModL – a language that provides high-level functions and features while having a familiar look and feel for users with experience programming in C. Also allows scripting to develop "wizards" or self-modifying models. Evolutionary Optimizer employs powerful enhanced evolutionary algorithms to determine the best model configuration.

GOLDSIM – GoldSim Technology Group, http://www.goldsim.com – Free Evaluation and Student version, Win

Uses the same approach based on an extendable library of icons ("hierarchical containers") for a variety of processes. The user controls the sequence of events, and can superimpose the occurrence and consequences of discrete events onto continuously varying systems. Other features include: particularly strong stochastic, Monte Carlo simulation component to treat uncertainty and risks inherent in all complex systems; embedded optimization, sensitivity analyses (e.g. tornado charts, statistical measures); external dynamic links to programs and spreadsheets; and direct exchange of data with ODBC-compliant databases. Models can be saved as player files. There are several extension modules (e.g. for Contaminant Transport using solvers for PDE, financial analysis, etc.).

SIMULINK – The Mathworks, http://www.mathworks.com/products/simulink/index. html – Free trial and web demo, Win/Mac/UNIX

Built on top of MATLAB (see below). Provides an interactive graphical environment and a customizable set of block libraries, which can be extended for specialized applications. More power, but harder to master. Can generate C code for your models, which can be further embedded into other applications. Based on the same concept of expandable libraries of predefined blocks, with an interactive graphical editor for assembling and managing block diagrams, with functionality to interface with other simulation programs and incorporate hand-written code, including MATLAB algorithms. Has full access to MATLAB for analyzing and visualizing data, developing graphical user interfaces, and creating model data and parameters.

Pros: Power, versatility, flexibility, expandability.

Cons: In a way the pros become their cons, since after investing much time to fully master these systems it is most likely that they will become your "hammer" for the future. Besides, when becoming wedded to proprietary software there is always a risk of running into limitations that will be hard to overcome.

Modeling languages, libraries and environments

Compilations of model languages, libraries appropriate to specific applications, and software environments are even more general, rely less on some embedded assumptions about the model structure and the logic of computations, but require more programming efforts.

Spreadsheets

The well-known spreadsheets are probably the most widely known software applications that can also help build quite sophisticated models. **Microsoft Excel** is by far the best-known and widely used spreadsheet. However, there is also **Lotus 123**, which actually pioneered the spreadsheet concept and is now owned by IBM, or the open-source **Open-Office** suite. Both offer very similar functionality. The other option is to use Google spreadsheets, which are found on the web and can be shared among several developers, who can then access and update the document from anywhere around the world using just an Internet browser.

The basic functionality that comes with spreadsheets is that formulas can be programmed using some very simple conventions. For dynamic models these formulas can be reiterated, using a TIME column, and using the results of previous calculations (rows) to generate the values for the next time step.

Pros: These tools are free or almost free, since they come as part of Microsoft Office, which is more or less standard these days, or can be downloaded as part of Open-Office, or can be used over the Internet with Google. Another advantage is that many users already know how to use them.

Cons: Spreadsheets can quickly get very cumbersome as model complexity increases, especially if you are trying to add dynamics to it. There is no good GUI for modeling, so models may be hard to present and visualize. Only the simplest numeric methods can feasibly be implemented (say, Euler for ODE).

Mathematical solvers

There are several specialized mathematical packages designed to help solve mathematical problems. As such they can be useful for modeling, since, after all, models are mathematical entities which need to be solved. These packages are not very helpful in formulating models. In this regard they are as universal as spreadsheets, but unlike spreadsheets, which are quite well known and intuitive to use, the mathematical packages have a steep learning curve and require learning specialized programming languages. On the benefit side, the computing power and versatility of mathematical methods is unsurpassed.

MATLAB –The MathWorks, http://www.mathworks.com/products/matlab/ – Free trial version, Mac/Win/Unix

This is a high-level technical computing language and interactive environment for algorithm development, data visualization, data analysis and numeric computation. It is faster to master MATLAB than C or Fortran, but it certainly requires a major investment of time. Includes mathematical functions for linear algebra, statistics, Fourier analysis, filtering, optimization (including genetic algorithms), and numerical integration; 2D and 3D graphics functions for visualizing data; tools for building custom graphical user interfaces; and functions for integrating with external applications and languages, such as C, C++, Fortran, Java, COM, and Microsoft Excel. May be a great tool to analyze models, but offers little help in conceptualizing and building them. There are sister products, such as Simulink (see above) or Simscape that are designed to handle the modeling process.

MATHEMATICA –Wolfram Research, http://www.wolfram.com/products/mathematica/index.html – Free web seminars and demos, Mac/Win/Linux/Unix

The software integrates numeric and, importantly, symbolic computations. It provides automation in algorithmic computation, interactive document capabilities, powerful connectivity, and rich graphical interfaces in 2D and 3D. It is based on its own advanced programming language, and it needs time and effort to master this. Has no specific tools to support modeling *per se*, but can be very useful to solve, run and analyze already built models. Can be very useful to study individual functions that are used in your model – for example, to test how parameters impact the functional response (see, for example, http://www.wolfram.com/products/mathematica/newin6/content/DynamicInteractivity/FindSampleCodeInTheWolframDemonstrationsProject.html).

Pros: Mathematical power that is hard to match.

Cons: Steep learning curve, requires a solid mathematical background.

Environments

Up to 80 percent of a modeling code may support various input/output functionality and interfaces with data and other programs. It makes perfect sense to build software packages that would take care of these data-sharing and communication procedures, so that modelers can focus on the actual formalization of processes and systems. There are numerous modeling environments developed to support modeling and to increase model functionality.

OpenMI – OpenMI Association, http://www.openmi.org/openminew/ – Open source, platform-independent

OpenMI stands for Open Modeling Interface and Environment, a standard for model linkage in the water domain. OpenMI avoids the need to abandon or rewrite existing applications. Making a new component OpenMI-compliant simplifies the process of integrating it with many other systems. It provides a method to link models, both legacy code and new ones. OpenMI standardizes the way data transfer is specified and executed; it allows any model to talk to any other model (e.g. from a different developer) without the need for cooperation between model developers or close communication between integrators and model developers. Based on Java and .NET technology, currently OpenMI has some 20+ compliant models in its library.

SME – UVM, http://www.uvm.edu/giee/IDEAS/lmf.html – Open source, Mac/Linux/Unix

The Spatial Modeling Environment (SME) links Stella with advanced computing resources. It allows modelers to develop simulations in the Stella user-friendly, graphical interface, and then take equations from several Stella models and automatically generate C++ code to construct modular spatial simulations and enable distributed processing over a network of parallel and serial computers. It can work with several GIS formats, and also provides a Java viewserver to present results of spatial simulations in a variety of graphic formats.

SAMT – ZALF, http://www.zalf.de/home_samt-lsa/ – Open source, Linux

Spatial Analysis Modeling Tool (SAMT) is a modeling system with some GIS features, designed to help with spatial analysis. It is an open system that links to different models (especially fuzzy-models, neural networks, etc.). It can also link to a general-purpose modeling language DESIRE.

Pros: Added functionality to other models and modeling tools.

Cons: Hardly any, since modeling environments mostly serve other modeling paradigms rather than imposing any of their own upon the user. In most cases, it is the next level of modeling that may require quite good modeling and systemic skills. Usually, user and developer groups are quite limited and are very much driven by enthusiasm. Therefore, future development and support may be quite uncertain.

Agent-based tools

Agent-based modeling requires more complicated formalism to describe the behavior and dynamics of individual agents and their spatial distribution and behavior.

Perhaps for this reason there are no "drag-and-drop" and "click-and-run" software packages available so far. All software tools in this area are designed around some programming language. It can be either versions of high-end full-fledged programming languages such as C++ or Java, or a simplified language such as Logo. However, it still requires some programming to get the model to run. All packages have links to GIS data, though some make a special effort to emphasize that. This connection usually goes in one direction, and is provided by routines that import data from raster GIS (ArcView, ArcGIS) and make it available for the modeling tools.

Swarm – Swarm Development Group, http://www.swarm.org/wiki/Swarm_main_page – Open source, any platform

This is a collection of software libraries, written in Objective C, originally developed at the Santa Fe Institute and since then taken up as an open-source project with developers all over the world. Swarm is a software package for multi-agent simulation of complex systems. It is specifically geared toward the simulation of agent-based models composed of large numbers of objects. EcoSwarm is an extension library of code that can be used for individual-based ecological models (http://www.humboldt.edu/~ecomodel/index.htm).

Repast – ROAD (Repast Organization for Architecture and Development), http://repast.sourceforge.net/ – Open source, any platform

Repast (REcursive Porous Agent Simulation Toolkit) is an agent-based simulation toolkit originally developed by researchers at the University of Chicago and the Argonne National Laboratory. Repast borrows many concepts from the Swarm toolkit. It is different in its multiple pure implementations in several languages (Java, C#, .Net, Python) and its built-in adaptive features, such as genetic algorithms and regression. Includes libraries for genetic algorithms, neural networks, random number generation, and specialized mathematics, has built-in systems dynamics modeling capabilities, has integrated geographical information systems (GIS) support.

MASON – George Mason University, http://cs.gmu.edu/~eclab/projects/mason/ – Open source, any platform

MASON Stands for Multi-Agent Simulator Of Neighborhoods ... or Networks ... or something ... the developers are not sure. It contains both a Java model library and an optional suite of visualization tools in 2D and 3D. It can represent continuous, discrete or hexagonal 2D, 3D or Network data, and any combination of these. Provided visualization tools can display these environments in 2D or in 3D, scaling, scrolling or rotating them as needed. Documentation is limited.

Cormas – Cirad, http://cormas.cirad.fr/indexeng.htm – Freeware

Programming environment to model multi-agent systems, with focus on natural-resources management. It is based on VisualWorks, a programming environment which allows the development of applications in SmallTalk programming language and is freely available from a third-party website.

OPENSTARLOGO – MIT, http://education.mit.edu/starlogo/ – Open source, Mac/Win

A programmable modeling environment for exploring the behaviors of decentralized systems, such as bird flocks, traffic jams and ant colonies, and designed especially for use by students. It is an extension of the Logo programming language, which allows control over thousands of graphic individuals called "turtles" in parallel. Comes with a nice interface, making it user-friendly and ready to use. Some basics of the Logo language are simple to learn, and users can start modeling in less than an hour.

NETLOGO – Uri Wilensky (Northwestern University), http://ccl.northwestern.edu/netlogo/ – Freeware, Mac/Win/Linux

NetLogo, a descendant of StarLogo, is a multi-platform general-purpose complexity modeling and simulation environment. The design is similar to StarLogo; it also has a user-interface. It is written in Java and includes APIs so that it can be controlled from external Java code, and users can write new commands and reporters in Java. It comes with hundreds of sample models and code examples that help beginners to get started. It is very well documented, and also has a systems dynamics component.

Pros: These systems offer perhaps the only possible way to identify emergent properties that come from interaction between agents. Most of the applications are open source, which creates infinite possibilities for linkages, extensions, and improvements.

Cons: Require programming skills, therefore may take a considerable time to learn.

Wrap-up of software

It should be noted that there are hundreds and maybe thousands of software packages that can be related to modeling, and by no means can we overview even a small fraction of them. My goal here was to look at some representative examples and try to put them in some order. Clearly, for beginners, especially those with no or few quantitative and programming skills, it makes more sense to start at the easier end of the spectrum and explore some of the existing models or modeling systems. They will take care of much of the tedious model organization and make sure that the model is consistent, they may help with unit conversions and logic of computations, and they will immediately offer some numeric methods to run a simulation. In some cases they may actually work as is, "off-the-shelf," for some applications that are repeated frequently for similar systems. As tasks become more complex, there will probably be the need to move higher up the diagram in Figure 2.16, and explore some of the more sophisticated modeling tools and methods.

Most of the software tools, like living organisms, go through life stages: after they are born, they develop, reach maturity, and then sometimes decline and die. It is hard to predict what the future of many of these products will be, especially when they are corporately owned and depend upon the dynamics of world markets. In this regard, well-developed open-source products promise more continuity, but even they can fade away or be replaced with something better. This is what we are now seeing with Swarm, which tends to morph into other products, such as Repast. Similarly, SME has been hardly developing over the past few years. In the proprietary world, there does not seem to be much progress in the ModelMaker development. Similarly, Stella seems to be relying on its previous success and has not shown much improvement over the past several years. There are also the models that are offered by federal agencies, which are free to download and use, but for which the source code is closed and proprietary. Except for the price factor, there is no big difference between these and the closed commercial products; in both cases users have very little to say about the future development of the software and entirely depend on some obscure decision-making process and funding mechanism, either in the corporate world or by the government.

The bottom line is that we need to keep a close eye on all these systems, and be flexible enough to migrate from one to another. The "hammer–nail" syndrome should be avoided. No modeling software is universal; there are always systems that could be better modeled using a different formalism and different mathematics and software. If you confine yourself to only one modeling system, you may start to think that modeling is only what the software is offering. In reality, there are numerous different approaches, and all of them may be worth considering when deciding how to model the system of interest.

Models built using open-source software are most desirable, since they can be modified to meet particular needs of various applications. Moreover, they can be tested and fixed if errors are found. While commercial proprietary software comes as closed "black boxes", where you can never be sure what's inside, open-source models are open, and the source code can be viewed and modified. On the other hand, commercial products tend to be better documented and supported. One rule of thumb is that if a project has involved a great deal of brainpower and enthusiasm, go for open source; if there is good funding for the project, go for commercial products.

Modeling is iterative and interactive. The goal is frequently modified while the project evolves. It is much more a process than a product. It becomes harder to agree on the desired outcomes and the features of the product. This certainly does not help when choosing the right software package to support modeling efforts. There is also a big difference between software development and modeling, and software engineers and modelers may have different attitudes regarding software development. For a software engineer, the exponential growth of computer performance offers unlimited resources for the development of new modeling systems. Models are viewed by software engineers merely as pieces of software that can be built from blocks or objects, almost automatically, and then connected, perhaps even over the web, and distributed over a network of computers. It is simply a matter of choosing the right architecture and writing the appropriate code – the code is either correct or not; either it works or crashes. Not so with a research model. Instead, scientists would say that a model is useful only as an eloquent simplification of reality, and needs profound understanding of the system to be built. A model should tell more about the system than simply the data available. Even the best model can be wrong, and yet still quite useful if it enhances our understanding of the system. Moreover, it often takes a long time to develop and test a scientific model.

As a result of this difference in point of view and approach, we tend to see much more rapid development of new languages, software development tools, and code- and information-sharing approaches among software engineers. In some cases, new software packages appear faster than their user community develops. In contrast, we see relatively slow adoption of these tools and approaches by the research modeling community. The applied modeling community, driven by strict deadlines and product-oriented, may be even more reluctant to explore new and untested tools, especially since such exploration always requires additional investment for acquisition, installation and learning. The proliferation of modeling software, as in the case of systems dynamics modeling tools, may even be considered an impediment, since if there were only one or two modeling tools generally accepted by all then these could be used as a common modeling platform, as a communication tool to share and distribute models. With so many different clones of the same basic approach we get a whole variety of dialects, using which it may be harder to find common ground.

In this book we will mostly be using Stella for our demonstrations. Stella's success is largely due to its user-friendly graphic interface (GUI) and a fairly wise marketing program that mostly targets students and university professors. Stella helps to illustrate a lot of modeling concepts. I do not intend to endorse or promote Stella in any way; it is no better than the other software packages available. It is just a matter of my personal experience and the legacy code that is available. Therefore, you will need to get at least a trial or Player version of Stella to be able to do the exercises and study the models that are presented in this book and can be downloaded from the book website. Doing this in alternative packages is an option that is only encouraged. Some systems dynamics tools described above offer tools to read and run Stella models. For example, Madonna will take Stella equations and, with some minimal tweaking, will run them – actually many times faster – and offer some additional exciting features. We will see how this works in Chapter 4.

The basic mathematical formalism and the interface conventions used in all these packages are quite similar – so once you have mastered one of them, it should be quite easy to switch to another if you are looking for certain special features. If you are unfamiliar with Stella, some limited instructions are available below. As mentioned above, the GUI in Stella is extremely user-friendly and the learning curve is gradual, so it should not take long for you to be able to use it for the purposes of this course.

A very quick introduction to Stella and the like

Before you start learning Stella or some of the other modeling packages described above, please ensure that you realize there are different modeling paradigms used in these packages and in certain respects this makes it hard to compare them. In this book we are mostly studying dynamic models, so we will be looking for software that can help us with this kind of modeling. The dynamic feature means that the system that we study changes over time and that there are variables that evolve. This also means that there are certain limitations and certain conventions. There are systems that are not very well suited to modeling with Stella and the like. If we simply want to use Stella for certain calculations we may probably do it, but this may not be the best way to go – using, say, Excel may be much simpler. For example, certain standard economic systems, which are usually formulated in terms of some equilibrium state, may be hard to define in Stella, unless we move away from the equilibrium and consider the transition processes as well. Stella has very limited capabilities for statistical analysis. If a simple empirical model is all you need, you may be better off with a statistical software package, which would also be much better for analyzing uncertainty and generating more sophisticated statistics.

Keep in mind that Stella and some other dynamic modeling packages assume the so-called "stock-and-flow" formalism, where the system is to be described in terms of reservoirs, called stocks, which are connected with pipes that carry flows. Stocks are therefore always measured in terms of certain quantities of material, energy, biomass, population numbers, etc., while flows are always rates of material transferred per unit of time, or energy passed per unit of time, etc. So when using Stella we start with identifying the state variables, which will be called stocks, and then figuring out what makes these variables increase (because there are flows of material or energy coming in), or decrease (because there are flows that go out). While all sorts of formulas can be used to define the flows, the stocks can be changed only by the flows; they cannot be calculated in any other way.

Stella opens with a graphic menu that contains a number of icons.

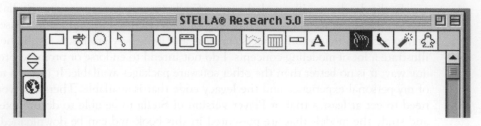

The first four are the main building blocks for your model. Figure 2.17 describes what they are.

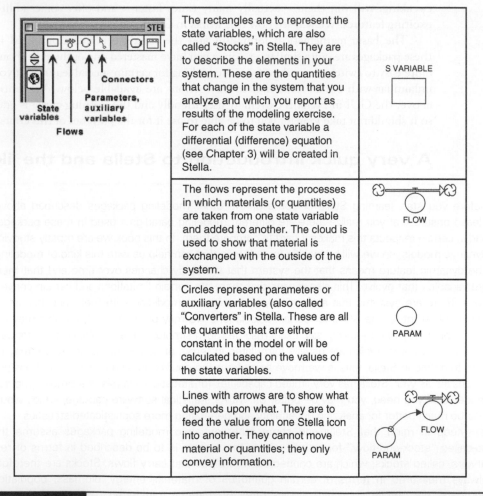

Figure 2.17   Main building blocks in Stella. Note that they are the same in Madonna, Simile and others.

To build a model using these tools, first click on the State variable icon and then somewhere in the window, where this variable will be located on the diagram. The variable appears with a "Noname" name. Click on this title and type the name that you want this variable to have.

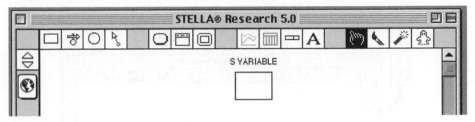

Next, click on the Flow icon and choose where you want to draw the flow that will go into or from the state variable. Then hold on the mouse button and drag the cursor from where the flow starts to where it goes. If you want a state variable to receive a flow or to be drained by a flow, make sure that the State variable icon is highlighted as you put the cursor on top of it. If it does, then it will be attached to the flow; if it does not, then it will not be associated with the flow – and this might create a problem in the future if you do not notice that there is a cloud placed somewhere on top of your State variable or right next to it. You will be thinking that the flow is there, while in reality it is not.

As you drag the flow from one element to another, you can make it angle if you hit the Shift key. This is useful to keep the diagram tidy and clear. Then give a name to the "Noname" flow similarly to the above.

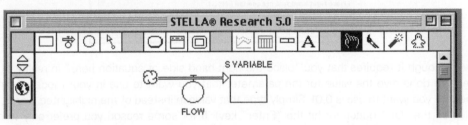

You may now need to add an auxiliary variable. In a similar way to above, choose the circle icon and place it somewhere on your diagram. Give it a name. If you click and hold on any of the names of the elements in the diagram, you can drag the name around the icon that it belongs to. This is useful to keep your diagram tidy.

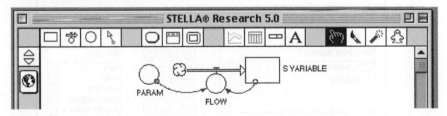

Finally, use the connector arrows to link variables, flows and state variables. Just as with flows, after choosing the connector icon you click the mouse on the origin of information and then drag the arrow to connect it with the recipient of the information.

By drawing this diagram, you have already formulated one differential (or rather difference) equation that will go into your model:

$$S\_VARIABLE(t) = S\_VARIABLE(t - dt) + FLOW*dt$$

Next, you need to specify the actual size of FLOW and the initial condition for S_VARIABLE. Before you go any further, switch the model from the so-called "Mapping Mode" to the "Modeling Mode." To do this, click on the button on the left-hand side that shows the globe. You may notice that "?" will appear on all elements in the diagram that need further definition.

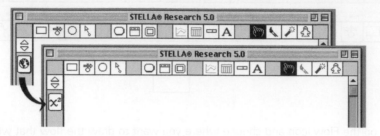

If you double click on the parameter icon, this dialogue window appears:

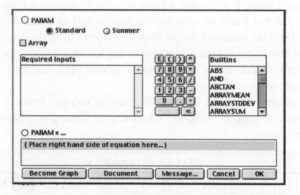

Even though it requires that you "place the right-hand side of equation here," in reality all you need to do is give the value for the parameter that you want to use in your model. Suppose the rate you want to use is 0.01. Simply type that value in instead of the highlighted words, and click on the "OK" button (or hit the "Enter" key). If for some reason you prefer only to click, you can also use the numerical pad offered in the dialogue box. You can also use some arithmetic expressions, like 1/100, or choose some of the built-in functions that are listed there.

Next we double click on the flow icon and open this dialogue box:

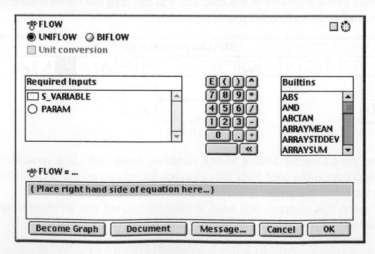

Now indeed you are to "place the right-hand side of equation here." The required inputs are listed above, and if you click on any of them they will be copied into the equation field. Here, describe the flow as a product of the available stock in S_VARIABLE, and the rate coefficient that has already been specified in PARAMETER: PARAMETER*S_VARIABLE goes into the equation field.

Click "OK" and similarly double click on the State variable icon. This opens the dialogue window to specify the initial conditions.

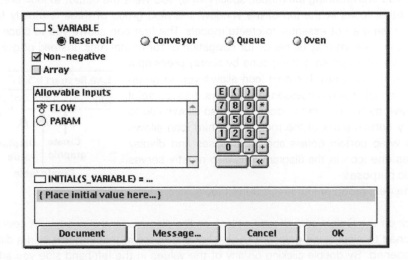

The list of "Allowable inputs" may be somewhat confusing, because it is only rarely that you will need any of the flows or other state variables to specify the initial conditions. In most cases you will simply type a constant – say, 10. Or you may store this constant as a parameter and then refer to it in this box.

Note that there is a check-box that says "Non-negative." By default, all state variables come checked as non-negative. This can be quite misleading. You want the flows in your model to make sense and to work in such a way that the stocks do not get depleted and negative. By clamping them with this check-box, you lose track of what is really happening in your model. You may well be generating some totally crazy behavior that is supposed to make a stock negative; however, you will not even notice that if that variable is clamped. It is recommended that you keep this box unchecked, unless you really know what you are doing and have a clear understanding of the effect of various flows on the state variable.

Now that all the question marks have gone, your model is ready to run. You can also check out the equations that you have generated by clicking on the downward arrow in the upper left-hand corner:

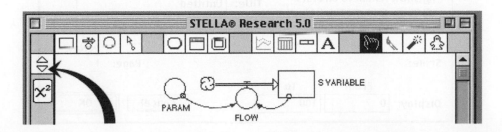

The equation here describes the exponential growth model:

☐   S_VARIABLE(t) = S_VARIABLE(t - dt) + (FLOW) * dt
      INIT S_VARIABLE = 10
      INFLOWS:
⇥   FLOW = PARAM*S_VARIABLE
◯   PARAM = 0.1

Before you start running the model, specify what you want the output to look like. Returning to the other icons at the top of the window, the next group of three is mostly for convenience; these are not essential to create models. The first one allows you to place buttons in the diagram, which may be handy for navigation or for model runs. Instead of going through menus, you can get something done by simply pressing a button in the diagram. The next icon allows you to group certain variables and processes into sectors; this is useful to achieve more order in the diagram. It also allows you to run only certain parts of the model. The third icon allows you to wrap certain details about processes and display them as one icon in the diagram; this also mostly serves esthetic purposes.

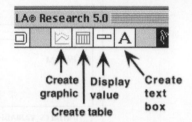

The next group of four is the Output Tools.

- Click on the Graph icon, choose a place on the diagram, and click again. A new window is opened to display the graphs. Double click anywhere in that window, and a dialogue box is opened. By double clicking on any of the values in the left-hand side you add them to the list of variables on the right-hand side. These will be graphed (no more than five per graphic). You can check out the different buttons and options available. The most useful is the scaling, which is controlled by highlighting (clicking once) a variable among those selected and then clicking on the arrow to the right of it. You can then change the minimal and maximal values for this variable that will define the scale in the graph.

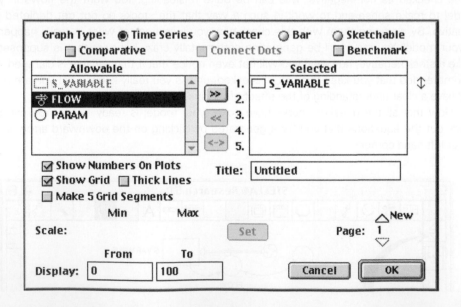

- In a very similar way, you can define a Table using output values that variables take during the course of the simulation. Choose the Table icon, click anywhere in the diagram, and a Table window opens; double click anywhere in that window, and a dialogue box opens.

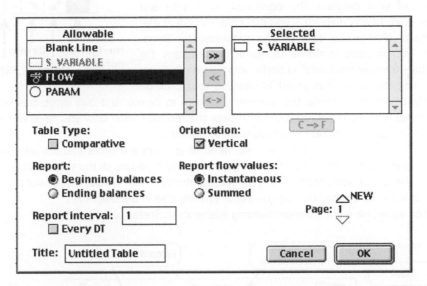

- Once again, choose the variables that you want to display and specify some of the characteristics of your Table. Similarly, you can generate output for an individual variable. In this case, you will get the value that it attains by the end of the simulation.
- As a result, for the model being built it is possible to design the output and run the model by choosing "Run" from the Run menu, or pressing Command-R (on a Mac) or Ctrl-R (in Windows). In this case, the graphic of exponential growth is displayed.

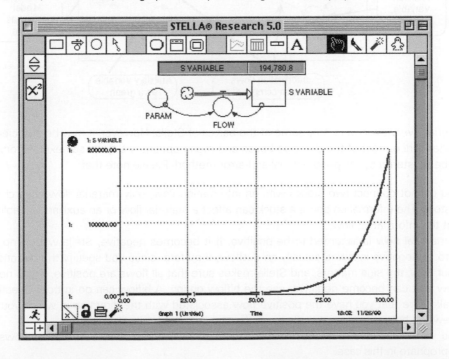

Finally, look at the Editing Tools presented by the fourth group of icons.

• The navigation cursor (the hand) is the one you will mostly be using to open and close windows and to arrange elements of your diagram. The paintbrush icon is to add some color to the diagram – you can color individual elements or change the color of your graphics. The 'dynamite' icon is used to delete things in the diagram. Be careful – there is no "undo" in Stella until version 8; if you blow something up, it is gone! To use this tool, click on the dynamite icon, choose the element you want to delete, and click on it; this will highlight what is to be blown up. Do not release the mouse button until you have verified that what is highlighted is really what you want to get rid of!

The "ghost" is very useful when you need to connect elements that are very far apart in the diagram. In that case, the diagram gets too busy if you do all the connections directly. Click on the ghost icon, then on any of the elements in the diagram you want to ghost. A copy of it will be created that can be put anywhere else in the diagram.

Once again, here are the main building elements in Stella:

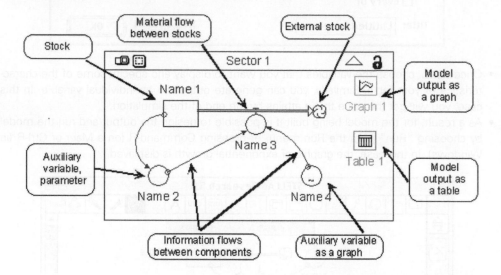

This brief overview covers only some of the basics of Stella. However, it may be sufficient as a starter, since most of the dialogue boxes and menu options are quite self-explanatory and may be mastered by the good old trial-and-error method. Please note that:

• You cannot connect two stocks with an information flow; only material flows can change a stock. The information about a stock can affect a material flow or an auxiliary variable but not the information flow.
• A material flow is assumed to be positive. If it becomes negative, Stella will clamp it to zero. A negative inflow is actually an outflow; therefore, since you specify the direction of your flow, the sign matters, and Stella makes sure that all flows are positive. If you need a flow that can become negative, use the biflow option. A biflow can go in both directions. Make sure that you have the positive flow associated with the direction in which you first drew the biflow. The negative flow will then go in the opposite direction.
• You cannot connect auxiliary variables with material flows. Only information flows are appropriate in this case.

Also please note the following rules of good style when building your models:

- Try to keep your diagram tidy. Avoid long connectors and confusing names, and avoid criss-crossing flows. The easier it is to read your diagram, the less errors you will make and the more appealing it will be to anyone else who needs to understand the model.
- There is no such thing as too much documentation. Every variable or flow in Stella has a document option, which is useful to record your ideas and comments about what you are modeling and the assumptions you are making. It is extremely important both for the model developer and the model user.

As mentioned above in our review, general-purpose spreadsheet software is a simple alternative to Stella and the like. For example, Excel may be used to build many models considered in this book (see Figure 2.18). You can download this model (Model_Of_Exponential_Growth) from the book website, and experiment

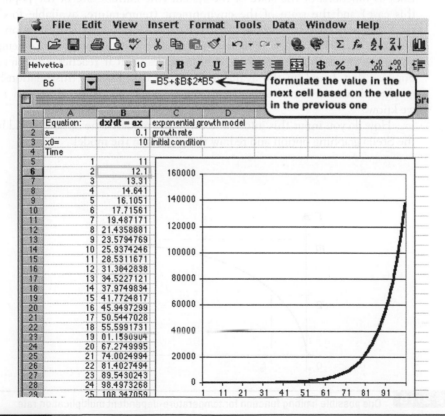

Figure 2.18   An example of a spreadsheet model for an exponential growth system (see Chapter 5 for more detail).

with the parameter and the formalization of the equation. It becomes quite difficult, if possible at all, to do this modeling for numerical methods other than Euler. Also, it gets very complicated as the model structure becomes more complex. Check out another example (Predator_Prey_Model), which is a two-variable model of predator–prey dynamics. It is still doable, but not as much fun as in the systems dynamics software.

2.3   Model formalization

The model formalization stage requires that each of the processes assumed in the conceptual model in a qualitative form be described quantitatively as a mathematical formula, logical statement or graph. This is what you do in Stella when you double click on any of the flow icons and get the dialogue box that invites you to specify the right-hand side of the equations. Choosing the right mathematical description to represent your qualitative ideas about a process may be quite tricky and ambiguous.

At this stage, describe how you envision the rates of flows between various variables. Suppose you are describing the growth rate of a bacterial population. The variable is the population size. There are two processes associated: the birth rate and the mortality rate. You need to decide how to describe both of these processes as functions of the state of the system (the current size of the population in this case) and the state of the environment (temperature, available food, space, etc.). Suppose that it is known that the reproduction rate is a function of temperature, such that at low temperatures the divisions are rare, and as temperature grows the number of divisions steadily increases until it reaches a maximal value. Suppose that, based on the available data, you can describe this relationship by a graphic shown in Figure 2.19. In this case, m is the maximal growth rate that we know.

How do you input this information into the model? One option would be to use the Stella graphing option and redraw the graphic in the model.

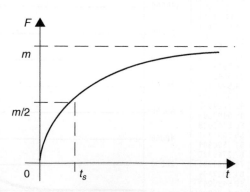

Figure 2.19   One possible limiting function for temperature-dependent multiplication rate.

In the Stella model that you have started to build, rename the variables to reflect the system that is being considered now.

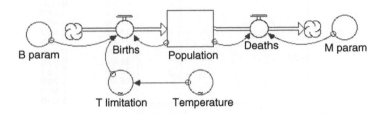

Now you have the stock that represents the population (measured here in biomass units instead of numbers). The size of the stock is controlled by two flows: the inflow is the births, similar to before, but now there is also the outflow, which is the deaths. In addition to simply having the births proportional to the size of the population, the temperature limitation is introduced by inserting the T_limitation function in the equation:

$$Births = B\_param * T\_limitation * Population$$

T_limitation should be a function of temperature, as described above. To use the graphic function in Stella to define it, double click on the T_limitation parameter to open the regular dialogue box that has temperature listed as the required input. Note the "To Graphical Function" button at the bottom left.

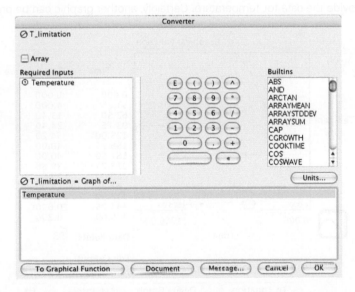

If you click on this button, you will see another panel, which is designed specifically to input a graphic that is to define what value this T_limitation parameter is to return, depending upon the value that the temperature parameter will feed in.

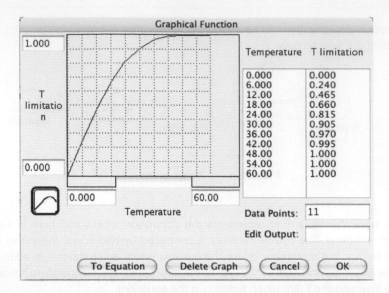

In this case, a function is designed that will change between 0 and 1 (like most of the limiting functions), while the temperature values are anticipated to be in the range between 0 and 60. Here the units for temperature are degrees Celsius, while the output of the limiting function itself is dimensionless – it will be a modifier for the birth rate that will slow down growth to zero when temperatures are low, and will have optimal growth values at temperatures close to 50°C. Perhaps this is good enough for bacterial growth. The actual values are then either typed as numbers in the table provided in the dialogue box, or produced when you draw the graphic using your cursor. Now the limiting function is finished with, but it is still necessary to figure out how to provide the data for temperature. Certainly, another graphic can be produced:

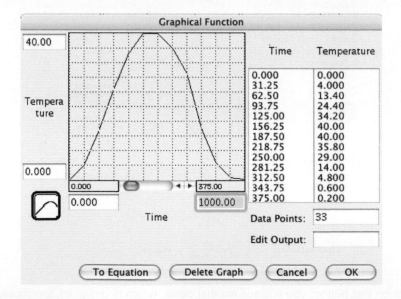

Here, a timeseries for temperature is presented, using a built-in called TIME to describe how temperature is changing in with time in the model. Some real climatic data can be copied from an Excel spreadsheet or text editor and then pasted into the table in the graphical function, or the graph can again be drawn using the cursor. In this case, a series of annual temperature cycles is defined where temperature varies from 0 to 40°C over a time period of about 365 days. There are data for about 3 years (or 1000 days), which can be seen in full in the graphic if the "ALT" key on the keyboard is pressed and held:

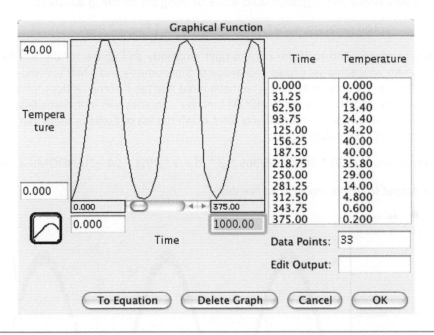

This graphical function is a nice feature, but it has one major drawback. If modification of the function is required to reflect some newly acquired knowledge, it is necessary to go into the graphic and manually redraw it. Suppose that you want to change the curvature, so that the optimal birth rate is attained faster, or suppose that the maximal birth rate should be increased – in all these cases, every time the graphic needs to be redrawn. This may become quite boring. As an alternative to the graphic representation of the data, you can assume a function that would generate the kind of response that you need. For example, for the temperature dependency shown above the function

$$F = \frac{mt}{t_s + t}$$

can be used, where m is the maximum birth rate and t_s represents the temperature at which the birth rate is half the maximal. This function happens to be known as the Michaelis–Menten function, widely used to model growth kinetics and population dynamics. Now there are two parameters that can easily modify the shape of the function. By changing m it is possible to raise or lower the asymptote, the maximal value to which the function tends. By moving t_s, the function can be made to grow faster (smaller t_s) or slower (larger t_s). All these modifications are made without any need to redraw any graphics, but simply by changing a parameter value.

Similarly, the timeseries for temperature can be defined as a formula instead of a graph. Of course, if you are using real climatic data it makes perfect sense to stick to it and import it into the graphic function, as described above. However, if the timeseries is hypothetical, it might be easier to have a formula to present it. For example, it is possible to generate the dynamics very similar to the graphic used above by using the following equation:

$$Temperature = 20 * SIN(TIME/365 * 2 * PI + 3 * PI/2) + 20$$

It does take some effort to figure out the right amplitude and phase for the SIN function; however, even with some very basic knowledge of trigonometry and a few trial-and-error runs in Stella, this can be done. It can even be made more realistic if some random fluctuations in temperature are added, using the RANDOM function – another built-in in Stella (just like SIN, PI or TIME – all these can be found if you scroll down the list of Stella built-ins that appear in any parameter of flow dialogue box):

$$Temperature = 20 * SIN(TIME/365 * 2 * PI + 3 * PI/2) + 24 + RANDOM(-4.4,0)$$

The output from this model looks like this:

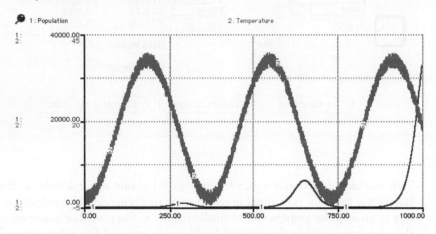

It may be hard quickly to find the right function to represent the kind of response that you have in mind for a particular process. It helps a lot if you know the behavior of some basic mathematical functions; then you can put together the right response curves by combining certain functions, the behavior of which you know. Figure 2.20 contains a collection of some useful functions that may be used as building blocks to describe various processes.

Note that for the numerical realization of the model you will need to provide actual values for all the parameters that are used in the functions. Therefore, the fewer parameters a function uses, the easier it will be to find all the values needed. It also helps a lot if the parameters used to describe a function have an ecological meaning. In that case, you can always think of an experiment to measure their value. For example, in the temperature function considered above, m can be measured as the birth rate at optimal conditions, when temperature is not limiting. Similarly, t_s can be estimated as the temperature value at which birth is approximately equal to half the maximal. Both these parameters can be measured and can be then used in the model. This is one of the basic differences between process-based and empirical

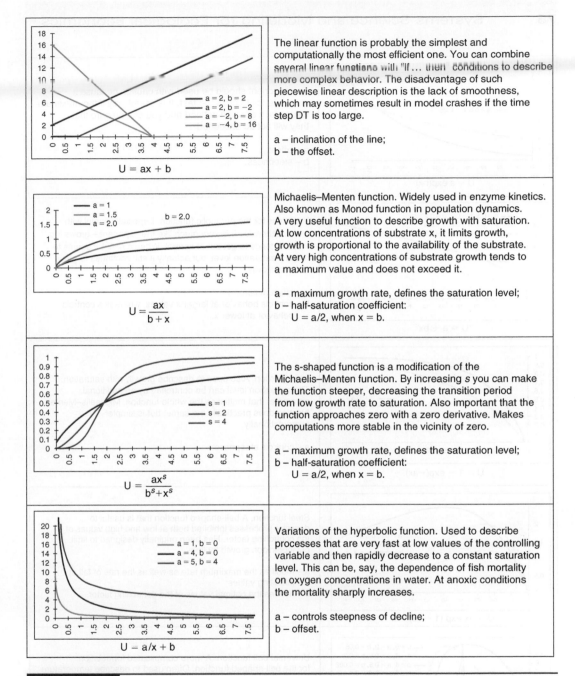

The linear function is probably the simplest and computationally the most efficient one. You can combine several linear functions with "If ... then" conditions to describe more complex behavior. The disadvantage of such piecewise linear description is the lack of smoothness, which may sometimes result in model crashes if the time step DT is too large.

a – inclination of the line;
b – the offset.

$$U = ax + b$$

Michaelis–Menten function. Widely used in enzyme kinetics. Also known as Monod function in population dynamics. A very useful function to describe growth with saturation. At low concentrations of substrate x, it limits growth, growth is proportional to the availability of the substrate. At very high concentrations of substrate growth tends to a maximum value and does not exceed it.

a – maximum growth rate, defines the saturation level;
b – half-saturation coefficient:
$U = a/2$, when $x = b$.

$$U = \frac{ax}{b + x}$$

The s-shaped function is a modification of the Michaelis–Menten function. By increasing s you can make the function steeper, decreasing the transition period from low growth rate to saturation. Also important that the function approaches zero with a zero derivative. Makes computations more stable in the vicinity of zero.

a – maximum growth rate, defines the saturation level;
b – half-saturation coefficient:
$U = a/2$, when $x = b$.

$$U = \frac{ax^s}{b^s + x^s}$$

Variations of the hyperbolic function. Used to describe processes that are very fast at low values of the controlling variable and then rapidly decrease to a constant saturation level. This can be, say, the dependence of fish mortality on oxygen concentrations in water. At anoxic conditions the mortality sharply increases.

a – controls steepness of decline;
b – offset.

$$U = a/x + b$$

Figure 2.20   A list of formulas to facilitate your choice of the mathematical expressions that can properly describe the processes in your model.

Certainly there are many others that you might find more appropriate for your particular needs. To check out how a function performs with different parameters and to choose the parameters that will best suit your particular needs, you can input the function into, say, a spreadsheet program such as Excel, and build graphs with various parameters. Another application that is especially useful for these purposes is the Graphing Calculator. It comes as part of the Mac OS-X, and probably there are also versions for Windows. It should be noted that in most cases using a formula with parameters is preferred, rather than inserting a graphic, to describe the process in your model. One significant advantage is that certain parameters that change the form of the curve can be easily used to study the sensitivity of your model to these sorts of changes. Similarly, changing graphics is a much more tedious job and more difficult to interpret.

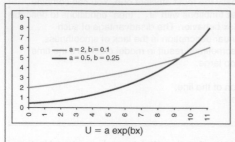

$$U = a \exp(bx)$$

The exponent should be used with caution because it always tends to grow too fast. It is useful only if the maximum values for x are well defined and you can be sure that they will not be exceeded.

a – offset;
b – steepness.

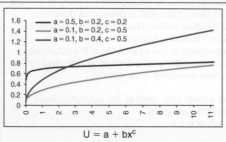

$$U = a + bx^c$$

Variations for the parabolic function. Especially useful with $c < 1$. Otherwise it is very much like the exponent – grows too fast and tends to get out of control. When $c \ll 1$ it seems to reach a saturation level, but actually it still continues to grow but very slowly.

a – offset at zero;
b – controls behavior at larger x values, whereas s controls behavior at lower x.

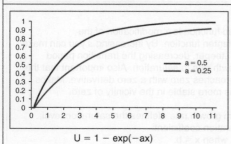

$$U = 1 - \exp(-ax)$$

Ivlev function. Also used to describe growth with saturation. The saturation level can be controlled by an additional parameter that multiplies the whole function. Michaelis–Menten function does practically the same, but is simpler computationally.

a – controls steepness.

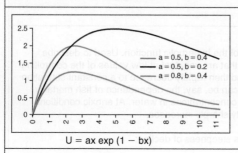

$$U = ax \exp(1 - bx)$$

Steel function. A bell-shaped function that is useful to describe processes inhibited both at low and high values of the controlling factor. Has been originally designed to limit phytoplankton growth by light.

a – modifies the maximum rate as well as the rate of fall at inhibiting values;
b – defines the optimal values of the controlling factor.

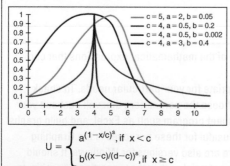

$$U = \begin{cases} a^{(1-x/c)^s}, & \text{if } x < c \\ b^{((x-c)/(d-c))^s}, & \text{if } x \geq c \end{cases}$$

Universal bell function. A more complicated formulation for the bell-shaped function. Often used to describe temperature limitation. Disadvantage – many more parameters to define. Advantage – much more flexibility and control over behavior. Can describe pretty much any bell function form.

a – value at zero;
b – value taken when x = d;
c – optimal value, where the function is maximal;
s – controls the steepness, when $s > 1$ the range of optimality can be made really big.

Figure 2.20   (Continued)

models. In process-based modeling, it is always possible (at least hypothetically) to design an experiment to measure the value of a parameter. In empirical models, parameters usually do not have any ecological meaning.

From this point on, it is assumed that you have a working knowledge of Stella and can run Stella models. Try to do Exercise 2.1 to test your knowledge of Stella. If you have problems with this exercise, try reading Chapter 3 first and then come back to it.

There may be numerous situations where the model produces dynamics that are unexpected, counter-intuitive or simply weird. In some cases that is really the system behavior discovering, which leads to new understanding and new research to explore and explain this behavior. However, in some cases the strange behavior is an artifact of model flaws or bugs. That is why you need to perform some diligent testing of the model before moving to conclusions about exciting system dynamics. However, before we start further analyzing our models, let us have a closer look at what is going on behind the scenes, inside Stella and other similar software packages.

Exercise 2.1

This exercise requires that you debug some Stella models. There are six models that have been built for particular systems; they run, but their output does not quite meet expectations. The models need be modified in such a way that they better represent the respective systems. You will need to download the Stella models from the book website, then analyze and debug them.

A model of a car dealership

You are about to open a car dealership, and plan your operations. The number of customers is random, and sales occur with a certain probability. The market analysis that you performed has given you an estimate of this probability, so you know what to expect. You also know that new stock is delivered with a delay. You need to place an order and then you get the cars shipped in several days. The size of your lot is also limited; you only have space to store a certain number of cars. You build the model and see that with the given parameters the stock goes to zero at certain times, meaning that you will not be able to sell cars to your potential clients. How can this be avoided if: (a) the delivery time is no less than 10 days? (b) the size of the warehouse is limited by 100 units? (c) both of the above are true – that is, the cars can be delivered only 10 days after they are ordered and there is space only for 100 cars in your lot?

A simple population growth model

You are modeling a population that has a certain growth rate and mortality rate. At this time it shows exponential growth, with the population increasing to infinity. Of course, if you change the parameters somewhat you can get the population to decrease to extinction. However, neither of these trends is what we observed in reality. Modify the model in such a way that the population will grow to 400 and stabilize at about that size. Can this be achieved by changing parameter values only, or do the functions that describe flows need to be modified?

A socio-economic model

The system is a human population, the growth of which also depends upon the level of "development." The larger the population, the higher level of development it can reach, since there are more people who can participate in the production process. The development in turn stimulates population growth, since there is better health care, lower child mortality, etc. As development increases, fewer babies die, therefore the greater the population becomes. Is there anything wrong with this conceptual model? What modifications could make the system "sustainable" (i.e. keep both variables within certain limits, avoiding extinction and runaway growth to infinity)?

A model of the biosphere

Here, we are modeling CO_2 and biomass on this planet. CO_2 is released into the atmosphere as fossil fuels are burnt. Forests then can uptake this CO_2 from the atmosphere. The forests are cut at a certain rate. They also decompose, adding to the pool of CO_2. Currently, the model displays infinite growth of forests, which does not look realistic. How can we modify the model to stabilize it (that is, make it retain the biomass of forests and CO_2 within certain limits)?

A model of body weight

You gain weight because you eat; you lose weight when you exercise. However, the more you exercise, the greater your appetite is and the more likely you are to eat. Also, the more you weigh, the greater your appetite will be. So the weight grows and grows. Can this pattern be changed? What do you need to change in this model to make your weight stabilize at a certain level and stay there?

A model of a river

A river is represented by five identical reaches. The water collected from the catchment area drains into river reach no. 5, and then travels down the river. The precipitation pattern is random, and results in sharp oscillations in the river level in the reaches. What changes to the parameters and/or functions are necessary to stabilize the system so that (a) there is no accumulation of water in the river, and (b) the oscillations in the water depth are smoothed out? Modify the model to describe the travel of one pulse of water through the river system.

Further reading

You will find a discussion of some of these basic modeling concepts in almost any book on modeling. If you want to explore some details in application to ecological modeling take a look at Jørgensen S.E., and Bendoricchio, G. (2001). *Fundamentals of Ecological Modelling*, 3rd edn. Elsevier Science; *or for more general modeling applications with a focus entirely on Stella, see* Hannon, B. and Ruth, M. (2001). Dynamic Modeling, 2nd edn (with CD). Springer-Verlag.

To read more about the modeling process see Jakeman, A.J., R.A. Letcher and J.P. Norton (2006). Ten iterative steps in development and evaluation of environmental models. *Environmental Modelling & Software*: 21(5): 602–614.

The Bonnini paradox was discussed in Starbuck, W.H. and Dutton, J.M. (1971). The history of simulation models. In: J.M. Dutton and W.H. Starbuck (eds.), *Computer Simulation of Human Behavior*. Wiley, p. 14.

Qu, Y. and Duffy, C.J. (2007). A Semi-Discrete Finite-Volume Formulation for Multi-Process Watershed Simulation. Water Resource Resolution. doi:10.1029/2006WR005752 – *This shows how non-structured triangulated grids can be used for watershed modeling.*

To get a better feeling of what a model is, try playing a computer game. My favorite for this purpose is Simcity – see http://simcity3000unlimited.ea.com/us/guide/. *The game, indeed, is a simulation model, which you run while changing some of the parameters. Try to figure out the spatial, temporal and structural resolutions that went into this computer model.*

3. Essential Math

3.1 Time

3.2 Space

3.3 Structure

3.4 Building blocks

> *The full set of dynamic equations of a given physical system presented in one of the approximate forms, along with the corresponding boundary conditions and with the algorithm for the numerical solution of these equations – inevitably containing means from a finite-difference approximation of the continuous fields describing the system – form a physico-mathematical model of the system.*
>
> A.S. Monin

SUMMARY

Many models are based on some mathematical formalism. In some cases it may be quite elaborate and complex; in many others it is straightforward enough and does not require more than some basic high-school math skills to understand it. In all cases it can help a lot if you know what the mathematics are that stand behind the model that you build or use. Most of the systems dynamics models that we use in this book are based on ordinary differential of difference equations. Some basics of those are explained in this chapter. We look at how models can tend to equilibrium conditions, and explore how these equilibria can be tested for stability. If the spatial dimension is added, we may end up with equations in partial derivatives. We will see how the advection and diffusion processes can be formalized. Finally, in the structural domain we may also find models that will be structurally robust and stable. Such models are preferable, especially when there is much uncertainty about model parameters and processes.

Keywords

Discrete vs continuous, initial conditions, ordinary differential equations, state variables, difference equations, exponential growth, time-step, numerical method, Euler method, Runge–Kutta method, equilibrium, stable or unstable equilibrium, box models, compartmental models, continuous models, advection, diffusion, equations in partial derivatives, rigid and soft systems, structural stability.

* * *

Modeling and systems analysis first appeared as mathematical disciplines. Reciprocally, much of modern mathematics has originated from models in physics. Until recently, a solid mathematical background was a prerequisite of any modeling effort. The advent of computers and user-friendly modeling software has created the feeling that mathematical knowledge is no longer needed to build realistic models, even for complex dynamic systems. Unfortunately, this illusion results in many cases in faulty models that either misrepresent the reality entirely or represent it only in a very narrow domain of parameters and forcing functions, while the conclusions and predictions that are made are most likely to be presented as being quite general and long lasting.

This does not mean that modeling cannot be done unless you have a PhD in mathematics or engineering. Many of the software packages that are currently available can indeed help a lot in the modeling process. They can certainly eliminate most of the programming work needed. It is important, however, for the modeler to know and understand the major mathematical principles that are used within the framework of those packages, otherwise the models will be prone to error. David Berlinski offers some noteworthy examples of how models can be misused, misunderstood, and in error when the mathematics is ignored.

Let us take another look at the model that we have developed in Stella. Open the model and click on the little arrow pointing downwards in the upper left corner. (In more recent versions of Stella the interface has been changed and you have separate tabs on the left of the window for the interface, the model and the equations.)

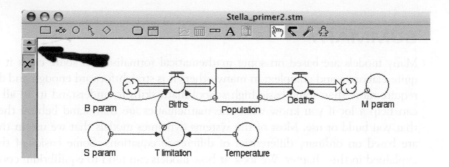

What you get is a list of equations:

Population(t) = Population(t - dt) + (Births - Deaths) * dt
    INIT Population = 10
    INFLOWS:
        ⟿ Births = B_param*T_limitation*Population
    OUTFLOWS:
        ⟿ Deaths = M_param*Population
○ B_param = 0.1
○ M_param = 0.05
○ S_param = 12
○ Temperature = 20*SIN(TIME/365*2*PI+3*PI/2)+24+RANDOM(-4.4,0)
○ T_limitation = Temperature/(S_param+Temperature)

So these will be the actual equations that the model will be solving. The Graphic User Interface has really just one purpose: to formulate these equations and then display the results of solving them. What are these equations, and how do they work?

As we have seen in the previous chapter, a system may be considered in three dimensions: *temporal, spatial* and *structural*. In the temporal dimension we decide how the system evolves in time; in the spatial dimension we research the spatial organization of the system; and in the structural dimension we define the complexity of the model. For each of these facets mathematics are used in modeling.

3.1 Time

Most computer models operate in discrete time. The time is represented as a sequence of snapshots, or states, which change momentarily every given time interval. The major question to be answered when considering this temporal evolution of a system is, what will be its state at time t if its state is known at the previous time $t - 1$? If we know how the system changes state, then we can describe its dynamics once we know the initial state of the system. Suppose we have a population of five cells and each cell divides into two over one time-step – say, 1 hour. Then after 1 hour we will have 10 cells, since each cell is to be replaced by two; after 2 hours there will be 20 cells; after 3 hours there will be 40 cells, and so on.

This is a verbal model of a system. Let us formalize it or describe it in mathematical terms. Let $x(n)$ be the number of cells at time-step $n = 1, 2, \ldots$ Then the doubling process can be described by

$$x(n + 1) = 2x(n) \qquad (3.1)$$

If we provide the initial condition $x(0) = a$, we can calculate the number of cells after any n time-steps:

$$x(n) = a2^n \qquad (3.2)$$

This is a simple *model of exponential growth*. The nice thing about the mathematical formalism is that it provides us with a general solution. Instead of doing iterative (i.e. repeating) calculations to find out the number of cells after, say, 100 divisions, and redoing these calculations if instead of five initial cells we were to consider six of them, we can immediately provide the result based on the general solution (3.2).

However, this model can only describe systems that are very well synchronized in time, where all the cells divide simultaneously and similarly. This is quite rare for real populations, where divisions occur all the time, and therefore the process is not so discrete. In this case it makes sense to assume a different model that we formulate in terms of growth rate. Suppose that each cell produces one new cell once an hour. This is more-or-less equivalent to the above model, but now we can remove

How did we get from (3.1) to (3.2)? Certainly, if $x(n + 1) = 2x(n)$, then similarly $x(n) = 2x(n - 1)$ and $x(n - 1) = 2x(n - 2)$. Substituting, we get: $x(n) - 2x(n - 1) = 2 \cdot 2x(n - 2)$, and so on, $x(n) = 2 \cdot 2 \ldots 2x(0)$. Keeping in mind that $x(0) = a$ and that $2 \cdot 2 \ldots 2(n$ times$) = 2^n$, we get the result in (3.2).

the synchronization. In two hours one cell will produce two cells, and in half-an-hour a cell will produce half of a new cell. This may not make sense for an individual cell, but with no synchronization this makes perfect sense for a population of, say, 100 cells. It simply means that in half-an-hour 50 new cells will be produced. We can then reformulate (3.1) as:

$$x(t + \Delta t) = x(t) + x(t) \cdot \Delta t \tag{3.3}$$

We have substituted t for n to show that time no longer needs to change in integer steps. Δt is the time increment in the model. Note that if $\Delta t = 1$, model (3.3) is identical to model (3.1):

$$x(t + 1) = x(t) + x(t) \cdot 1 = 2x(t) \tag{3.4}$$

However, if we run the same model with $\Delta t = 0.5$, we get a different result:

$$x(t + 0.5) = x(t) + x(t) \cdot 0.5 = 1.5x(t)$$

Then similarly

$$x(t + 1) = x(t + 0.5) + x(t + 0.5) \cdot 0.5 = 1.5x(t + 0.5).$$

Substituting from the above, we get:

$$x(t + 1) = 1.5 \cdot 1.5x(t) = 2.25x(t),$$

which is different from what we had for $\Delta t = 1$ in (3.4). We see that when we change the time-step Δt, we get quite different results (see Figure 3.1). The more often we update the population of cells, the smaller the time-step in the model, the faster the population grows. Since new growth is based on the existing number of cells, the more often we update the population number, the more cells we get to contribute to further growth.

Indeed, let us take a closer look at Figure 3.1 and zoom in on the first two steps (Figure 3.2). We start with a certain initial condition – say, 2. We decide to run the model with a certain time-step – say, $\Delta t = 1$. According to (3.3), the next value at $x(1) = x(0) + x(0) \cdot \Delta t = 2 + 2 \cdot 1 = 4$. At time 0 we define (3.3) for the first time-step, and we know that during this period of time nothing is supposed to change in the equation. Only when we get to the next point in time do we re-evaluate the variables in (3.3). Now we change the equation and diverge from the straight bold line that we once followed: $x(2) = x(1) + x(1) \cdot \Delta t = 4 + 4 \cdot 1 = 8$.

If we had chosen to run the model with a time-step $\Delta t = 2$, then we would have stayed on the course longer (the broken line in Figure 3.2), and then $x(2) = x(0) + x(0) \cdot \Delta t = 2 + 2 \cdot 2 = 6$. See the difference?

Alternatively, if we had chosen to run the model with a time-step of $\Delta t = 0.5$, we would have corrected the trajectory already after the first half-step taken (the dashed line in Figure 3.2), and then later on every half-step we would have been correcting the course. As a result, by the time we got to time 2 we would have got to a different value, a substantially different one compared with the case with $\Delta t = 2$. *So the smaller the time-step we use, the more often we correct the course, the less the computational error.*

Note that equation (3.3) is similar to the kind of equations that we saw previously, generated in Stella. Remember when we were clicking the little triangle in

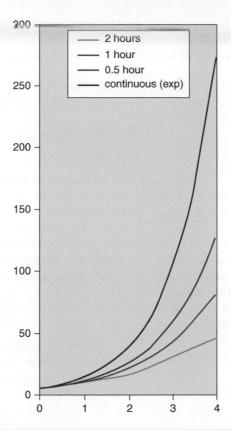

Figure 3.1   Exponential growth model.

A 1-hour time-step corresponds to model (2) dynamics. A 2-hour time-step results in slower growth, and a 1/2-hour time-step produces a steeper curve. The exact solution to the differential equation in case of continuous time produces the fastest growth.

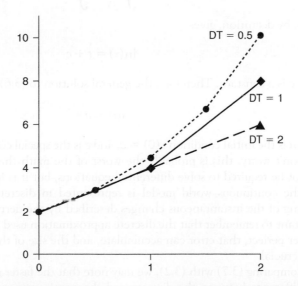

Figure 3.2   A zoom-in to get a better view of the first several steps of the exponential growth model.

the upper left-hand corner to get the equation view? The only obvious difference is that instead of Δt they use dt in their equations. Actually there is no big difference, because usually dt is what is used in mathematics when Δt becomes very, very small, $\Delta t \to 0$.

What will be the system dynamics if we try to update the population numbers permanently, as often as possible, continuously? Let us rewrite (3.3) as

$$\frac{x(t) - x(t - \Delta t)}{\Delta t} = x(t) \tag{3.5}$$

How did we get from (3.5) to (3.6)? What is dx/dt? This is just a mathematical notation that represents $(x(t + \Delta t) - x(t))/\Delta t$, where Δt is approaching zero, dx is the increment in the variable x and dt is the increment in time. We see that dx/dt has the units of a rate (km/h, kg/d, kcal/s, individuals/y, etc.). Note that it is the dx/dt that we define with the flows in Stella.

Allowing Δt to become smaller and smaller, $\Delta t \to 0$, we get the instantaneous increment in the population or the rate of change. When $\Delta t \to 0$, the left-hand side of (3.5) is called a *derivative* of x. In this case time is no longer discrete, it is assumed to change in infinitesimally small increments and we get an ordinary *differential equation*:

$$\frac{dx}{dt} = x \tag{3.6}$$

There is a wealth of analytical and numerical methods available to solve ordinary differential equations, especially simple ones. In our case we can rewrite (3.5) as:

$$\frac{dx}{x} = dt$$

Integrating both sides of this equation, we get

$$\int \frac{dx}{x} = \int dt$$

which, by definition, gives

$$\ln(x) = t + c$$

where c is a constant. Therefore, the general solution to (3.6) will be:

$$x(t) = ae^t \tag{3.7}$$

where a is the initial condition $x(0) = a$, and e is the special constant $e = 2.71828\ldots$

Don't worry; this is probably the worst of the math that we will see here. You will not be required to solve differential equations, but it is important to understand how the continuous world model is represented in discrete terms, and what the meaning of the instantaneous changes described by the derivative dx/dt is. It is also important to remember that the discrete approximation used in the computer models is never perfect, that error can accumulate, and the size of the time-step used (Δt) is really crucial.

Comparing (3.7) with (3.2), we may note that the faster growth is because $e > 2$. The difference between the discrete and the continuous model is quite large. The ongoing updates of the population in the continuous model give a rather small change to the model formalization: 2.72 in (3.7) is not that much larger than 2 in (3.2). However, the exponent that is used in the model blows up this difference tremendously. The time-step used in the model becomes a crucial factor. This is something to remember: *models with rapidly changing variables are extremely sensitive to the size of the time-step used.*

So what does this mean in terms of our little Stella model? As we have seen above, there is a *dt* in the equations file. So now we know what it is all about. There is also a way to change this *dt*. Click on the "Run" menu, and then choose "Run Specs." This will open a dialogue box that contains the time specifications for your model run. "From" will specify at what time you start the simulation, "To" tells when to end. "DT" is the time-step to use in the simulation.

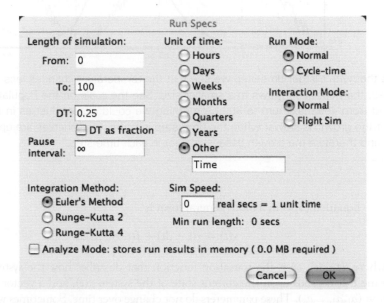

Let us start with DT = 0.25 and run the model. The result should look something like this:

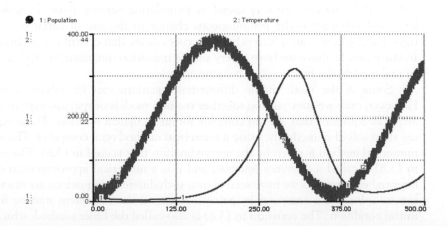

If we now change the time-step to DT = 2 we will get a similar picture, but not exactly the same:

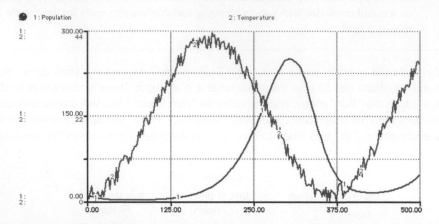

Notice that with a larger time-step we see that the temperature changes less frequently, and, besides, the population grows to a smaller size. See the scale on the Population axis: it has changed from 400 maximum to 300. Something we could expect: just as in Figure 3.2, we see that the growth is slower when the time-step is larger, the variables are updated less frequently, and therefore the growth base is smaller in each time-step.

Equation (3.3) in a more general form is

$$x(t) = x(t - dt) + f(t, x(t), a)dt \tag{3.8}$$

where $f(t, x(t), a)$ is the transition function that describes how the system changes at time t. It depends upon the current state of the system $x(t)$, and a vector of parameters $a = (a_1, a_2, \ldots, a_n)$. These parameters do not change over time. Sometimes we assume that the parameters are hidden and write simply $f(t, x)$. As a differential equation, this will be:

$$\frac{dx}{dt} = f(t, x, a)$$

Differential equations are very useful in formulating various dynamic models. The left-hand side dx/dt is the instantaneous change in the size of variable x. On the right-hand side, we can specify what the processes are that contribute to this change. In the example above we have a very simple transition function: $f(t, x(t), a) = x(t)$. In real models, the function can be quite complex.

Some of the more simple differential equations can be solved analytically. However, once we start putting together realistic models of systems, very quickly we arrive at equations that are too complex for an analytical solution. These equations are then solved numerically, using a numerical method on a computer. The simplest numerical method is given by the approximation that is used in (3.8). The equation in (3.8) is called a *difference equation*, and it is a numerical approximation of a differential equation. As we have seen above, such difference equations are discrete and can be solved on a computer by going through all the time-steps starting from the initial condition. The equation in (3.8) is also called the Euler method, which is the simplest way to find a numerical solution for a differential equation. This method also creates quite large errors, especially when the right-hand side is changing fast.

There are also other numerical methods that are way more efficient than Euler's method. One of the most widely used methods is the Runge–Kutta method. This method also takes the value of x at time t and calculates the value of x at time $t + dt$. However, in the simple version of this method, instead of assuming that the transition function $f(t, x)$ is the same over the whole period dt, as in Figure 3.2, we approximate the value of the function at mid-point $t + dt/2$ and use this value to improve the result. Instead of the equation

$$x(t + dt) = x(t) + f(t, x(t))dt$$

we use the equation:

$$x(t + dt) = x(t) + f\left(t + \frac{dt}{2}, x\left(t + \frac{dt}{2}\right)\right)dt = x(t) + f\left(t + \frac{dt}{2}, x(t) + \frac{k_1}{2}\right)dt,$$

where $k_1 = f(t, x(t))dt$. Note that here we jump to the next time point at $t + dt$ using the forecasted value at mid-point rather than the value at the beginning point, t. Also note that any improvement in accuracy in this method comes at a price: we have to calculate the transition function f twice to make the move to the next time-step.

Why not run the Euler method with halved time-step, instead? The amount of calculations will be about the same: same two calculations of the transition function. The formula would be:

$$x\left(t + \frac{dt}{2}\right) = x(t) + f(t, x(t))\frac{dt}{2}$$

$$x(t + dt) = x\left(t + \frac{dt}{2}\right) + f\left(t + \frac{dt}{2}, x\left(t + \frac{dt}{2}\right)\right)\frac{dt}{2}.$$

Substituting, we get

$$x(t + dt) = x(t) + f(t, x(t))\frac{dt}{2} + f\left(t + \frac{dt}{2}, x\left(t + \frac{dt}{2}\right)\right)\frac{dt}{2}$$

This is pretty close to the above, although not exactly the same. Comparing the two you can see what the invention of Runge and Kutta actually was.

Exercise 3.1

1. Consider a population where cells divide into three every time-step. What mathematical model will describe this population?
2. Formulate the model with triple growth in terms of a differential equation and try to solve it. Is the difference between the discrete and continuous models larger or smaller than in the double-growth model?
3. Suppose you have $10,000 in your bank savings account with an annual interest rate of 3%. Build a model to calculate your interest earnings in 5 years. What will be the difference in your earnings if the interest is calculated monthly instead of annually? How much does the bank make on your account by calculating the interest monthly, instead of doing it continuously?

Let us take another look at our population model. Suppose we get rid of the temperature function, and have a simple exponential growth model with an initial condition of Population = 10, birth rate B_rate = 0.1 and death rate M_rate = 0.05.

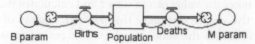

B param   Births Population   Deaths   M param

A simple analytical calculation like that we did above gives us the formula $x(t) = ae^{kt}$, where a is the initial condition (10 in our case) and k is the growth coefficient (which in our case is equal to B_rate − M_rate = 0.05. Running the model for, say, 100 days gives us a population of 1,484. Now let us run this model in Stella with the Euler method. Note the list of available methods in the Run Specs … panel:

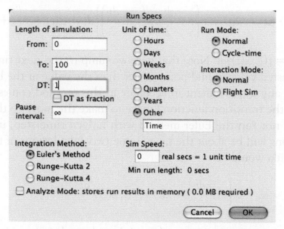

The result we get is 1,315 – quite a substantial difference. Let us click on the Runge–Kutta 2 method and rerun the model. The result is 1,481, which is getting quite close to the analytical solution. Running the same model using the Euler method and the time-step DT = 0.5 requires the same number of calculations for the right-hand side $f(t, x)$; however, the result is 1,395, which is nowhere nearly as good as the Runge–Kutta model run. With some other models, the difference can become even more dramatic. For instance, running the predator–prey model (see Chapter 5) we get qualitatively different dynamics of the variables than when running the Euler and Runge–Kutta methods. While the Euler method results in increasingly large oscillations (1), the Runge–Kutta results in a more or less cyclic dynamics (2):

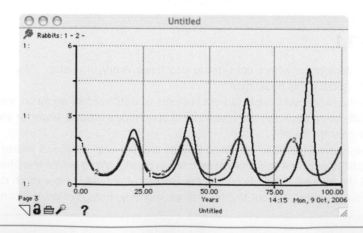

The more complicated and most often used version of the Runge–Kutta algorithm uses a weighted average of several approximated values of $f(t, x)$ within the interval $(t, t + dt)$. The formula known as the "fourth order Runge–Kutta formula" is given by

$$x(t + dt) = x(t) + \left(\frac{dt}{6}\right)(k_1 + 2k_2 + 2k_3 + k_4)$$

where

$$k_1 = f(t, x(t)), \quad k_2 = f\left(t + \frac{dt}{2}, x(t) + \left(\frac{dt}{2}\right)k_1\right), \quad k_3 = f\left(t + \frac{dt}{2}, x(t) + \left(\frac{dt}{2}\right)k_2\right),$$
$$k_4 = f(t + dt, x(t) + dt\ k_3)$$

In our population model, if we choose the Runge–Kutta 4 method we get the population of exactly 1,484 after the 100 days of a run. That is a perfect match with the analytic solution. Quite outstanding!

To run the simulation, we also start with the initial condition, at t_0, $x_0 = x(t_0)$ and find $x_1 = x(t_0 + dt)$ using the formula above. Then we plug in x_1 to find $x_2 = x(t_1 + dt) = x(t_0 + 2dt)$, and so on. Once again we pay a price for the improved accuracy of calculations: now we have to calculate the transition function four times.

The Runge–Kutta algorithm is known to be very accurate and behaves well for a wide range of problems. However, like all numerical methods, it is never perfect and there are models where it fails. One universal rule is that the smaller the time-step, no matter what method we use, the better the accuracy of the simulation. *To ensure that you are getting the right result with your numerical method, you may want to keep decreasing the time-step until you do not see any difference in the results that you are generating.* There are some adaptive step-size algorithms that do exactly that automatically. Other algorithms are also available, such as the Adams method or Bulirsch–Stoer or predictor–corrector methods, that can be way more efficient for some problems, especially when very high accuracy is essential. Just remember that there is always a price to pay for higher accuracy. The smaller the time-step, the longer it takes to run the model. The more accurate the method, the longer it takes to run the model. However, sometimes one method is simply better for a particular type of a model – it runs faster and gives better accuracy. So it always makes sense to try a few methods on your model and see which one works best.

We have already seen that the size of the time-step chosen for the numerical solution of the model can significantly change the output produced. Let us consider another example. Suppose we are modeling a stock of some substance that is accumulated due to a flow coming in and is depleted by an outflow:

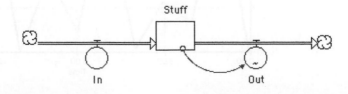

The equations for this model will be:

☐ Stuff(t) = Stuff($t - dt$) + (In − Out) * dt
INIT Stuff = 0
INFLOWS:
⇨ In = 1
OUTFLOWS:
⇨ Out = GRAPH(Stuff)
(0.00, 0.09), (0.1, 0.63), (0.2, 1.06), (0.3, 1.32), (0.4, 1.47), (0.5, 1.56), (0.6, 1.62),
(0.7, 1.67), (0.8, 1.72), (0.9, 1.75), (1, 1.77)

The inflow is constant, whereas the outflow is a function of the substance accumulated. It may be described by a simple graphic function of the form:

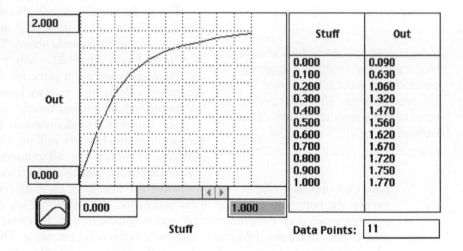

Stuff	Out
0.000	0.090
0.100	0.630
0.200	1.060
0.300	1.320
0.400	1.470
0.500	1.560
0.600	1.620
0.700	1.670
0.800	1.720
0.900	1.750
1.000	1.770

Data Points: 11

If we first run the model with DT = 1 using the Euler method, we get a very bumpy ride, and an oscillating trajectory:

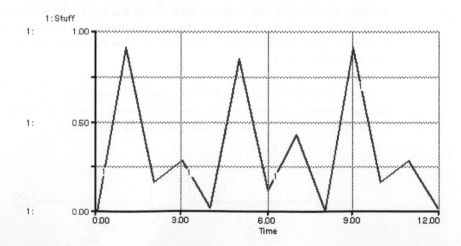

The very same model but with a smaller time-step of DT = 0.25 produces entirely different dynamics:

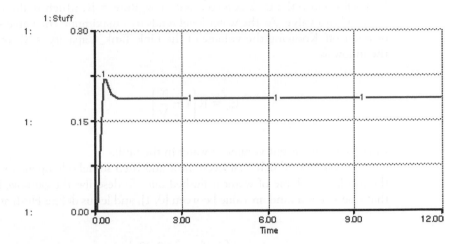

Finally, if we switch to the Runge–Kutta, fourth-order method, we get very smooth behavior, with the trajectory reaching saturation level and staying there:

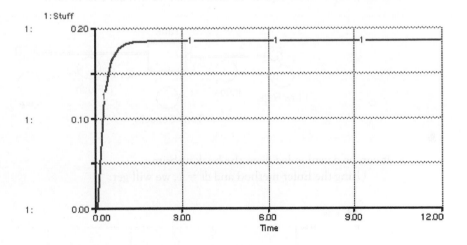

The very same model produces entirely different dynamics by simply changing the time-step assumed. Clearly you do not want to run your model with too small a DT, since it will require more computational time and may become more difficult to analyze properly. However, too large a DT is also inappropriate, since the results you produce may be entirely wrong.

Here is yet another example that shows that DT matters and that it is always important to remember the equations that are solved to run your model. Quite often in models we want to do something to the entire amount stored in one of the variables. For example, at certain times we need to deplete a reservoir, and then start filling it all over again. Or we may be looking at an age-structured population, when after reaching a certain age the entire population moves from one stage (say, eggs) to another stage (say, chicks).

Let us consider a simple model of a flush tank that we all use several times a day. Water flows into the tank at a constant Flow_Rate = R, which is altered by a floater attached to a valve. As the water level tends to a maximum, the valve shuts the flow of water off. Knowing the volume of the tank Tank_Capacity = V, we can describe the inflow as

$$F_{in} = R\left(1 - \frac{T}{V}\right)$$

where T is the current volume of water in the tank.

The outflow is such that every now and then somebody opens the gate and all the available volume of water is flushed out. To describe the outflow, let us assume that Use = u is a random value between [0, 1], and let us define Flush as:

$$F_{out} = \begin{cases} 0, \text{ if } & u < 0.99 \\ T, & \text{otherwise} \end{cases}$$

If we just put these equations into Stella, we will get this model:

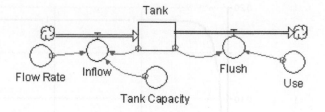

It can be downloaded from the book website.

Using the Euler method and $dt = 1$, we will get:

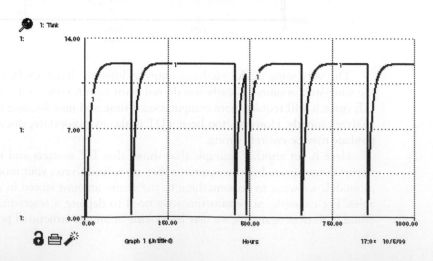

which looks exactly how we wanted. Every now and then the tank is emptied, and then it is gradually refilled. Suppose now that instead of $dt = 1$ we wish to get a more accurate solution and make $dt = 0.5$. Now, the output looks quite different:

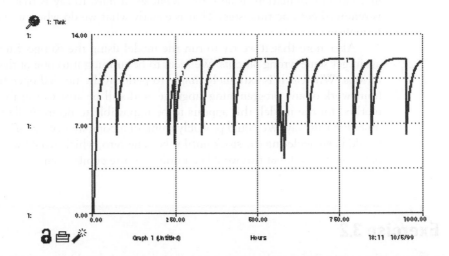

Not quite as expected. The tank does not get emptied any more. What has happened? Let us look at the Stella equations for this model:

⬜ Tank(t) = Tank(t - dt) + (Inflow - Flush) * dt
INIT Tank = 0
INFLOWS:
⇥ Inflow = Flow_Rate*(1-Tank/Tank_Capacity)
OUTFLOWS:
⇥ Flush = IF Use>0.999 THEN Tank ELSE 0
◯ Flow_Rate = 0.1
◯ Tank_Capacity = 12
◯ Use = RANDOM(0,1,14)

It is clear that, contrary to what we intended, the outflow is not T, but $T \cdot dt$. That is why dt started to modify the model output so dramatically. It should be remembered that whenever a flow is described in Stella or another similar package, it is then multiplied by dt when it is inserted into the real equations to be solved. Therefore, if it is actually the entire stock that you want to move, you should describe the flow as T/dt. Then when it is inserted into the equations, the dt gets cancelled out and we can really flux the entire amount as it was intended.

Therefore, the correct Stella equations should be:

⬜ Tank(t) = Tank(t - dt) + (Inflow - Flush) * dt
INIT Tank = 0

INFLOWS:
⇥ Inflow = Flow_Rate*(1-Tank/Tank_Capacity)
OUTFLOWS:
⇥ Flush = IF Use>0.995 THEN Tank/dt ELSE 0
◯ Flow_Rate = 0.9
◯ Tank_Capacity = 12   **Note this change!**
◯ Use = RANDOM(0,1,14)

All this is quite clear if we do a strict check of all the units in the model formalization. The flows should have units of amount per time. If that is kept in mind, then it is obvious that specifying the outflow in terms of just the state variable, as in the first model formulation, is an error. What we *wanted* to say is that the entire stock is removed per one time-step. That is exactly what we do when we use the Tank/dt formalism.

Also, note that if we try to run the model using the Runge–Kutta method, we again get some wrong outputs. We seem to be running into one of those Stella limitations. Whereas it is extremely simple to perform the needed operation within the framework of any programming language, and still be able to run the Runge–Kutta method for the model, this appears to be impossible to do in Stella in a direct and unambiguous way. We could probably think of a combination of "if" statements that would keep depleting the stock until it became zero, which would work for any computation method, but this would certainly be quite cumbersome.

Exercise 3.2

Can you reformulate the model in such a way that the stock would be drained out properly under any computational method and any time-step DT? As you may have noticed, when we switch to other methods, the outflow starts to deplete the stock, but then, apparently, does not have enough time to take it all out. How can this be fixed?

Note that the temporal issues that we are discussing here are different from the choice of the temporal resolution that we were deciding in Chapter 2. In that case, we were thinking about the timescales of the processes that we were to consider – whether we should observe the system once every day, or every hour, or every year. There, we were considering how fast the system changes, and how often we would need to update the model and report results to keep track of the change. Here, we are mostly concerned with the accuracy of simulations; we are looking at the behavior of the model equations and deciding how big an error we can afford in the model we build.

Let us consider another system. Suppose x is the amount of dollars in your bank account. This changes with time, so we should write $x(t)$. Every month you deposit your paycheck in the account, adding p ($/month). You are on a fixed salary, so p is a constant. Note that p is a rate; therefore the units are per time. In addition there is the monthly interest rate k (1/month), which is also constant, that adds kx ($/month) to your account. Your average monthly expenses of q ($/month), which you also keep constant, should be subtracted from your account. As a result, you get a model to describe the dynamics of your bank account:

Always mind the units when formulating your models. It can save you a lot of trouble in debugging.

$$\frac{dx}{dt} = kx + p - q \qquad (3.9)$$

It is important to make sure that the units on the left and right-hand sides of the equations match.

In this case the left-hand side is the variation in your account, measured in $ per month. We make sure that the flows on the right-hand side are presented in similar units. For example, it is important to remember that the interest rate k is monthly, and should therefore be recalculated from the more frequently used annual interest rate.

When $dx/dt = 0$, there is no change in the variable. If the inflows and outflows are balanced, the variable is at equilibrium; it does not change because nothing is added to it and nothing is taken away. By setting $dx/dt = 0$, we can calculate the equilibrium conditions in our model. From (3.9) we get:

$$kx + p - q = 0,$$

$$x^* = \frac{q - p}{k}. \tag{3.10}$$

If you make $(q - p)/k$ your initial condition: $x(0) = (q - p)/k$, there will never be any increase or decrease in the value of the variable; your account will remain unchanged. A nice guideline to balance your account! However, what will happen if your initial condition is slightly larger or smaller than the equilibrium (3.10)?

In model (3.9) if we are even slightly below the equilibrium: $x < (q - p)/k$ then $dx/dt < 0$. The derivative is negative when the function is decreasing. Therefore, for values less than the equilibrium equation, (3.9) takes us further away from it and we will be getting decreasing values for the account (Figure 3.3). Similarly, if we start even slightly above the equilibrium, then $x > (q - p)/k$ and $dx/dt > 0$. Now the derivative is positive, so the function grows, and therefore again we start moving away from the equilibrium. The farther we move away from the equilibrium, the larger dx/dt gets, the farther it takes us away from the equilibrium. This positive feedback sets us on a path of exponential growth. The equilibrium state is unstable. If you take one step away from the equilibrium, even a very small one, you will slide further away from it. Small deviations from the equilibrium will only increase with time.

Here is another example. Suppose a population of woozles lives on a small island that has enough grass to support only A woozles. If there are more woozles than

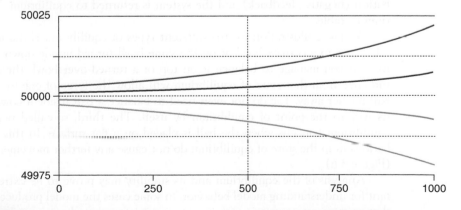

Figure 3.3 Unstable equilibrium. Small displacements from steady state result in increasing divergence from it.

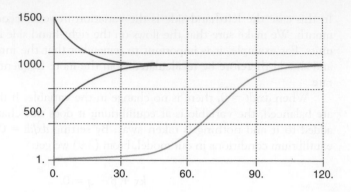

Figure 3.4 Stable equilibrium. When system is perturbed from steady state $A = 1000$, it returns to it.

A, they die off from hunger. We can use the following formalism to describe this system:

$$\frac{dx}{dt} = kx\left(1 - \frac{x}{A}\right) \tag{3.11}$$

Here, k is the growth rate of woozles and A is the carrying capacity of the island. As x approaches A, the multiplier $(1 - x/A)$ effectively slows down the growth rate dx/dt, making it zero when $x = A$. If somehow x becomes larger than A, this same multiplier makes the growth rate negative, providing that the population size decreases until it reaches the size $x = A$ (Figure 3.4).

We see that $x = A$ is an equilibrium point. There is yet another equilibrium point, where $dx/dt = 0$. This is $x = 0$. From Figure 3.4, we readily see that when $0 < x < A$ the derivative is positive and therefore x grows. If x could be negative (not the case in our system, but it could be if the same formalism was used for a different system), then $dx/dt < 0$ and therefore x further decreases, tending to $-\infty$. The equilibrium $x = 0$ is clearly unstable. On the contrary, as we can see when the system is perturbed from the equilibrium $x = A$, the sign of the derivative is opposite to the sign of the perturbation (negative feedback) and the system is returned to equilibrium. This equilibrium is stable.

A classic illustration for the different types of equilibrium is the movement of a ball put in a convex bowl or on the same bowl turned upside down (Figure 3.5). Even if you manage to balance it on top of a turned-over bowl, the slightest disturbance from that state of equilibrium will allow the force of gravity to move the ball further away. You do not even need to balance a ball inside a bowl; it will find its way to the point of equilibrium by itself. The third, so-called neutral type of equilibrium happens when the ball is placed on a flat surface. In this case, perturbations from the state of equilibrium do not cause any further movement of the ball (Figure 3.6).

Analysis of the equilibrium and its stability may prove to be extremely important for understanding model behavior. In some cases the model produces trajectories that seem to converge to a certain state, no matter what changes are made to model parameters. In that case, chances are that the trajectory is at equilibrium and there

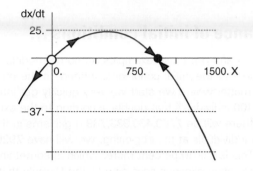

Figure 3.5   Derivative *dx/dt* as function of *x* (*A* = 1000).

Figure 3.6   Three types of equilibrium: A. unstable, B. stable, C. neutral.

is no way you can change model dynamics by modifying parameters. If that behavior is unwelcome, it is the model structure or formalism that needs to be changed. However, it is only by analytical study that you can find the equilibrium states and analyze their stability. Numerical analysis can only speak in favor of or against stability of a certain state.

When we represent a system using several variables, we write equations for each of them and get systems of equations that need be solved simultaneously. That is when the analytical methods generally start to fail. In most cases, only some very simple systems of ordinary differential equations (ODE) can be solved when there are more than two variables. With only two variables, already, the analytical solution may be calculated only for some special cases. However, even for those systems of ODE that cannot be solved analytically the analysis of equilibria may be possible and quite useful.

Exercise 3.3

1. Consider a population of wizzles which, unlike woozles, can multiply only when their numbers are larger than a certain minimal value *a*. It there are less wizzles than *a*, they cannot find a partner for mating and the population dies off. The carrying capacity of the island where wizzles live is also *A*. Obviously, *A* > *a*. What model would describe the population dynamics of wizzles on this island?
2. Are there any equilibria in the wizzles model? Are they stable?

The importance of initial conditions ...

... can be easily seen from these examples. In some models, the outcome of simulation is almost independent of the initial conditions. Indeed, in the model with exponential growth, for example, no matter where we start we very quickly get into very, very big numbers. Let us say we have 100 individuals that double according to the exponential growth model. After 25 generations, there will be 7,200,489,933,739 organisms in the population. If we start with 10 percent more individuals at the beginning, we will have 7,920,538,927,112 organisms after 25 generations. This is still 10 percent more, which is something we could expect bearing in mind that the population expands according to the formula that we derived above: $P = P_o e^t$, where P_o is the initial condition and t is time. However, when you get into the very, very big numbers these 10 percent do not seem to matter as much, and the graphs of the two results look very similar: you still get a very, very big number ...

Similarly, in the last example that has a stable equilibrium for the population of woozles. In this case, the ultimate population is always the same, no matter where we start. The initial conditions can be 10 or 1000 percent off; still we will end up at the same equilibrium point.

The system in the bank account example, however, displays a radically different type of behavior. In this case, the end result is entirely determined by the initial conditions. If we start with a little over $50,000, we are destined for prosperity. Our account will grow indefinitely. However, if we are among the disenfranchised and have just a little less than $50,000 to start with, we will inevitably end up in total poverty.

This is not to imply that our real economic system is really described by such a simple model, but if we think about it, some of the major features of wealth accumulation are pretty well captured by it. The results we observe seem to resonate with the well-documented fact that "the rich get richer, and the poor get poorer." In reality, there are certainly many more processes involved that make the picture more obscure, creating possibilities for perturbations that bounce certain individuals out of the trend. However, common sense will tell us that most of the additional processes will actually serve in strengthening this system behavior. Indeed, when wealth also leads to more power (as we will see in Chapter 7), this power will be then used to create more wealth. This is another positive feedback, which adds to the positive feedback already embedded into the economic system that we described using the bank account model. (Can you figure out what the feedback is that we are describing?)

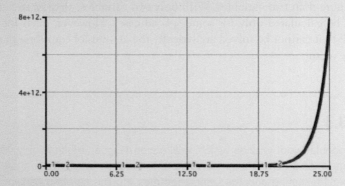

On the other hand, I wonder if this is also one of the reasons why the mainstream economy is so addicted to exponential growth as the only possible vision of viable economic system performance. Indeed, in an exponentially growing system it is much easier to hide the importance of initial conditions. As a participant of the exponentially growing system, even with very little initial wealth, as we have seen above, one can always expect to exponentially increase this wealth and become as rich as some of the other participants – even those who were way ahead at the beginning. Well, maybe not exactly *as* rich, but very, very rich. This future becomes much more unlikely if the economy is not growing exponentially, and therefore the role of initial conditions becomes much more exposed and obvious. This is the end of the "American Dream" for many – actually, for most.

3.2 Space

How can we represent space? This is the next decision to be made when creating a model. Once again, we go back to the goal of the model to decide how much detail about the spatial extent and spatial resolution of the system is needed. Both must be taken into account when choosing the appropriate mathematical formalism.

Box models

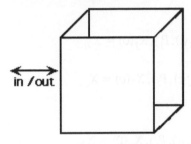

In box models, everything is assumed to be spatially the same. There is no difference between various spatial locations. This may be because the system is indeed spatially homogeneous, or if we want to deal only with variables that are averaged across space, or if the only data we have about the system are measured at only one point. In these cases, the system can be represented by just one point (box) in space. For this system, as we have seen above, we have only the temporal dynamics to take care of – and a system of ordinary differential equations with initial conditions can easily handle that:

$$\frac{dX}{dt} = F(X(t), P)$$

$$X(0) = X_0$$

Here, X is the vector of the state variables $X = (x_1, x_2, \ldots, x_n)$, P is the vector of parameters used in the model $P = (p_1, p_2, \ldots, p_k)$, X_0 is the vector of initial conditions $X_0 = (x_{10}, x_{20}, \ldots, x_{n0})$ and F is the vector of functions that describe the interactions within the system, as well as the in- and outflows that connect the boxed system with the outside world, $F = (f_1, f_2, \ldots, f_n)$. So the equation above actually stands for n equations, written out for each of the x_i: $dx_i/dt = f_i(x_1, \ldots, x_n, p_1, \ldots, p_k)$.

The actual size of the system modeled is not a criterion for the choice of the spatial representation. For example, the famous Forrester–Meadows models of the biosphere represent our whole globe as one box.

Compartmental models

When the system is not entirely uniform over space, but the spatially different components are large enough and are homogeneous within, then we can represent this system by a compartmental model. In this case a separate box model represents each

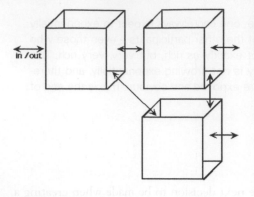

spatially homogeneous component, also called a compartment. These are then linked together by flows of material or energy. In effect, a compartmental model is a number of box models joined together. For example, this is how we might want to present a small stratified lake, where the upper part of the lake (epilimnion) is separated from the deep waters (hypolimnion) because of the temperature gradient. The warmer water stays on top and gets well mixed by wind-induced currents, making this upper layer spatially uniform. However, the currents are strong enough to mix only a certain portion of water; the rest of the cooler water is not involved in the turnover and remains somewhat separated from the epilimnion. It makes sense to represent each of these spatial units as separate box models and link them by certain fluxes, such as the sedimentation process of material across the boundary of the two compartments.

Each of the box models may be described by a system of differential equations with initial conditions:

$$\frac{dX_1}{dt} = F_1(X_1(t), P_1), X_1(0) = X_{01}$$

$$\frac{dX_2}{dt} = F_2(X_2(t), P_2), X_2(0) = X_{02}$$

$$\ldots$$

$$\frac{dX_n}{dt} = F_n(X_n(t), P_n), X_n(0) = X_{0n}$$

Here, once again, X_i is the vector of the state variables in compartment i, P_i is the vector of parameters used in the model in compartment i, X_{0i} is the vector of initial conditions for compartment i.

As a discrete interpretation, similar to that which Stella generates, we get a system of differential equations:

$$X_1(t) = X_1(t - dt) + F_1(X_1(t - dt), P_1)dt$$

$$\ldots$$

$$X_n(t) = X_n(t - dt) + F_n(X_n(t - dt), P_n)dt$$

These are then linked by flow equations:

$$X_i(t) = X_i(t - dt) + \left(\sum_{j \neq i} X_j(t - dt) \cdot Q_{ji} - \sum_{j \neq i} X_i(t - dt) \cdot Q_{ij} \right.$$
$$\left. + \sum_{j \neq i} D_j \cdot (X_j(t - dt) - X_i(t - dt)) \right) dt$$

Here, we distinguish between two types of flows: advection and diffusion. Advection describes motion caused by an external force (such as gravity, which causes sedimentation). Q defines the advective flux. Q_{ji} represents the flows that flow into the ith compartment from the surrounding jth compartments; Q_{ij} are the flows that flow out of the ith compartment. Diffusion is defined by the gradient or difference between the

amounts of material in different spatial locations. D_j is the diffusion coefficient for the
jth compartment. Note that when the values in the surrounding j compartments are
higher than inside compartment i: $(X_j(t - dt) > X_i(t - dt))$, we get a positive term and
the amount of material in i, $X_i(t)$, is increasing. Otherwise, it will be decreasing, and
the diffusion will be taking material out of this compartment.

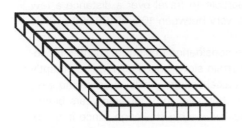

Continuous models

If there is much variability in space and this variability is an important factor for the goals of our analysis,
then the system is spatially heterogeneous, and we
need to make the compartments as small as possible.
When the size of the compartments tends to zero, we
get an equation in partial derivatives. Now we have
two independent variables: time t, as before, and a
spatial coordinate, z. In the one-dimensional case,
the model is formulated as follows:

$$\frac{\partial X}{\partial t} = \frac{\partial}{\partial z}\left(D_z \frac{\partial X}{\partial z}\right) + Q_z \frac{\partial X}{\partial z} + F(t, X, P)$$

X is the vector of the state variables. This is
similar to that which we were considering above
for the differential equations that appear when the
time increment approaches zero. In that case, we
were specifying the initial conditions to fire up our
model. In this case, in addition to initial conditions:

$$X(0, z) = X_0(z)$$

we need to provide the boundary conditions, or
the initial conditions in the space dimension:

$$X(t, z_0) = X'(t).$$

Note that now the initial conditions are
a function of the spatial coordinate z, and the
boundary conditions are functions of time, t. D_z
describes the diffusive flux while Q_z presents the
advective flux. F is the function of local ecological processes, similar to the one used in the box
model presentation, and P is a vector of ecological
parameters in this function.

> So what is this rounded ∂ all about? We
> just got over the dX/dt, now we have
> $\partial X/\partial t$. Moreover, there is also $\partial X/\partial z$.
> Well, this is what happens when we
> realize that our state variable depends
> upon two independent variables, time
> and space, simultaneously. In this case,
> X is a function both of time, t, and a
> spatial coordinate, z. While before we
> were looking at instantaneous change
> of our function as the time increment
> was approaching zero, now we can look
> at similar change in two dimensions –
> temporal and spatial. Therefore, the
> instantaneous increment when consid-
> ered in one dimension is called a par-
> tial derivative and is shown as ∂X. Then
> $\partial X/\partial t$ is the change in time, and $\partial X/\partial z$ is
> the change in space. It becomes even
> more interesting if we consider more
> than one spatial coordinate. Then we
> get a function of even more independ-
> ent variables.

Time and space scales ...

... are somewhat related. In most cases we observe that systems with larger spatial scales
(sizes) have longer temporal scales. It is interesting to note how in cosmic scales time and
space become combined into a unit such as a light year, which is actually the distance that is
covered if traveling with the speed of light for 1 year – 1 light year = 9.4605284×10^{15} meters,

which is quite a distance! Something similar is found in the subatomic world. As we have seen, the tiny particles of the microworld have lifetimes of less than a millionth of a second. This is extremely short in the human timescale. However, their size is also very small and they travel at lightning-fast velocities. To make more sense of this comparison, physicists have come up with a measure called a "particle second" – a unit of time equal to 10^{-23} seconds, which is the time needed for a particle to travel over a distance a few times its own size. Various particles have lifetimes that vary between 10 and 100,000 particle seconds.

Such resolutions do not make any sense if we are considering geological change, movement of continents or rising of mountains. However, certain slow processes may be abruptly interrupted by fast and violent fluctuations. Slow geological change yields to an earthquake, when in minutes and hours we see more disturbance than over the thousand years before that. Modeling processes that occur on a variety of scales is a big challenge, since it is prohibitively hard to represent the slow processes at the scale of the rapid ones. However, if we ignore the singularities completely we may miss some really important changes and transformations in the system.

When sizes don't differ that much, there is no exact relationship between temporal and spatial scales. For instance, snails register their environment once in every 4 seconds. Even though humans are larger, they can do a better job. For us, the world around us changes approximately once every 1/24th of a second. This resolution of ours is what defines the rate of change of snapshots in movies that we watch. If we do it less frequently, we see how the motion becomes discontinuous; figures start to move in jerks. If we do it faster, we will not see the difference. We can actually insert another frame and we will not register it. They say that there is a method of manipulating people, called the "25th frame." This is when a 25th frame is inserted and the movies run at 1/25th of a second. This single 25th frame can be entirely out of context, and humans to not consciously register it. However, apparently it affects our subconscious and the information finds its way to the right parts of the brain, influencing our opinions and decisions.

This would not be possible for a fly, which scans the environment 20 times faster than we do. A fly would stare at the 25th frame for long enough to realize that something totally out of context was being displayed. On the other hand, a snail would never even see this frame. Moreover, if you move fast enough, in 4 seconds you can pick a snail from the ground and put it in your basket. For the snail this kind of transformation will occur instantaneously; it will never know how it got from one place to another. These considerations are important when choosing the right resolutions for your models.

Let us take a look at a couple of examples of how this kind of equation can be derived.

Modeling advection

Consider the distribution of a certain constituent in space and time. Suppose we are looking at only one spatial dimension, perhaps a long pipe or a canal. To account for spatial heterogeneity we will assume that the whole length of the canal can be divided into equal segments, each Δx long. The concentration of the substance in each segment will be then a function of both time and length: $C(t, x)$. When Δx is large enough we can think about this model as a compartmental one and use some of the formalism described above.

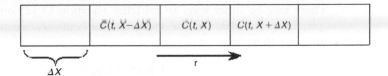

We also assume that there is a certain velocity of flow in the canal, r, and that it is constant.

Let us now define the concentration of the contents in any given segment at time $t + \Delta t$, assuming that we know the concentration there at time t. Since calculating concentration may be confusing, let us write the equation for the total amount of material in segment x at time $t + \Delta t$:

$$C(t + \Delta t, x) \cdot \Delta x = C(t, x) \cdot \Delta x + \underbrace{C(t, x - \Delta x) \cdot r \cdot \Delta t}_{\text{Brought in from segment above}} - \underbrace{C(t, x) \cdot r \cdot \Delta t}_{\substack{\text{Moved out to segment} \\ \text{below}}}$$

Rearranging the terms, we get:

$$C(t + \Delta t, x) \cdot \Delta x - C(t, x) \cdot \Delta x = C(t, x - \Delta x) \cdot r \cdot \Delta t - C(t, x) \cdot r \cdot \Delta t$$

Dividing both sides by $\Delta x \Delta t$:

$$\frac{C(t + \Delta t, x) \cdot \Delta x - C(t, x) \cdot \Delta x}{\Delta x \cdot \Delta t} = \frac{C(t, x - \Delta x) \cdot r \cdot \Delta t - C(t, x) \cdot r \cdot \Delta t}{\Delta x \cdot \Delta t}$$

Or, cancelling Δx on the left-hand side and Δt on the right-hand side:

$$\frac{C(t + \Delta t, x) - C(t, x)}{\Delta t} = r \cdot \frac{C(t, x - \Delta x) - C(t, x)}{\Delta x}$$

Now if we let $\Delta x \to 0$ and $\Delta t \to 0$, we get the well-known advection equation as a partial differential equation:

$$\frac{\partial c}{\partial t} = -r \frac{\partial c}{\partial x}.$$

In discrete notation, the equation for concentration at the next time-step is:

$$C(t + \Delta t, x) = C(t, x) - \frac{[C(t, x) - C(t, x - \Delta x)] \cdot r \cdot \Delta t}{\Delta x} \tag{3.12}$$

If we know the concentration at the previous time-step, we can calculate the concentration at the next time-step. To be able to use this equation at any (x, t), we still need to define two more conditions. First, we need to know where to start – what was the distribution of material along the canal at the beginning, at time $t = 0$. That will be the initial condition:

$$C(0, x) = c_0(x)$$

Besides, if you look at equation (3.12) you may notice that to solve it for any t we need to know what the concentration at the left-most cell is, where $x = 0$. That is the boundary condition:

$$C(t, 0) = b(t)$$

There may be other ways to initialize equation (3.12) on the boundary. For example, instead of defining the value on the boundary, we may define the flow, assuming, say, that

$$C(t, 0) = C(t, 1)$$

This will be a condition of no flow across the boundary, and will also be sufficient to start the iterative process to solve equation (3.12).

Modeling diffusion

Let us now consider diffusion as the driving force of change in the concentration in our system. The force that makes the substance move in this case is the difference between concentrations in adjacent segments. It is also good to remember that in this discrete approximation we are actually dealing with points on a continuum, in this case a line 0x. The concentrations that we are considering are located at these points. We are dealing with average concentrations for the whole segments, and are assuming that these averages are located at these nodes. Therefore, if there is no outside force to move the material, it would be reasonable to assume that the farther away the points we consider are, the less material can be moved between them by the concentration gradient.

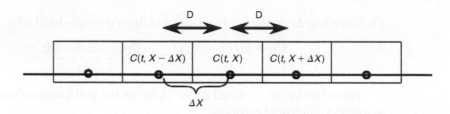

Just as before, let us define the concentration of material in any given segment at time $t + \Delta t$, assuming that we know the concentration there at time t. The equation for the total amount of material in a segment at time $t + \Delta t$ is:

$$C(t + \Delta t, x) \cdot \Delta x =$$
$$C(t, x) \cdot \Delta x + \frac{C(t, x - \Delta x) - C(t, x)}{\Delta x} \cdot D \cdot \Delta t + \frac{C(t, x + \Delta x) - C(t, x)}{\Delta x} \cdot D \cdot \Delta t$$

In this equation, $\dfrac{C(t, x + \Delta x) - C(t, x)}{\Delta x}$ is the empirically derived equation for the diffusive flux between two adjacent segments. D is the diffusion coefficient that characterizes the environment, the media; it tells us how fast diffusion can occur in this kind of media.

After some rearranging we get:

$$\frac{C(t + \Delta t, x) - C(t, x)}{\Delta t} = D \frac{\dfrac{C(t, x - \Delta x) - C(t, x)}{\Delta x} - \dfrac{C(t, x) - C(t, x - \Delta x)}{\Delta x}}{\Delta x}$$

Once again, if we let $\Delta x \to 0$ and $\Delta t \to 0$, we get the well-known diffusion equation as a partial differential equation:

$$\frac{\partial c}{\partial t} = D\frac{\partial^2 c}{\partial x^2}$$

In discrete notation, the equation for concentration at the next time-step becomes:

$$C(t + \Delta t, x) = C(t, x) + \frac{[C(t, x - \Delta x) - 2C(t, x) + C(t, x + \Delta x)]D \cdot \Delta t}{\Delta x^2} \qquad (3.13)$$

If we know the concentration at the previous time-step, we can calculate the concentration at the next time-step. Just as in the advection example, to calculate this equation at any (x, t) we need to define the initial condition:

$$C(0, x) = C_0(x)$$

As for the boundary conditions, in this case we will need two of them. We cannot use equation (3.13) to calculate the value both on the left-hand side boundary $C(t, 0)$ and on the right-hand side boundary $C(t, N)$, where N is the number of the maximal segment that we consider. Therefore, we need two boundary conditions:

$$C(t, 0) = b_1(t); \quad C(t, N) = b_2(t).$$

Similarly, there may be other types of boundary conditions, such as:

$$C(t, 0) = C(t, 1), \quad C(t, N - 1) = C(t, N).$$

This will be a condition of no flow across the boundaries.

3.3   Structure

Consider a community of two competing species that eliminate one another. We can describe this system by the following two ODEs:

$$\frac{dx}{dt} = -by$$

$$\frac{dy}{dt} = -ax \qquad (3.14)$$

where a and b are hunting efficiencies of species y and x respectively. This model can be resolved analytically:

$$\frac{dx}{dy} = \frac{by}{ax},$$

$$ax\,dx = by\,dy,$$

$$ax^2 - by^2 = const$$

A good way to look at system dynamics, especially in case of two variables, is to draw the phase portrait, which presents the change in one variable as a function of the other variable. Figure 3.7 presents the phase portrait for model (3.14). It can be seen that the two populations eliminate each other following a hyperbola. The initial conditions define which trajectory the system will follow. In any case, one of the two species gets eaten up first, while the other species remains. If the initial condition is on the line equation $\sqrt{(ax)} = \sqrt{(by)}$, then the two populations keep exterminating each other at infinite length, tending to complete mutual extermination. If the initial conditions are below this line, then y is exterminated and x persists. If the initial conditions are above this line, then y wins. Models like those considered above may be called *rigid* (Arnold, 1997); their structure is totally defined. In contrast to a rigid model (3.14), a *soft* model would be formulated as:

$$\frac{dx}{dt} = -b(x, y)\, y$$

$$\frac{dy}{dt} = -a(x, y)x \tag{3.15}$$

where $a(x, y)$ and $b(x, y)$ are certain functions from a certain class. It may be shown that for most functions $a(x, y)$ and $b(x, y)$ the phase portrait of system (3.15) is qualitatively similar to the one in system (3.14) (Figure 3.8). One of the species is still exterminated, but the threshold line is no longer straight.

An important feature of model (3.15) is its structural stability. Changes in functions $a(x, y)$ and $b(x, y)$ that describe some features of the populations do not change the *overall qualitative behavior* of the system. Since in most cases our knowledge about the objects that we model is not exact and uses a good deal of qualitative description, soft models are more reliable for predicting the system dynamics. Unfortunately, there are very limited analytical methods to study the structural stability of models. The only way to analyze structural stability in broader classes of models is to run extensive sensitivity analysis, varying some functions and relations in the model as well as changing parameters and initial conditions.

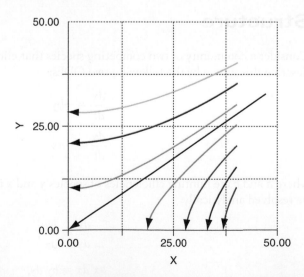

Figure 3.7   Phase portrait for the model of mutual extermination.

Structural analysis of models requires quite sophisticated mathematics. Even for a simple model like that above, analysis of its structural stability lies way beyond the scope of this book. In general, Table 3.1, from von Bertalanffy (1968), shows that there is a very small domain of mathematical models that can be analyzed by analytical methods.

Most of the real-world models turn out to be non-linear, with several or many equations. Besides, most of the systems are spatially distributed, which almost precludes analytical methods of analysis. However, there are numerous examples of quite successful and stimulating analytical studies that have led to new theories and new understanding. Physics especially has an abundance of this sort of model. Probably this is why most of the mathematics that is used in modeling came from physical applications.

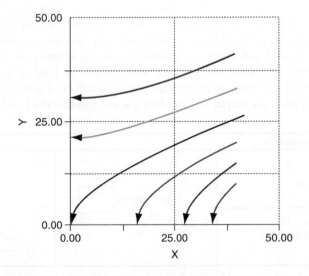

| **Figure 3.8** | Phase portrait for the soft model of mutual extermination. |

Table 3.1	Mathematical models that can be analyzed by analytical methods					
	Linear equations			**Non-linear equations**		
Equation	One equation	Several equations	Many equations	One equation	Several equations	Many equations
Algebraic	Trivial	Easy	Essentially impossible	Very difficult	Very difficult	Impossible
Ordinary differential	Easy	Difficult	Essentially impossible	Very difficult	Impossible	Impossible
Partial differential	Difficult	Essentially impossible	Impossible	Impossible	Impossible	Impossible

Ecology, social sciences and economics have yet to develop adequate mathematical methods of analysis. Up till now, most of the models in these sciences have been numerical, analyzed by means of computer simulations.

3.4   Building blocks

Let us consider some of the main types of equations and formulas that you can encounter in dynamic models (Figure 3.9). If you have a good feel for how they work, you can put together quite sophisticated models using these simple formalizations as building blocks. While, indeed, complex non-linear models are notorious for springing surprises, for their unexpected behavior, it is always nice to have some level of control regarding what is going on in the model. Knowing some of the math behind the equations and formulas in a modeling software package such as Stella will add some predictability to how your model may behave. Knowing how some of the very simple formalizations perform as stand-alone modules will help you to construct models that will be better behaved and easier to calibrate. Certainly, interaction of these processes will create new and uncertain behavior, which it will be hard or impossible to predict in some cases. However, in many other cases you will be able to have a pretty good expectation of what the output will be when you put together the building blocks.

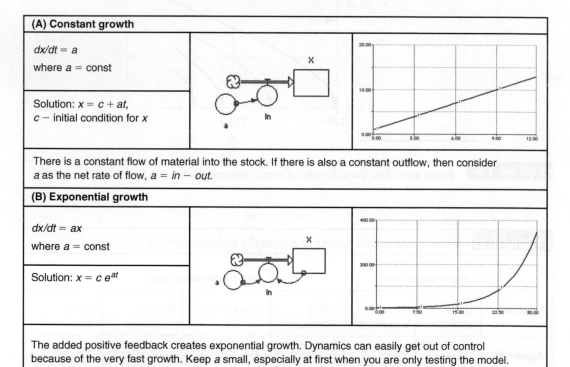

(A) Constant growth

$dx/dt = a$

where a = const

Solution: $x = c + at$,
c − initial condition for x

There is a constant flow of material into the stock. If there is also a constant outflow, then consider a as the net rate of flow, $a = in - out$.

(B) Exponential growth

$dx/dt = ax$

where a = const

Solution: $x = c\,e^{at}$

The added positive feedback creates exponential growth. Dynamics can easily get out of control because of the very fast growth. Keep a small, especially at first when you are only testing the model.

Figure 3.9   Growth: (A) Constant growth; (B) Exponential growth; (C) Growth with saturation; (D) Growth with peaking; (E) Delayed response.

(C) Growth with saturation

$dx/dt = ax - bx^2$

where a and b = const

Solution: $x = \dfrac{ace^{at}}{a + bc\,(e^{at} - 1)}$

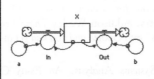

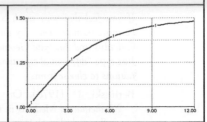

The exponential growth is now dampened by exponential decline. At smaller populations the linear function (ax) dominates. As numbers increase the parabola (bx^2) overwhelms and shuts down growth. The solution is the so-called *logistic* equation. Note that the model is identical to the model with carrying capacity: $ax - bx^2 = ax(1 - bx/a)$. The carrying capacity in this model is then a/b. When $x = a/b$ the growth is zero and the model saturates.

(D) Growth with peaking

$dx/dt = ax - bx^{st}$

where a, b, and s = const

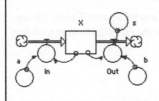

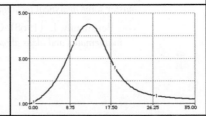

A simple way to make the model peak and then decline is to have a variable exponent in the outflow part and make this outflow grow with time. In this case again at first the outflow is very small and the system grows. Later on the outflow becomes dominant and gradually reverses the dynamics eventually getting the system down to zero. Used less often than the first three blocks but still may be handy.

(E) Delayed response

$dx/dt = ax - bx^2(t - \Delta t),$

where a and b = const
Δt — is time delay

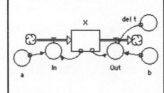

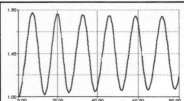

A powerful way to get pretty confusing results. In this model of saturated growth (see above) we assumed that mortality is controlled by the population size several time-steps ago. This may be if we assume that mortality is due to a disease and the disease has an incubation period of Δt. If $\Delta t = 1$ we still have a saturation. If $\Delta t = 2$ we suddenly run into oscillations as shown in graph. With $\Delta t > 2$ we have a population peak and collapse somewhat similar to the dynamics in the previous block. The delay function should be always used with caution, since it can easily destabilize your model.

Figure 3.9   (Continued)

Further reading

If you feel that your math is too flaky you may want to refresh it. Any textbook in calculus will be more than enough. Try this one for example: Thomas, G.B., Finney, R.L. (1989). Elements of Calculus and Analytic Geometry. Addison-Wesley.These days you can also find a lot on the web. Just type "differential equations" into Google and you will get quite a few links with pretty good explanations to choose from.

Berlinski, D. (1978). *On Systems Analysis. An Essay Concerning the Limitations of Some Mathematical Methods in the Social, Political, and Biological Sciences.* MIT Press – *This does a really good job explaining why mathematics can be quite important for building good models. Berlinski may be overly critical of some of the classic modeling treatises, including books of Bertalanffy and Meadows, however most of his criticism makes a lot of sense. It is important to remember that models are more than mathematical objects, and that in some cases they may be useful even with flawed or inadequate mathematics.*

Vladimir I. Arnold has been stressing the difference between soft and rigid models in his 1997 presentations. His classic book: Arnold, V. I. (1992). *Ordinary differential equations.* Springer-Verlag – *Can be recommended for those who want to get a better understanding of modeling with ODE's and master some analytical techniques.*

von Bertalanffy, L. (1968). *General System Theory.* George Braziller – *Contains some important mathematics and ideas about the building blocks in modeling.*

4. Model Analysis

SUMMARY

There are many ways in which a model can be analyzed and tested, and some of them have become more-or-less standard for the trade. There may be many unknowns or assumptions that go into the model. Sensitivity analysis is a way to figure out how important these assumptions are and what effect they may have on the model performance. Sensitivity can be tested by disturbing a model component that is not known for certain (a parameter, a function, a link), and then seeing how this disturbance propagates through the model structure and how different the results that come from the disturbed model are. A second standard analysis is performed to see how closely the model can be made to reproduce the experimental data (qualitative and quantitative). This is model calibration. The model parameters are modified to minimize the difference between model output and the available data. Finally, other tests can be conducted to validate the model and verify its performance. This analysis includes different methods, ranging from diligent debugging of software code and mathematical formalizations to comparisons with independent data sets, and extensive scenario runs.

Keywords

Uncertainties, parameters, initial conditions, critical parameters, inverse problem, data model, error model, Theil's index, R^2 index, weighted average, empirical model, trendline, process-based modeling, objective function, minimization, trial and error, optimization, Madonna software, curve fitting, open systems, CLIMBER model, validation, verification, scenario, credibility.

* * *

Choosing variables and connecting them with flows and processes is not enough to build a model. Actually, this is just the beginning of the modeling process. By identifying the variables and formalizing the processes that connect them, in Stella or in any other modeling tool, only one possible description of the system is created. We still need to make sure that this description really describes the system, and then try to use the model in a meaningful way to generate additional knowledge about the system. Why else model at all?

This stage of testing and working with the preliminary model built is called *model analysis*. If the model is a mathematical formalization – say, a system of ordinary differential equations – we may try to solve the equations. If this is possible, we get a functional representation for all model variables and can pretty much say what they will be at any time or place, and see clearly how different parameters affect them. However, as previously indicated, the chances are quite slim that we will get an analytical solution. We may still try to analyze the phase plane of the model variables and derive some general understanding of the model behavior – perhaps by testing for equilibrium conditions, or trying to identify when variables grow and when they decline. The more results we can derive from this analytical analysis the better, because all the analytical information we obtain is general and it describes the system behavior for all kinds of parameter values that we may insert into the model – not just the single set of parameters that we use when we run the model numerically on the computer.

4.1   Sensitivity analysis

If no analytical analysis is possible, we have to turn to numerical methods. Using Stella, in order to see how the model performs we need to "Run" it. By doing this, we numerically solve the system of difference equations that Stella has put together based on the diagram and process formalizations that we have formulated. A numerical solution of a model requires that all parameters take on certain values, and as a result is dependent on the specified parameter values. The result of a model run is dependent on the equations we choose, and the initial conditions and parameters that are specified. Some parameters do not matter much; we can vary them quite significantly, but will not see any large changes in the model dynamics. However, other parameters may have a very obvious effect on the model performance. Even small changes in their values result in dramatically different solutions.

Analyzing model performance under various conditions is called *sensitivity analysis*. If we start modifying a parameter and keep re-running the model, instead of a single trajectory we will generate a bunch of trajectories. Similarly, we can start changing the initial conditions or even some of the formalizations in the process descriptions. By comparing the model output, we get an idea of the most essential parameters or factors in the model. We also get a better feeling of the role of individual parameters and processes in how the model output is formed, what parameters affect what variables, and within which ranges the parameters may be allowed to vary. This is very important because, in contrast to an analytical solution where we could find an equation relating model output to the input parameters, with numerical models we do not have any other way to learn what the connection is between the various parameters and the model output, except by rerunning the model with different parameter values. Whereas in the analytical solution we can use a formula that clearly shows how a parameter affects the output, in case of numeric runs we know nothing about what to expect from the output when a parameter changes.

In Stella, there is a method of making estimates for model sensitivity. Choose "Sensi Specs…" in the Run menu. A window will open that will allow you to set up your sensitivity test.

The following steps will be required:

1. Double click on the parameter that you want to test for model sensitivity. It will be moved to the right pane.
2. Highlight the parameter in the right pane.

3. Set the number of parameter values that you wish to test for.

4. Choose how you want the parameter to change its value.

5. Make sure you click the Set button to fill in the table on the right, where parameter values will be automatically calculated to run the model.

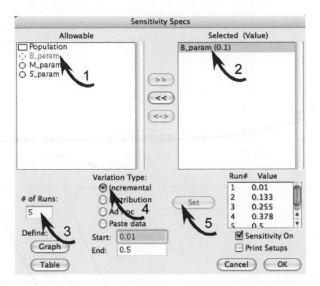

If you now click "OK," the model will run several times in a row for the different values of the parameter chosen. Before you do that, you need to prepare your output. Make sure you create a "Comparative" graph to see the difference in the output that you will be generating. For example, in the model that we were building above, if we start changing the Birth Rate parameter we will produce a family of curves, which show that the model is quite sensitive to changes in this parameter.

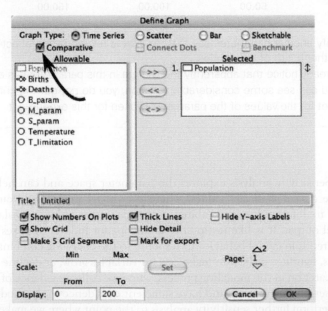

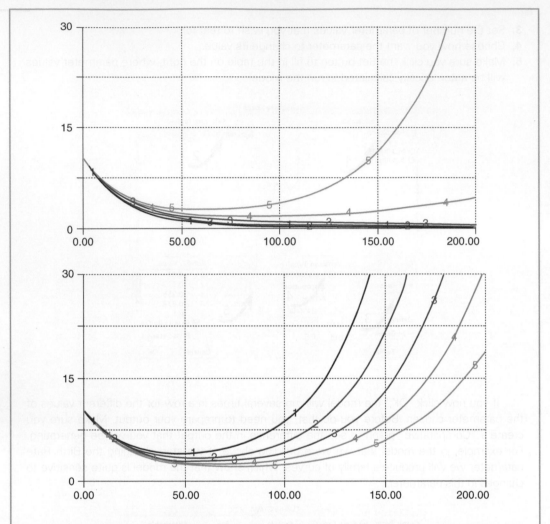

If you modify another parameter, say the one that is related to the effect of temperature, you will get another bunch of trajectories.

You may already notice that apparently the change in this parameter has a less prominent effect. While you can see some considerable variation, you do not get the curve to decline to zero – at least not for the values of the parameter chosen for this experiment.

Sensitivity analysis explores the parameter space and can help us identify some of the *critical parameter* values, where the model might, for example, crash or run away to infinity. Every combination of parameter values translates into a specific model output. It is like testing the landscape for hidden surprises and trying to capture trends in model behavior in response to the changing combinations of parameter values, figuring out how to make certain variables grow, or decline and at what time.

Later on in the modeling process, when we collect evidence of the model actually representing the system, and have sufficient confidence in the model performance, we can perform further sensitivity analysis to the point where we make conclusions about the sensitivity of the original system to certain processes and factors. It will then help identify those processes that should get the most attention in experimental research, and which may become important management tools if we intend to modify the system behavior to match certain criteria.

A full sensitivity analysis of the model is quite a difficult task, since changing parameters one by one and in combinations may produce entirely different results, especially with non-linear equations. However, even a partial analysis that looks at only some parameters and their combinations is certainly better than nothing. It will also be of great help for the next step of model analysis, which is calibration.

4.2   Model calibration

The next thing we need to do when analyzing the model is to compare its output with the other data that are available about the system. In many cases we may have better data about the dependent variables in the model than data regarding the independent variables or parameters. For instance, US Geological Survey (USGS) provides quite extensive data sets for water flows measured over a network of river gauges. For a stream hydrology model that is to produce river flow dynamics, we will most likely have quite good information about the flows but poor data about the hydrologic coefficients, such as infiltration, transpiration and evaporation rates, etc.

By solving an *inverse problem* we will be determining the values of parameters such that the model output will be as close as possible to the observed data. This process of model refinement in attempt to match a certain existing data set is called *model calibration*. We compare the model output to the data points, and start changing the model parameters or structure in order to get a fit that is as close as possible. Suppose that in our example there are certain measurements of the population size changing in time. As we compare the model output to the data points, we may start modifying the B_param or the S_param parameters in order to get as close a fit as possible. It is like tuning a radio by slowly turning the dial on the radio and listening to the sound of it until the best reception possible is found – or, when cooking, you add a certain amount of salt, then try the food, then add a little more until you get the taste you like. However, it may be hard to take the salt out if you add too much. Not so in modeling – you can do whatever you want with model parameters.

How is it that we are solving an *inverse problem*? Suppose we have a simple linear equation: $y = ax + b$. This will be our direct problem: if we know the values of a and b for any x value, we can calculate the value for y. As a result, we get a graph that is a straight line describing the linear functional response defined by the equation. Suppose now that we do not know what the values of a and b are, but we do know what the graph looks like – that is, we know that there are two points with coordinates (x_1, y_1) and (x_2, y_2) that sit on this graph. (Since we can draw only one line passing through two given points, the two coordinates that are defined above should be enough to define the graph.) So how do we figure out the equation for the line that will pass through these two points? Let us solve the inverse problem.

We can write that $y_1 = ax_1 + b$ and $y_2 = ax_2 + b$. Now a and b are the unknowns. Solving this system of linear equations, we immediately get:

$$b = y_1 - ax_1; \quad y_2 = ax_2 + b = ax_2 + y_1 - ax_1$$

It follows that

$$a = \frac{y_2 - y_1}{x_2 - x_1}$$

and

$$b = \frac{y_1 - x_1(y_2 - y_1)}{x_2 - x_1} = \frac{y_1 x_2 - x_1 y_2}{x_2 - x_1}$$

By solving the inverse problem, we have identified the parameters of our equation based on the graph of the observed function. That is pretty much exactly what we are doing in the calibration effort, except here we have the luxury of an analytical solution, which is quite rare in real models.

Sensitivity analysis may have already informed us which parameters need to be modified to produce a certain change in model trajectories. Now we actually change them in such a way that the trajectory of the model output matches the data "closely enough." How closely? This depends on our level of confidence in the data we have and upon the goals of our study. It also depends on the model that we built. Sometimes we find it very difficult to produce the change in the model output that is needed to get the trajectories closer to the data points. Sometimes it is simply impossible, and we have to find other ways to fix the model, either digging into its structure, or realizing that we have misinterpreted something in the conceptual model or in the time or space scales. Modeling is an iterative process, and it is perfectly fine to go back and re-evaluate our assumptions and formalizations.

Note that the data set used for calibration, in a way, is also a model of the process observed. The data are also a simplification of the real process, and they may also contain errors; they are never perfect and, besides, they have been collected with a certain goal in mind, which does not necessarily match the goal of the newly built numerical model. We may call these monitoring results an experimental model or a *data model*. In this process of calibration we are actually comparing two models and modifying one of them (simulation) to better match the other (data).

This model comparison will be important at all stages of model analysis – here, when we calibrate the model, and further on when we test it. When comparing models, it makes sense to think of a measure of their closeness, or a measure of the fit of the simulation model to the data model. We may call this measure the *error model*. There may be very different ways to represent this error, but they all have in common one feature, which is that they represent the distance between the data model and the formal model. These error models may be qualitative and quantitative. The very simplest qualitative error model is "eyeballing," or visual comparison. That is when we simply look at the graphs and decide whether they are close enough or not. In doing this visual comparison, we may focus on various features of the model. For example, it may be important that the peaks are properly timed and match well, in which case we will be paying special attention to how the maximums and minimums compare. Alternatively, we may be more concerned about the overall comparison, and instead look at the trend (growth vs decline) or the average values in the dynamics. We may not even know what the data are, and calibrate only for the range of possible variation.

Qualitative comparison may become difficult as we close on our target, getting the model output almost identical to data. We may be still improving the results somewhat, but we can no longer distinguish the gains by simply staring at the graphics. Another case is when we get a better match between output and data in one time range for one set of parameters, but achieve a better match with a different time range for another set of parameters. Which parameters do we choose then? In these cases, visual comparisons can fail. Quantitative mathematical formulas can then become useful. One simple formula for the error model is:

$$E = \sum_{i=1}^{n} \frac{(x_i - y_i)^2}{y_i^2} \tag{4.1}$$

where x_i are the data points and y_i are the values in the model output that correspond in time or space to the data points. Note that this formula tracks the relative proximity of the two models – that is, for larger values we allow larger errors. The smaller the error, E, the better the model calibration. This index is quite similar to Theil's measure of forecast quality:

$$E_t = \frac{\left[\sum_{i=1}^{n} (x_i - y_i)^2\right]^{1/2}}{\left[\sum_{i=1}^{n} y_i^2\right]^{1/2}} \tag{4.2}$$

In some cases, we may be concerned only with the average values over certain time periods. Then we can compare the mean values:

$$E_a = \frac{\left|\sum_{i=1}^{n} x_i - \sum_{i=1}^{n} y_i\right|}{n} = \frac{\left|\sum_{i=1}^{n} (x_i - y_i)\right|}{n} \tag{4.3}$$

Very often, the metric used to compare the models is the Pearson moment product correlation coefficient,

$$r = \frac{n\sum_{i=1}^{n} x_i y_i - \sum_{i=1}^{n} x_i \sum_{i=1}^{n} y_i}{\sqrt{\left[n\sum_{i=1}^{n} x_i^2 - \left(\sum_{i=1}^{n} x_i\right)^2\right]\left[n\sum_{i=1}^{n} y_i^2 - \left(\sum_{i=1}^{n} y_i\right)^2\right]}} \tag{4.4}$$

or the R^2 value, which is equal to r^2. This correlation coefficient is good for matching the peaks. Note that unlike the above error models, where the best fit came with the minimal value of E, here the best fit is achieved when $r^2 = 1$.

These formulas become more cumbersome if we calibrate for several variables at once. In the simplest case, we can always take an average of error models for individual state variables:

$$E^* = \frac{\sum_{j=1}^{k} E_j}{k}$$

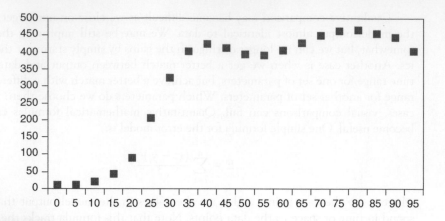

Figure 4.1 Experimental model of the microbial system.

where E_j are the individual errors calculated using equations (4.1)–(4.3) or similar metrics. In some cases, calibration with regard to one variable may be more important than the fit for the other ones. For example, when we have the best data model for a certain variable but very approximate information about the others, we would want to make sure that we calibrate more to the reliable information, while the importance of other data sets may be secondary. In this case, we may want to introduce certain weights into the formula so that particular variables get more attention in the comparison:

$$E_w^* = \frac{\sum_{j=1}^{k} w_j E_j}{k}$$

where w_j are the weights associated with k different variables,

$$\sum_{j=1}^{k} w_j = 1$$

The error model is then affected most of all by the variable that has the higher weight. This means it is more efficient to get the error down for that variable as far as possible, since the total error then gets reduced the most.

Let us consider an example. Suppose that we have been running an experiment in the lab measuring the growth of a batch of microorganisms over a period of 100 hours, taking a sample every 5 hours. We then use a spreadsheet program to store the results and to present them in a graphic format (Figure 4.1). Also suppose that we are measuring a certain limiting factor – say, temperature, or substrate availability – that describes how suitable the lab environment is for the growth of the organisms that we are observing (Figure 4.2). We are normalizing this measured value to bring it within a range of [0,1]. This can be done if we divide all the data by the maximum observed value.

Let us build a model of the system. Suppose we are not interested in the structure of the system and want to build an empirical, "black-box" model.

Empirical model

The output that we have consists of the data about the number of organisms. The input is time, and the information about the temperature in the environment. One simple empirical model can be created immediately in a spreadsheet program. For example, in Excel it is called "adding a trendline to the graph."

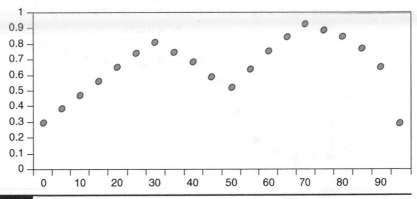

Figure 4.2 Experimental model of temperature in the microbial system.

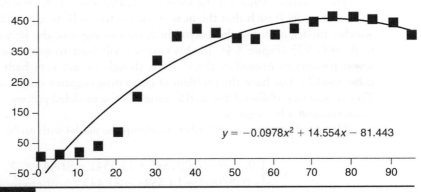

$$y = -0.0978x^2 + 14.554x - 81.443$$

Figure 4.3 A trendline as a black-box model that uses time as input.

In this case, the only input information that is used is time. The model is the equation of the line, which is a polynomial of order 2:

$$y = -0.0978x^2 + 14.554x - 81.443$$

As we can see in Figure 4.3, the trendline does a pretty good job of representing the model results, though there is obviously a difference between the model output and the data points available. Note that Excel labels the independent variable x, while in our case it should rather be t for time. By adjusting some of the parameters in the model, we may make the model output closer to or further away from the data points measured in the experiment. Actually, this is exactly how Excel came up with this equation. It took a general form of a second-order polynomial and started to tweak the three coefficients. We can see how this works if, instead of "Adding the trendline" in the Chart menu, we set up a general form of polynomial and use the "Solver" option in the "Tools" menu. We will then be able actually to see how the values of the three coefficients will be modified while Excel will be optimizing something to get the two curves to match as closely as possible.

This process of tweaking the model parameters in an attempt to get a better representation of the data available is the calibration of the model. In our case, the coefficients of the polynomial are the unknown model parameters that have been varied in an attempt to get the polynomial trendline as close as possible to the data points.

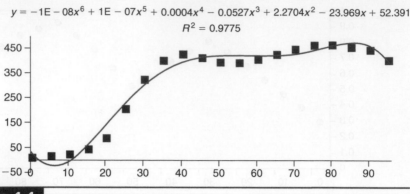

$$y = -1E - 08x^6 + 1E - 07x^5 + 0.0004x^4 - 0.0527x^3 + 2.2704x^2 - 23.969x + 52.391$$
$$R^2 = 0.9775$$

Figure 4.4   A fifth-order polynomial as a black-box model that uses time as input.

The R-squared value for the model described above is $R^2 = 0.9232$. Recall that this error model is such that the fit is getting better as R^2 is approaching 1. If we use another model, a sixth-order polynomial, we can improve the R^2 value and raise it to $R^2 = 0.9775$ (Figure 4.4). In this case we will have to guess the best values for seven parameters instead of three. Even though we get very high R^2 values from these models, they have the problem of generating negative output at certain times. This should be prohibited due to the nature of the modeled process – the population numbers cannot be negative.

The simplest way to avoid this is to clamp the model with an "if" statement:

$$y = \begin{cases} 0, & \text{if } -0.0978x^2 + 14.554x - 81.443 < 0 \\ -0.0978x^2 + 14.554x - 81.443, & \text{otherwise} \end{cases}$$

This would be then our empirical model, where the numeric coefficients are the calibrated values.

There are other statistical tools that are available in Excel (such as the Solver or the Goal Seek tools) or in other packages that may be further used for a refinement of our calibration. We may also try to bring in the other available data set – that is, temperature – and run multiple regression for time and temperature to try to improve further our empirical model; however, this will require more sophisticated statistical tools than Excel, unless we formulate our own equation and use the Solver to minimize the error model.

In any case, what is important is that, when building these empirical models, we entirely rely on the information that we have in the data sets. We come up with some type of equation and then quite mechanically adjust the parameters in an attempt to reproduce the data as well as possible. All the information we know about the system is in the data. It may be somewhat risky to use the same model in different conditions – for example, when the temperature is consistently 5° lower. Temperature has not been included in this model at all, and clearly the results will be totally off if it changes.

Process-based model

Instead of further exploring the empirical model, let us try to build a process-based model for the microbial system that we are studying. We will draw on some of our understanding of population growth, consider some of the processes that may be involved, and describe them in the model. This brings up a whole different paradigm of modeling, where, in addition to the information contained in the data sets, we bring in other information available from similar studies conducted before on similar systems, or from general ecological theory, or from mass-conservation laws, or simply from common sense.

For the microbial system that we are considering, just as for any other population, the processes of growth and death are most likely playing an important role. Perhaps we can try to describe the life of the whole population in terms of these two processes. The simplest model of population growth can be then presented by the following Stella equations:

Population(t) = Population(t − dt) + (Growth − Mortality) * dt

INIT
Population = 10
INFLOWS:
Growth = GrowthRate*Lim_factor*Population*(1 − Population/C_Capacity)

OUTFLOWS:
Mortality = MortalityRate*Population
C_Capacity = 500
GrowthRate = 0.6
MortalityRate = 0.15.

We can also insert the values for the limiting temperature factor as a graphic:

Lim_factor = GRAPH(TIME)
(0.00, 0.305), (10.0, 0.47), (20.0, 0.65), (30.0, 0.815), (40.0, 0.7), (50.0, 0.505), (60.0, 0.745), (70.0, 0.93), (80.0, 0.86), (90.0, 0.71), (100, 0.00)

By looking at the data points we see that, after the initial period of rapid growth, the population size seems to saturate at a certain level. As we have seen above, there is a simple way to control growth in the model by introducing the Carrying Capacity, which represents the maximum number of organisms that can survive in the lab environment. With the parameters listed above, the model produces the following dynamics:

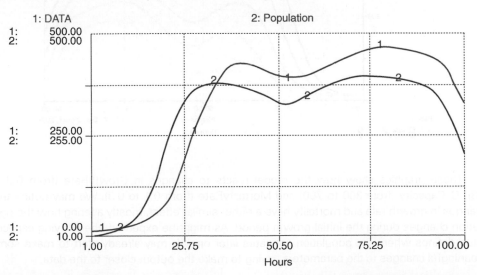

Curve (1) on the graph represents the experimental values that we have been observing, while curve (2) is the simulated behavior. Here, too, we see that there is a certain error or distance between the two models. The size of this error depends on the parameter values used in the model. Let us run sensitivity analysis for the three parameters in this model.

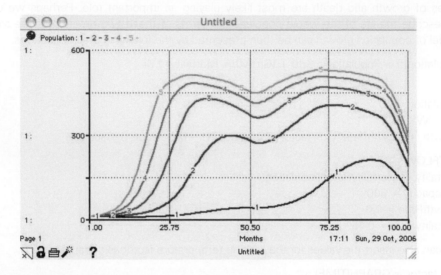

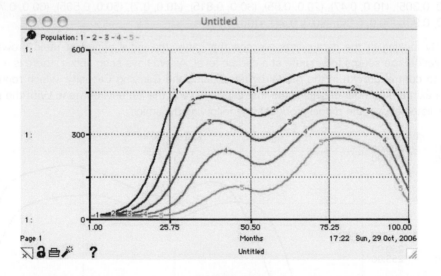

These graphics show how the model reacts to changes in GrowthRate (from 0.3 to 0.8), C_Capacity (from 300 to 700) and MortalityRate (from 0.1 to 0.3). We may notice that changes in growth rate and mortality have a rather similar effect, mostly altering how the population changes during the initial growth period. As might be expected, the carrying capacity value defines where the population saturates later on. We may already start to make some meaningful changes to the parameters, trying to make the output closer to the data.

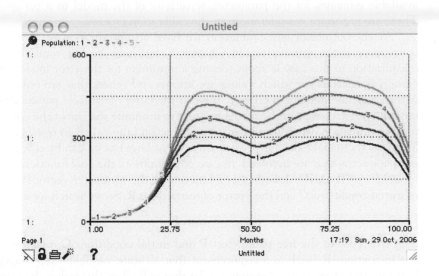

To keep track of our gains and losses, we can put together an error model. Described in terms of Stella equations, the error model might be as follows:

Error(t) = Error(t−dt) + (Er_In) * dt
INIT Error = 0
INFLOWS:
Er_In = Population − DATA)^2/DATA^2

This formula reproduces the metrics described above as the sum of squares E in (4.1). Notice that at each time-step we add another error term, which makes it equivalent to the summation that we see in (4.1). Keeping in mind the results of sensitivity analysis, we can now start to tweak some of the model parameters and see how this changes the distance between the data and population that is also measured by the error variable. Most likely the GrowthRate will need to go down a little to make the population grow slower, but the C_Capacity should probably go up to make it saturate at a higher level. That should bring the model output somewhat closer to the Data. This is an iterative trial-and-error process that may or may not get us to the perfect match.

You may have noticed that there is a difference in calibrating empirical and process-based models. In empirical models, we rely entirely on the information that we have in the data sets. We come up with some type of equation, and then quite mechanically adjust the parameters in an attempt to reproduce the data as well as possible. All the information we know about the system is in the data, and the parameters usually can take any values as long as the error model is minimal.

In process-based models calibration is different, since we are restricted by the ecological, physical or chemical meaning of the parameters that we change. Besides, there are usually some estimates for the size of the parameters; they are rarely precisely measured, but at least the order of magnitude or a range is usually known. Moreover, there are other factors that may play a role, such as confidence in the available estimates for the parameter, sensitivity of the model to a parameter, etc. These are important considerations in the calibration process.

At the bottom of any calibration we have an optimization problem. We will learn more about optimization in Chapter 8, but here we just want to note that optimization in this case is about seeking a minimum for the error model. We have certain parameters for which values are known and others that are only estimated within a certain domain of change. We call the latter ones "free" parameters. These are the ones to change in the model in order to minimize the size of the error. To perform optimization, we first formulate a *goal function* (also called an *objective function*). Then we try to make this function as little (or as large) as we can by changing different parameters that are involved. In case of calibration, the goal function is the error model $E = f(\mathbf{P}, \mathbf{C}, \mathbf{R})$, described as a function of the parameter vector $\mathbf{P}$, the vector of initial conditions $\mathbf{C}$ and the vector of restrictions $\mathbf{R}$. So we search for a minimum:

$$\min E$$

over the space of the free parameters $\mathbf{P}$ and initial conditions $\mathbf{C}$, making sure that the restrictions $\mathbf{R}$ (such as a requirement that all state variables are positive) hold. It is rare that there is a real system model that will allow this task to be solved analytically. It is usually a numerical procedure that requires the employment of certain fairly complicated software.

There are different ways to solve this problem. One approach is to do it manually, as we did above with the so-called trial-and-error method or educated-guess approach. The model is run, then a parameter is changed, then the model is rerun, the output is compared, the same or another parameter is changed, and so on. It may seem quite tiresome and boring, but actually this process is extremely useful in understanding how the system works. By playing with the parameters we learn how they affect output (as in the sensitivity analysis stage), but we also understand the synergetic effects that parameters may have. In some cases we get quite unexpected behavior, and it takes some thought and analysis to explain how and why the specific change in parameters had this effect. If no reasonable explanation can be found, chances are there is a bug in the model. A closer look at the equations may solve the problem: something may have been missed, or entered with a wrong sign, or some effect may not have been accounted for.

In addition to the educated-guess approach, there are also formal mathematical methods that are available for calibration. They are based on numerical algorithms that solve the optimization problem.

Some modeling systems have the functionality to solve the optimization problem and do the curve fitting for models. One such package is **Madonna**. One big advantage of Madonna is that it can also take Stella equations almost as is and run them within its own shell. Madonna also has a nice graphic user interface of its own – so it is as well for us to start putting the model together directly in Madonna, if we expect some optimization to be needed.

To do the parameter calibration for our Stella model in Madonna we will have to:

- Go to the Stella equations
- Save them as a text file (File -> Save As Text)
- Open the file from Madonna, using the Open command in the File menu

- (Alternatively you can "choose all" and "copy" the equations from Stella, and then "paste" them directly into an Equations window in Madonna; however, in this case you will have to remove all the "INFLOW:" and "OUTFLOW:" statements in the equations by hand)
- Define the control specs such as the STARTTIME, STOPTIME, and DT

The model is now ready to run in Madonna.

Running the same population model, built now in Madonna, we get the following output, which is – not surprisingly – identical to the Stella output:

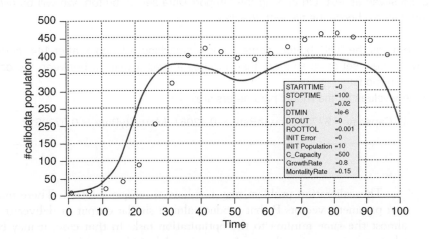

As we start running the model, the first thing we notice is that Madonna runs much faster than Stella. That is because in contrast to Stella, which interprets the equations on the fly, Madonna has a built-in compiler that first compiles our model and only then runs it. On some models, the difference is quite significant, up to orders of magnitude. This is especially essential for optimization, since all optimization algorithms require numerous model runs to be performed.

The next thing we need to do to calibrate our model is input the data into Madonna. This is done as part of the optimization dialogue, which in this case is called Curve Fitting. In the "Parameters" menu, we choose "Curve Fit...." A dialogue box will open:

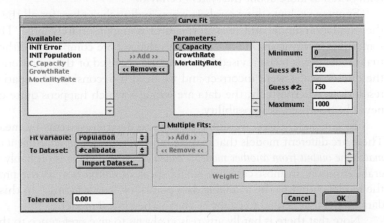

Here, we need to specify four items:

1. Choose the free parameters that can be changed for model calibration
2. For each parameter, identify the maximal and minimal allowed values, and two "guesses" – values in the domain of change that will be used to initialize the optimization process
3. Choose the state variable that we are calibrating – "Population" in this case
4. The data set to which we wish to calibrate the model – "#calibdata" in this case. The data set should be in a file, one value on a row, which can be generated, say, from Excel if the data are saved as Text. On clicking the "Import Data set…" button, we will be given the opportunity to choose the file with the data.

Now, if we press the "OK" button, some number crunching will begin; after 144 model runs we will get a new set of parameters that provides a much closer fit between the data and the simulation model.

The new values for the model parameters are:

C_Capacity = 577.3, GrowthRate = 0.42061, MortalityRate = 0.0760512

The calibration problem may not have a unique solution. There may be several parameter vectors **P** that produce almost similar output or deliver the same or almost the same minima to the optimization task. In that case, it may be unclear what parameters to choose for the model. Other considerations and restrictions may be used to make the decision. For instance, with C_Capacity = 600, GrowthRate = 0.5, MortalityRate = 0.1, we get a fit almost as good as that achieved with Madonna. Which of the two parameter sets should we choose for the model? Normally this decision is made based on the other information about the system that is available. For example, there may be some experimental data that would either identify a value for one of the rate coefficients, or at least put a range on them. Then we can see which of the calibrated values is in better agreement with these restrictions. In some cases this information may not be available, and there may be some uncertainty about the system. This can further drive our experiments with the system, or tell us more about the system behavior.

Suppose we have done our best when finding the values for all the parameters in the simulation model and yet still the error is inappropriately large. This means that something is wrong in one of the models that we are comparing. Either the conceptual model needs to be revised (the structure changed or the equations modified), or the chosen scales were incorrect and we need to reconsider the spatial or temporal resolution. Alternatively, the data are wrong – which happens quite often, and can never be dismissed as a possibility.

To conclude, there are different ways to describe systems by means of models. There are different models that may be built. *The process of adjustment of one model to match the output from another model is called calibration.* This is probably the most general definition. In most cases we would speak of calibration as the process of fitting the model output to the available data points, or "curve fitting." In this case, it is the data model that is used to calibrate the mathematical model.

Note that there is hardly any reason always to give preference to the data model. The uncertainty in the data model may be as high as the uncertainty in the simulation model. The mathematical model may in fact cover areas that are not yet presented in data at all. However, in most cases we will have data models preceding mathematical models, and, at least initially, assume that the data models convey our knowledge about the system.

Empirical models are entirely based on data models, they may be considered as "extensions" of the data models. They attempt to generalize the data available and present them in a different form. The process-based models, in addition to knowledge about the modeled system, may also employ information about similar systems studied elsewhere, or they may incorporate theoretical knowledge about processes involved. In a way, these process-based models can be even better than the data available for the particular system that is modeled. Therefore, we may hope that process-based models will be performing better outside of the data domains that were used for their calibration, and thus it may be easier to apply process-based models to other similar systems than to use empirical models, which would require a whole new calibration effort.

The calibration problem may not have a unique solution. There may be several parameter vectors **P** that deliver the same or almost the same minima to the optimization task. In this case, it may be unclear what parameters to choose for the model. Other considerations and restrictions may be used to make the decision.

Exercise 4.1

Download the calibration exercise model from the book website, or reconstruct it from the equations below. In this model the output once matched the data, but then someone came and changed some of the parameters. As a result, the model is no longer producing the right dynamics. Can you help to find the original parameter values? What you actually need to do is to calibrate the model to the curves that are now called Data. At least in this case you can be sure that the solution to the calibration task exists. Please remember that initial conditions are also a special case of parameters, so make sure you check them as well. Take a look at the comments in the equations. It always helps to understand what the equations are doing, and why certain formalism has been chosen to describe the system.

Algae(t) = (Algae(t − dt) + A_grow − A_mort) * dt
INIT Algae = 1
INFLOWS:
A_grow = c_a_grow*Nutrients*Algae*T_limit
{Algal growth is dependent upon the available Nutrients and is limited by temperature.}
OUTFLOWS:
A_mort = c_a_mort*Algae
{Mortality is proportional to the existing biomass}

Detritus(t) = Detritus(t − dt) + (A_mort + M_mort − D_decomp − M_d_grow − Out) *dt
INIT Detritus = 10
INFLOWS.
A_mort = c_a_mort*Algae
M_mort = c_m_mort*Macrophytes
{Detritus is produced by the dead Algae and Macrophytes}
OUTFLOWS:
D_decomp = c_decomp*Detritus

{Loss of detritus due to decomposition}
M_d_grow = c_md_grow*Detritus*Macrophytes
{A part is consumed directly by Macrophytes that grow on Detritus}
Out = 0.09
{A certain part is removed with a constant outflow}

Macrophytes(t) = Macrophytes(t − dt) + M_grow + (M_d_grow−M_mort) *dt
INIT Macrophytes = 3
INFLOWS:
M_grow = c_m_grow*Nutrients*Macrophytes
M_d_grow = c_md_grow*Detritus*Macrophytes
{Macrophytes can uptake both nutrients dissolved in water and detritus}
OUTFLOWS:
M_mort = c_m_mort*Macrophytes
{Mortality is a constant proportion of the biomass}

Nutrients(t) = Nutrients(t − dt) + (D_decomp + Load−A_grow−M_grow) *dt
INIT Nutrients = 0.2
INFLOWS:
D_decomp = c_decomp*Detritus
{Detritus is decomposed providing nutrients back into the water}
Load = c_load*Precipitation
{In addition nutrients are provided with surface runoff, which is generated by rainfall}
OUTFLOWS:
A_grow = c_a_grow*Nutrients*Algae*T_limit
M_grow = c_m_grow*Nutrients*Macrophytes

{These are the parameters that you may want to change. Also don't forget about the initial conditions}
c_a_grow = 0.06
c_a_mort = 0.07
c_decomp = 0
c_load = 1.2
c_md_grow = 0.002
c_m_grow = 0.012
c_m_mort = 0.03

{Temperature is defined as a function of time}
Temperature = SINWAVE(6,730)^2 + RANDOM(−4,4)*RANDOM(0,1)
{Growth rate of Algae is limited by this temperature dependent function}
T_limit = MAX(0, 0.05*Temperature*EXP(1−0.05*Temperature))

Algae_data = GRAPH (TIME)
(0.00, 1.00), (10.0, 0.49), (20.0, 0.26), (30.0, 0.17), (40.0, 0.12), (50.0, 0.09), (60.0, 0.09), (70.0, 0.09), (80.0, 0.14), (90.0, 0.42), (100, 1.28), (110, 1.61), (120, 1.30), (130, 1.08), (140, 0.85), (150, 0.79), (160, 0.59), (170, 0.38), (180, 0.38), (190, 0.36), (200, 0.4), (210, 0.48), (220, 0.88), (230, 1.41), (240, 1.76), (250, 1.40), (260, 0.96), (270, 0.69), (280, 0.74), (290, 0.89), (300, 1.18), (310, 0.96), (320, 0.6), (330, 0.37), (340, 0.23), (350, 0.15), (360, 0.1)

Detritus_data = GRAPH(TIME)
(0.00, 10.0), (10.0, 9.99), (20.0, 9.69), (30.0, 9.33), (40.0, 8.95), (50.0, 8.55), (60.0, 8.16), (70.0, 7.77), (80.0, 7.41), (90.0, 7.20), (100, 7.55), (110, 8.38), (120, 9.15), (130, 9.62), (140, 9.87), (150, 10.0), (160, 10.0), (170, 9.83), (180, 9.56), (190, 9.28), (200, 9.02), (210, 8.78), (220, 8.79), (230, 9.15), (240, 9.87), (250, 10.6), (260, 10.9), (270, 10.9), (280, 10.9), (290, 10.9), (300, 11.1), (310, 11.3), (320, 11.3), (330, 11.1), (340, 10.7), (350, 10.3), (360, 10.1)

Macrophytes_data = GRAPH(TIME)
(0.00, 2.00), (10.0, 1.70), (20.0, 1.78), (30.0, 1.92), (40.0, 1.94), (50.0, 1.92), (60.0, 1.89), (70.0, 1.88), (80.0, 1.94), (90.0, 2.39), (100, 2.92), (120, 2.58), (130, 2.39), (140, 2.20), (150, 2.14), (160, 1.97), (170, 1.74), (180, 1.74), (190, 1.71), (200, 1.74), (210, 1.81), (220, 2.10), (230, 2.31), (240, 2.38), (250, 2.21), (260, 1.99), (270, 1.81), (280, 1.82), (290, 1.90), (300, 2.08), (310, 2.03), (320, 1.87), (330, 1.76), (340, 1.88), (350, 2.40), (360, 2.99)

Nutrients_data = GRAPH(TIME)
(0.00, 0.2), (10.0, 1.05), (20.0, 2.61), (30.0, 2.07), (40.0, 1.87), (50.0, 1.82), (60.0, 1.58), (70.0, 2.12), (80.0, 2.87), (90.0, 4.93), (100, 2.40), (110, 1.34), (120, 1.18), (130, 1.09), (140, 1.92), (150, 1.02), (160, 0.82), (170, 1.72), (180, 1.57), (190, 1.95), (200, 1.64), (210, 3.52), (220, 2.28), (230, 2.36), (240, 1.29), (250, 0.98), (260, 0.44), (270, 1.99), (280, 2.39), (290, 1.72), (300, 1.70), (310, 0.85), (320, 0.79), (330, 1.75), (340, 3.42), (350, 3.97), (360, 3.05)

Precipitation = GRAPH(TIME)
(1.00, 0.00), (5.00, 0.41), (10.0, 0.00), (15.0, 1.47), (20.0, 0.00), (25.0, 0.00), (30.0, 0.00), (35.0, 0.05), (40.0, 0.00), (45.0, 0.3), (50.0, 0.09), (55.0, 0.00), (60.0, 0.00), (65.0, 0.4), (70.0, 0.00), (75.0, 0.00), (80.0, 0.01), (85.0, 1.40), (90.0, 0.97), (95.0, 0.00), (100, 0.00), (105, 0.156), (110, 0.03), (115, 0.1), (120, 0.13), (125, 0.00), (130, 0.15), (135, 0.00), (140, 0.00), (145, 0.00), (150, 0.02), (155, 0.00), (160, 0.33), (165, 0.00), (170, 0.89), (175, 0.17), (180, 0.00), (185, 0.20), (190, 0.29), (195, 0.00), (200, 0.00), (205, 0.38), (210, 0.17), (215, 0.00), (220, 0.00), (225, 1.10), (230, 0.00), (235, 0.00), (240, 0.22), (245, 0.00), (250, 0.00), (255, 0.00), (260, 0.00), (265, 0.34), (270, 0.00), (275, 1.20), (280, 0.00), (285, 0.00), (290, 0.39), (295, 0.00), (300, 0.00), (305, 0.00), (310, 0.00), (315, 0.60), (320, 0.00), (325, 0.00), (330, 0.00), (335, 0.41), (340, 0.00), (345, 0.23), (350, 0.00), (355, 0.01), (360, 0.00)

4.3   Model testing

Now we have a simulation model that represents the data set closely enough. Does this mean that we have a reliable model of the system, which we can use for forecast or management? Did we really capture the essence of the system behavior, do we really understand how the system works, or have we simply tweaked a set of parameters to produce the needed output? Are we representing the system and the processes in it, or, as in empirical models, are we only seeing an artifact of the data set used?

We build process-based models with the presumption that they describe the "guts" of the system and therefore are general enough to be reapplied in different conditions, since they actually describe how the system works. That would be indeed true if all the parameters in the process formulations could be measured experimentally and then simply substituted into the model. However, usually these data are non-existent or imprecise for many of the parameters.

The solution is to approximate the parameter values based on the data we have about the dynamics of state variables, or flows. That is the model calibration procedure. We are solving an inverse problem: finding the parameters based on the dynamics of the unknowns. This would be fine if we could really solve that problem and find the exact values for the parameters. However, in most cases this is also impossible and, instead, we are finding approximate solutions that come from model fitting. But then how is this different from the fitting we do when we deal with empirical models? In that case, we also have a curve equation with unknown coefficients, which we determine empirically by finding the best combination of parameters that make the model output as close as possible to the data.

The only difference is that instead of some kind of generic equation in the empirical models (say, a polynomial of some form), in process-based models we have particular equations that have some ecological meaning. These equations display certain behavior by themselves, no matter what parameters are inserted. A polynomial can generate pretty much arbitrary dynamics as long as the right coefficients are chosen. However, an equation of exponential growth will always produce an exponent, and, say, a classic predator–prey system (considered in the next chapter) will always produce oscillations, no matter what coefficients we insert. Of course, for some parameters they may crash even before generating any meaningful output, but otherwise the dynamics will be determined by the type of equations used, at least for a large enough range of coefficients. So we may conclude that, to a large extent, we are building a good model as long as we chose the right dynamic equations to describe our system.

On top of the basic dynamic equations we overlay the many other descriptions for the processes that need to be included in the model. These may be the limiting factors, describing the modifying effect of temperature, light or other external conditions. There may be some other details that we wish to add to the system. However, if these processes are not studied experimentally, and if the related coefficients are not measured, their role in the model is no different from that of the coefficients that we have in an empirical model. In both cases we figure out their values based on a time-series of model output; in both cases the values are approximate and uncertain. They are only as good as they are the best ones found; we can never be sure that a better parameter set does not exist.

So the bottom line is that there is a good deal of empiricism in most process-based models, and the more parameters we have estimated in the calibration process, the more empiricism is involved, the less applicable the model will be in situations outside the existing data range. How can we make sure that we have really captured the essence of the system dynamics, and can reproduce the system behavior beyond the domain that we have already studied?

To answer these questions, the model needs to undergo a process of vigorous testing. There is not (and probably never will be) a definite procedure for model testing and comparisons. The obvious reason is that models are built for various purposes; their goals may be very different. Moreover, these goals may easily change when the project is already underway. There is no reason why goal-setting should be left out of the iterative modeling process. As we start generating new knowledge and understanding with a model, its goals may very well change. We may start asking new questions and need to modify the model even before it has been brought to perfection.

Besides, ecological and socio-economic systems are open, which makes their modeling like shooting at a moving target. While we are studying the system and building a model of it, it is already evolving. It evolves even more when we start administering control, when we try to manage the ecosystem. As a result, models can very well become obsolete even before they are used to produce results. We are modeling the system as it was until a year ago, but during the last year, because of some external conditions (say, global climate change), the system has already evolved and the model is no longer relevant. The data set considered for calibration and collected during one period may not represent quite the same system as the one that produced the other data set that is intended for verification. We might be calibrating a model of one system, and then trying to verify the same model but for a different system.

Nevertheless, there are several procedures of model testing that became part of good modeling practice and should be certainly encouraged. Ironically, in various applications the names for these processes may be used interchangeably, which can only add to the confusion. "Model testing" is probably a more neutral and general term.

One way to test the model is to compare its output with some independent data set, which has not been used previously for model building and calibration. This is important to make sure that the model output is not an artifact of the model formalization, and that the processes in the model indeed represent reality and are not just empirical constructs based on the calibrated parameters. This process is called *validation* (or *verification*, in some texts). We start running the model for places and time periods for which we either did not have data, or have deliberately chosen to set the data aside and not use it for model calibration. We may have the luxury of waiting until the new data sets are acquired, making our predictions first and then comparing them with what we are measuring, or we may have enough data to afford to set aside some of them and pretend that we do not know these while constructing the model. Then, when the model is built and calibrated based on the remaining data, we will want to bring the other portion of data into light and see if we have equally well matched this other data set. This time we do not do any calibration and we do not tweak model parameters or functions, we only compare and estimate the error model. If the error is small, we may conclude that the model is good and have some confidence in applying the model for future predictions.

CLIMBER-2 is coupled climate model of intermediate complexity, designed for long-term climate simulations. The model has a coarse spatial resolution that can only resolve the continental-scale features and differentiate between oceanic basins. The atmosphere model is based on a statistical-dynamical approach, and does not resolve synoptic variability; instead, it parameterizes it as large-scale horizontal diffusion. The vertical structures of temperature, specific humidity and atmospheric circulation are parameterized. These parameterizations compute the vertical profiles of temperature, humidity and velocity that are used for calculating the three-dimensional advective, diffusive and radiative fluxes. These fluxes are computed using a multilevel radiation scheme (16 levels) that accounts for water vapor, ozone, CO_2 and the computed cloud cover. The ocean model is a multi-basin, zonally averaged model. The sea-ice model predicts ice thickness and concentration, and includes ice advection. In the hierarchy of models, this model is placed between spatially uniform energy balance models and general circulation models (GCMs). On a workstation, the model can be integrated for about 10,000 model years in a day.

The model has been used to analyze a variety of climatic situations. Here is one example. Many palaeoclimate records from the North Atlantic region show a pattern of rapid climate oscillations, the so-called Dansgaard–Oeschger (D–O) events, with a quasi-periodicity of 1,470 years for the late glacial period. Various hypotheses have been suggested to explain these rapid temperature shifts, including internal oscillations in the climate system, and external forcing, possibly from the Sun. There are well-known and well-pronounced solar cycles of 87 and 210 years, the so-called DeVries and Gleissberg solar cycles, but no one so far has detected a 1,470-year solar cycle.

Interestingly enough, when forced by periodic freshwater input into the North Atlantic Ocean in cycles of 87 and 210 years, CLIMBER simulated rapid climate shifts similar to the D-O events with a spacing of 1,470 years. This robust 1,470-year response time can be attributed to the superposition of the two shorter cycles, together with strongly non-linear dynamics and the long characteristic timescale of the thermohaline circulation. For Holocene conditions, similar events do not occur. It could therefore be concluded that the glacial 1,470-year climate cycles could have been triggered by solar forcing, despite the absence of a 1,470-year solar cycle. A frequency of 1,470 years is not found in the forcing; it is found only in the model response. The study is not aimed at suggesting a certain mechanism for solar influence on freshwater fluxes; this should be studied with more detailed and higher-resolution models. The simplified approach implies that the known solar frequencies are present in the hydrological cycle, and that translates into the D-O cycles by the complex and non-linear nature of the system.

Whenever the model can reproduce something that is observed but was not used to build the model, it speaks in favor of model validity. Moreover, in 2001 model results were published showing that Arctic and Antarctic temperatures cycle out of phase. There were some records from the South Atlantic Ocean and parts of Antarctica that show that the cold events in the North Atlantic were associated with unusual warming there (the "bipolar see-saw effect"). However, it was only in 2006 that the EPICA team (the European Project for Ice Coring in Antarctica) published their data that connect the Antarctic ups and downs of climate with the much greater ones of Greenland. Once again, the model predicted something that was not in the data used to construct it – providing more strong evidence in favor of its validity.

In reality, unfortunately, it rarely happens that there are enough data for thorough model validation. The temptation is too strong to use all the data available when building the model and, as a result, there are usually no good data remaining for a true validation. Besides, even when validation is undertaken, in most cases it proves to be less accurate than the calibration and therefore the researcher is likely to jump into model modifications and improvements to make the validation result look better. However, this immediately defeats the purpose of validation. Once we start using the validation data set for model adjustments, we have abandoned our validation attempts and gone back to further calibration.

Actually, this has become quite standard in many ongoing modeling projects, and is called data assimilation. Special procedures are designed to constantly update and improve models based on the incoming flow of new experimental data. This becomes crucial for complex open systems (which is most usually the case for ecological and socio-economic systems), which are always changing and evolving. As a result, the data set considered for calibration and collected during one period may not represent quite the same system as the one that produced the other data set that is intended for validation. We might be calibrating a model of one system, and then trying to validate the same model but for a different system.

Another important step in model analysis is *verification*. A model is verified when it is scrupulously checked for all sort of internal inconsistencies, errors and bugs. These can be in the equations chosen, in the units used, or in links and connections. There may simply be programming bugs in the code that is used to solve the model on the computer, or there may be conceptual errors, when wrong data sets are used to drive the model. Once again, there is hardly a prescribed method to weed these out. Just check and recheck. Run the model and rerun it. Test it and test again. There is no agreed procedure for model verification, especially when models become complex and difficult to parameterize and analyze. We just keep studying its behavior under all sorts of conditions.

One efficient method of model testing is to run the model with extreme values of forcing functions and parameters. There are always certain ranges where the forcing functions can vary. Suppose we are talking about temperature. We make the temperature as high as it possibly can be in a particular system, or as low as it can be, and see what happens to the model. Will it still perform reasonably well? Will the output stay within certain plausible values, or will the model crash? If so, we need to try to figure out why. Is it something that can be explained? If so, then probably the model can be still salvaged and we may simply need to remember that the forcing function should stay within certain allowed limits. If the behavior cannot be explained, we need to keep digging – most likely, there is something wrong.

Just as when we are testing a new car, the best way to find out how it performs is to force it. Step on the pedal, and let it run as fast as it can. See if something goes wrong, and where it might fail. The beauty of testing the model is that it is not wrecked when it goes wrong! If we force the car too hard, we will ruin it. With the model, we can do whatever we want to it – change all the parameters as much as we wish. If the computer does not overheat, we can always go back to previous parameter values, and the model will run again like new. However, we will collect some valuable information about what to expect from it, where the bugs and the features are, what we can let users do to it, and where we should add some limits to make sure they do not have surprises that we cannot explain.

Another important check is based on first principles, such as mass and energy conservation. It is important to make sure that there is a mass balance in the model, so that nothing gets created from nothing and nothing is lost.

Running scenarios is another great way to test a model. This step may already be considered as model use rather than just testing. A scenario in this context is a story about what can happen to the system. To define a scenario, we need to formulate all the forcing functions (say, patterns of climate, or pollution loading, or landuse patterns) and all the control parameters (say, management rules, or external global variables). In a way, we are modeling what the external forcings are to which the system will be reacting. For example, if we are considering a model of landuse change for an urban area, we can formulate a so-called "business as usual" scenario that will assume that all the existing development trends continue into the future: the population, the economy, the investments, etc. will continue to grow at the same rate, there will be no additional controls or limits introduced, or climatic perturbations, etc. These we feed into the landuse model and run it to generate patterns of landuse under this scenario.

We may then figure out a different scenario – perhaps a sustainable development plan. We will need to formulate this in terms of the model. This means that we translate the sustainable development plan into the parameter values and forcing functions that will most closely describe that. In a way, we model what we think will be a sustainable future. In our case we may assume that there is a control over population growth, so that certain birth-rate reductions are introduced. Furthermore, we will tie economic growth to the natural resources that are available in the area, and make the growth rate slow down as natural capital gets depleted. We can also include some rules for investments that would stimulate the green economy. As a result, we will get a different set of parameters that control the model, and the model run will now produce some different pattern of landuse as a result of this scenario.

Yet another scenario can be put together for devastating climatic conditions – say, a storm that will flood the area and destroy property and population. We will need to formulate some climatic conditions describing this storm. Once again, we are modeling certain conditions or forcings for the system. Note that scenarios are also models, coherent and feasible models of external conditions that will then drive the model of the system that we are studying.

Note that scenario runs are also powerful tools of model testing. In this case, we are likely to explore the unknown domains of model parameter values. We do not have the data about the model behavior that we might expect, but we do want the model to produce something qualitatively reasonable. If that does not happen, we may question the model validity and have some clues where to look for errors. For example, if a model of sustainable growth results in patterns of further urban sprawl, this would be a warning indicating that something is not working right in the model. We should take a closer look at the formalism we used, or perhaps at the parameter values that we calibrated.

The bottom line regarding all this testing is that there is no perfect model. It is hardly possible to get a perfect calibration, and the validation results will likely be even worse. No matter how long you spend debugging the model and the code, there will always be another bug, another imperfection. Does this mean that this is all futile? By no means! As long as we reach new understanding of the system, as long as the model helps to communicate understanding to others and to manage and control the system, we are on the right path and our efforts will be fruitful. *Any model that is useful is a good model.*

4.4　Conclusions

One obvious conclusion from all the above is that putting the model together is not just about establishing variables and connections and writing the equations for them. We also need to do a lot of number crunching, running the model many times. If the model is complex and requires a great deal of computer power to run it, we will be limited in the extent of testing and improving that can be done with the model. We will have to be prepared to do the job on our slow desktop (and spend more time), or we will need to find a more powerful super-computer (and spend more $$), or we will have to limit ourselves in the amount of testing and calibrating that we can do (and get a poorer model and less well-understood system). Yet another option is to go back to the model design stage and try to simplify the model.

There is a potential Catch-22 in this process. On the one hand, the more information about the system we can use in our model, the more processes we can include and the more detail about these processes we can formalize, the better our model should be and the more it should be able to tell us about the real system. On the other hand, the more complexity there is built into the model, the longer it will take to run it, the less testing we will be able to afford, the less understanding we will be able to derive from the model, and the less useful the whole exercise will be.

This is why we keep repeating *ad nauseam* that modeling should always be iterative. All the time during the modeling process we need to check our balances. Is the level of complexity justified? Are we maintaining control over the model, or is it becoming too complex in itself to be useful? Does the model complexity match the goals of the study? Do we really need all that? In some cases we may even go back, cutting down on model complexity, if at some stage we see that it is no longer justified. Suppose we run sensitivity analysis that tells us that large chunks of the model have very little if any impact on the overall systems dynamics. It should be safe to remove those components and parameterize them, put them into some empirical relationship and let it sit there as a black box.

However, this is also not a simple mechanical effort of trimming and cropping. Parameter sensitivity does not have to be linear, nice and smooth. There may be thresholds, there may be conditions when the model flips into an entirely different state, and a parameter or process suddenly becomes very crucial for the system behavior. It is important to watch for these thresholds, especially if we are parameterizing and removing certain processes, and to make sure that these model components behave the same over the whole domain of model inputs.

Oxygen is an important component of an aquatic ecosystem. Lack of oxygen may cause a fish-kill. The anoxia that is dangerous for fish will occur at O_2 levels of about 2 mg/l or less. If we were to model a fishpond, we might therefore want to include oxygen as a state variable and would probably add an oxygen forcing function for fish mortality: in this function, just as needed, the mortality would increase as oxygen concentrations fell below 2.

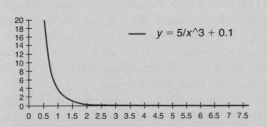

Suppose we build this model and start running it with the data that we have, and with the existing information we never get anywhere even close to hypoxia. We can safely run the model for oxygen concentrations well over 3 mg/l. This indicates that all the sensitivity analysis that we perform will unequivocally tell us that there is absolutely no sensitivity to oxygen-related parameters, and it will become very enticing to remove the relevant processes from the model. Shouldn't we make the model as simple as possible? And why would we want to have the redundancy of an extra state variable (O_2) and all the associated processes? However, if we do this and start "managing" the ecosystem by, say, increasing the amount of fish feed that we apply, we are very likely slowly to move the system towards the 2 mg/l threshold. The process and the parameters would then become very important once again, but they are no longer in the model, since we have "simplified" it. We are now running a model that is no longer valid, since it is ignoring some important processes in the system.

Rykiel (1996) defines model credibility as "a sufficient degree of belief in the validity of a model to justify its use for research and decision-making." In this case, the "sufficient" degree should be evaluated based on the relative risk associated with the decision to be made. This is probably a good way to frame it. Here, we include both the model goal and the model users in the evaluation process. Indeed, there is no use talking about some overall universal model validity; the model is valid only with respect to the goals that it is pursuing, and only the users of the model can define whether it suits their needs or not.

There is a good deal of concern about the uncertainties that are inherent in almost any modeling effort. Pretty much any stage of the modeling process is full of uncertainties. We start from the goals of the study and immediately we realize that there are different expectations that various users may have for a model. The goals are communicated in some linguistic form, in words, and this in itself is a model of a collection of thoughts or ideas about what we want. Such models already may be fuzzy, and may change as the mind, knowledge and ideas of people evolve. Especially when we are dealing with socio-economic processes that include people, their opinions, and priorities, we immediately enter a realm of huge uncertainty and much guesswork.

Very much like in quantum physics, where the mere occurrence of the experiment influences its results, so it is in social work, where, for example, by polling people and asking them a question we immediately bias the outcome by how we ask the question and by the simple fact of the question, which already can make people think differently from how they might have done without being exposed to the question.

"How do you value that forest?" Well, chances are the respondents never even noticed the forest and could not care less about its existence. However, now that they are asked about it, they may start thinking: "So why would they ask me? Actually yes, there is that forest. And I remember going there as a kid. Once. And it was pretty cool. And how am I going to look if I say that I don't care about this forest? No, probably I should say that I value it at least somewhat. And maybe actually there is value in it, or why would they ask otherwise?" We see that the response is already different from what is was supposed to be at first. The person quickly built a mental model, analyzed it and produced an answer, which in fact is still full of uncertainties, especially since we will never know what the real chain of thought was and what intermediate evolution the person's mind had gone through.

It does not get any better as we step up to the next stages of model building. As we have already seen, we hypothesize all sorts of things about a system when we model it. Besides, we need to simplify it, introducing even more uncertainties. And then of course there is all of the calibration process, when looking at the sensitivity test should be enough to realize that different parameters can result in a dramatically different model output. A model that does not have much sensitivity to its parameters, that is quite robust, will be adding less to the overall uncertainty than will a model that is very sensitive to certain parameters. Sensitive parameters then need to be measured with especially high accuracy, which may not be possible in some cases. Obviously, as models become more complex, overall uncertainty also grows very fast. In some cases, greater complexity can make the model more robust to variations in parameters; however, this normally comes at the expense of overall model controllability, when the complex model starts to operate as an entity in itself, and we approach the Bonnini paradox situation – that is, we replace the real-life complex system by another complex system – the model.

Still, we will model. There is simply no other better way to perform analysis and to produce synthesis. We have to find a way to simplify a complex system if we want to understand it. As long as we are ready to go back, to try again, to reiterate and test, test, test, we will eventually end up with a useful product. And if it is useful, it means that the model we have built is a good one.

Exercise 4.2

There are many interesting papers on the controversies of verification and validation that are worthy of your attention. Some of them are mentioned in the bibliography. Choose three papers from the list, or come up with your own sources, then write a short essay (700–1000 words) on what you think about model testing, verification and validation. Is it an important step in model development? How much attention does it deserve? Is it doable at all? How can we do it for various models and systems?

Modeling requires practice. Describe some of your own experiences in testing the model that you are developing for your own project. Submit your essay to the book website – the most interesting ones will be then posted there.

Further reading

Oreskes, N., Shrader-Frechette, K. and Belitz, K. (1994). Verification, validation and confirmation of numerical models in the earth sciences. *Science*, 263: 641–646 – *A rather philosophical paper that explores to what extent models are actually means of finding the "truth". It raises the issue of open systems and how modeling them is difficult and hardly possible to really verify.*

Rykiel, E.J. (1996). Testing ecological models: the meaning of validations. *Ecol. Modelling*, 90: 229–244 – *Another important paper that basically agrees with Oreskes that validation is not a procedure for testing scientific theories or testing the 'truth' of scientific understanding. The author argues that validation should not even be required for every modeling effort. Validation only means that the model is acceptable for its intended use and meets specified performance requirements. Also gives an excellent history of validation in ecological modeling.*

Haefner, J.W. (1996). *Modeling biological systems: principles and applications.* Chapman and Hall – *A very good textbook on biological modeling, which has an interesting chapter of model testing and calibration.*

Jakeman, A.J., Letcher, R.A. and Norton, J.P. (2006). Ten iterative steps in development and evaluation of environmental models. *Environmental Modelling & Software*, 21(5): 602–614 – *The authors argue that there is good practice in modeling, which should be followed as much as possible. Among other steps they describe how models should be tested.*

Beven, K. (1993). Prophecy, reality and uncertainty in distributed hydrological modelling. *Advances in Water Resources*, 16: 41–51 – *Focuses on uncertainty and describes the Generalized Likelihood Uncertainty Estimation (GLUE) technique in application to hydrological modeling.*

Loehle, C. (1987). Errors of construction, evaluation, and inference: A classification of sources of error in ecological models. *Ecol. Modelling*, 36: 297–314 – *This provides a review of various sources of error and uncertainty in ecological models, and puts them into three major classes: errors of construction, evaluation and inference.*

Villa, F., Voinov, A., Fitz, C. and Costanza, R. (2003). Calibration of Large Spatial Models: a Multi-stage, Multi-objective Optimization Technique. In: R. Costanza and A. Voinov (eds.), *Spatially Explicit Landscape Modeling.* Springer, pp. 77–118 – *Here you can find a Model Performance Index that was designed to calibrate models using several different metrics, both quantitative and qualitative. There is also a discussion on calibration of complex spatial models.*

To read about the CLIMBER climatic model and its applications see Ganopolski, A., Rahmstorf, S. (2001). Rapid changes of glacial climate simulated in a coupled climate model. *Nature* 409: 153–158, *where the CLIMBER model was first used to analyze the D-O cycle and its stability. The connection with the shorter solar cycles is explored in:* Braun H., M. Christl, S. Rahmstorf, A. Ganopolski, A. Mangini, C. Kubatzki, K. Roth, B. Kromer (2005). Possible solar origin of the 1,470-year glacial climate cycle demonstrated in a coupled model. *Nature, Vol* 438: 208–211. *To learn more about climate change and models that are used to understand it check out the blog at* http://www.realclimate.org/. *Their post at* http://www.realclimate.org/index.php/archives/2006/11/revealed-secrets-of-abrupt-climate-shifts/ *is about the story we described here.*

5. Simple Model, Complex Behavior

SUMMARY

Non-linear systems are those that can generate the most unusual and hard to predict behavior. A system of two species where one eats the other is a classic example of such non-linear interactions. The predator–prey model has been well studied analytically and numerically, and produces some very exciting dynamics. This simple two-variable model can be further generalized to explore systems of many species that are linked into trophic chains. Further complexity is added when these populations are considered spatially as so-called metapopulations.

Keywords

Lotka–Volterra model, non-linear systems, trophic function, equilibrium, phase plane, carrying capacity, Monod function, Kolmogorov theorem, periwinkle snail, even and odd trophic levels, Yellowstone wolves, Stella arrays, Simile, Spatial Modeling Environment (SME).

* * *

Two-state-variable systems have been honored with the most attention from mathematical modelers. This may be readily explained by the dramatically increasing complexity of mathematical analysis as the number of variables grows. As seen previously, it is only the simplest models that can be treated analytically. On the other hand, two state variables produce much more interesting dynamics than one variable, especially if there is some non-linearity included. Mathematically, such systems are more challenging and certainly more rewarding. All sorts of exciting mathematical results have come from analysis of these systems. In addition to advancing mathematics, analysis of these simplest two-state-variable systems has provided a wealth of results that may have important ecological implications and are certainly interesting in the art of modeling even in more general and complicated cases.

One of the first and also best-studied communities is the so-called "predator–prey" system, where organisms of one population serve as food for those of the other. Vito Volterra studied fish populations, and in 1926 formulated a model that turned out to be very insightful regarding the understanding of population dynamics. Alfred Lotka proposed the same model in 1925, so the model is sometimes known as the Lotka–Volterra model, or just the Volterra model, since it was he who did most of the mathematical analysis.

5.1 Classic predator–prey model

Suppose we are considering a predator–prey system, where rabbits are the preys and wolves are the predators. The conceptual model for this system can be presented by the simple diagram in Figure 5.1.

In this case we are not concerned with the effects of the environment upon the community, and focus only on the interactions between the two species. Let $x(t)$ be the number of rabbits and $y(t)$ be the number of wolves at time t. Suppose that the prey population is limited only by the predator and, in the absence of wolves, rabbits multiply exponentially. This can be described by the equation:

$$\frac{dx}{dt} = \alpha x \tag{5.1}$$

When the wolves are brought into play, they start to consume rabbits at a rate of $V = V(x)$, where $V(x)$ is the number of rabbits that each wolf can find and eat over a unit time. Naturally this amount depends on the number of rabbits available, x, because when there are just a few rabbits it will be harder for the wolves to find them than when the prey are everywhere. The form of the function for $V(x)$ may be different, but we may safely assume that it is monotone and increasing. Then the equation for rabbits will be

$$\frac{dx}{dt} = \alpha x - V(x)y \tag{5.2}$$

The growth of the wolf population is determined by the success of the wolves' hunting activities. It makes sense to assume that only a certain part of the biomass (energy) consumed is assimilated, while some part of it is lost. To account for that, we describe the growth of the wolf population as $kV(x)y$, where $0 < k < 1$ is the efficiency coefficient. The wolf population declines due to natural mortality, with μ being the mortality rate. As a result, we get a system of two ordinary differential equations (ODE) to describe the wolf–rabbit community:

$$\frac{dx}{dt} = \alpha x - V(x)y$$

$$\frac{dy}{dt} = kV(x)y - \mu y \tag{5.3}$$

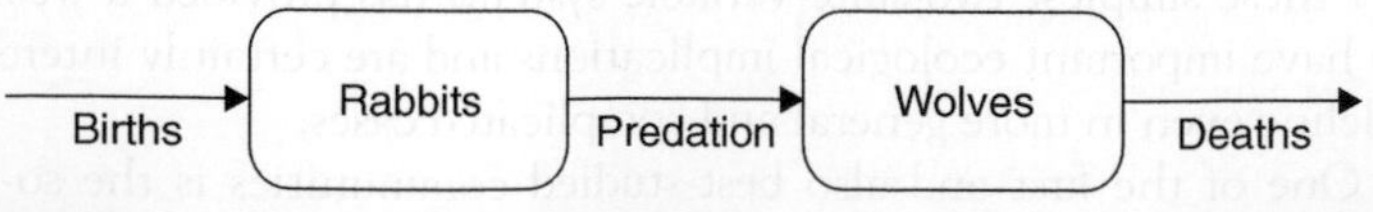

Figure 5.1 A simple conceptual model for a predator–prey system: wolves eat rabbits.

In the absence of rabbits, the wolf population exponentially decreases. $V(x)$ is called the trophic function, and it describes the rate of predation as a function of the prey abundance. The form of the trophic function is species-specific, and may also depend upon environmental conditions. Usually it grows steadily when the prey population is sparse, but then tends to saturation when the prey becomes abundant. Holling has identified three main types of trophic functions, as shown in Figure 5.2.

The first two types of the trophic functions (A, B) are essentially the same, except that in case B the function has a well pronounced saturation threshold. The third type of trophic function behaves differently for small values of prey densities. It tends to zero with a zero derivative, which means that near zero the trophic function decreases faster than the prey density. This behavior is found in populations that can learn and find refuge from the predator. For such populations there is a better chance to persist, because the predator cannot drive the prey to total extinction.

Volterra considered the simplest case, when the trophic function is linear. This corresponds to function B below the saturation threshold. The wolves are assumed to be always hungry, never allowing the rabbits to reach saturation densities. Then we can think that the trophic function is linear: $V = \beta x$. The classical Volterra predator–prey model is then formulated as:

$$\frac{dx}{dt} = \alpha x - \beta x$$

$$\frac{dy}{dt} = k\beta xy - \mu y$$

(5.4)

It can easily be seen that this system has two equilibria. The first is the so-called trivial one, which is when both the wolves and the rabbits are driven to extinction, $x = 0$, $y = 0$. There is also a non-trivial equilibrium when $x^* = \mu/k\beta$, $y^* = \alpha/\beta$. Obviously, if the community is at an equilibrium state, it stays there. However, the chances that the initial conditions will exactly hit the non-trivial equilibrium are null. Therefore, it is important to find out whether the equilibria are stable or not. For a simple model like this, some qualitative study of the phase plane may precede further analytical or numerical analysis of the model. In fact, we may note that when there are more rabbits than at equilibrium ($x > x^*$), the population of wolves decreases ($dy/dt < 0$). The opposite is true when $x < x^*$. Similarly, when there are more wolves than at equilibrium ($y > y^*$), the population of rabbits declines ($dx/dt < 0$); it grows

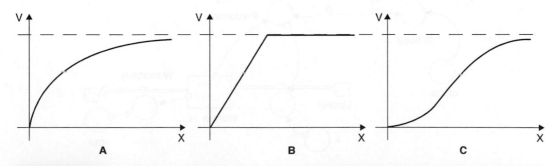

Figure 5.2 Different types of the trophic function, according to Holling.

when $y < y^*$. We may therefore break the phase plane into four areas and in each of them show the direction of the trajectory of the model solution (Figure 5.3).

This qualitative analysis already shows that there appears to be some cyclic movement around the equilibrium point. The trajectories are likely to wind around this point. There is still a chance that the point is stable, in which case we start circling around the equilibrium, gradually moving back into the center. However, this qualitative analysis only indicates that the model trajectories will loop around the non-trivial equilibrium, but it is not clear whether these loops form a spiral converging towards the equilibrium (point stable) or whether the spiral will be heading away from the center (point unstable). In any case, we may expect oscillations in populations of rabbit and wolf. Let us see what a simple Stella model can tell us about the dynamics in the predator–prey system (Figure 5.4).

You can either put together a model yourself for further analysis, or download it from the book website. The phase portrait very well matches our expectations. We do get the loop that behaves exactly as our qualitative analysis predicted. As expected, the model produces cyclic behavior, where an explosion in the rabbit population is followed by a peak in the wolf population. The rabbits are then wiped out, after

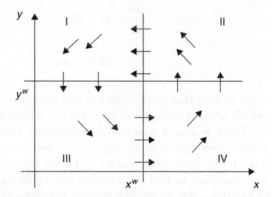

Figure 5.3   The direction of change on the phase plane for the Volterra model.
In I, both x and y decline; in II, x declines as y grows; in III, x grows and y falls; in IV, both x and y grow.

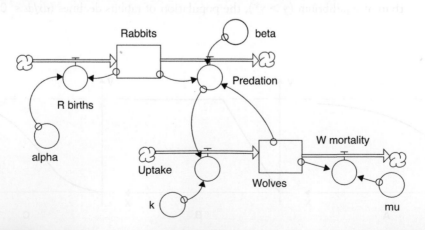

Figure 5.4   The Stella diagram for the predator–prey model.

which the wolves die from starvation, almost to extinction. When there are very few wolves left, the rabbits start to multiply again and the pattern recurs (Figure 5.5). If we run the model with the Euler method, we see that there is no trend towards the equilibrium in the center, and the amplitude of the oscillations gradually increases until the system crashes. However, if we switch to the Runge–Kutta fourth-order method, we find that actually we get a closed loop in the phase plain. Populations of both wolf and rabbit follow the same identical trajectory, going through the same pattern of oscillations (Figure 5.6). There is no convergence towards the equilibrium in the center, and neither is there a run-away from it, which we erroneously suspected at first when running the model with the Euler method.

However, unless we find an analytical solution we cannot be really sure that this will be the kind of behavior that we get under all conditions and combinations of parameters. Luckily, in the time of Vito Volterra there were no computers and he studied the model quite rigorously, analytically proving that the model trajectories always loop around the equilibrium point.

It may be noted that the initial conditions turn out to be very important for the overall amplitude of the cycle. Note that if all the parameters stay the same but the initial conditions are modified the system still produces a cycle, although its form may change quite dramatically. This is a somewhat unexpected result, showing that the current state of the system depends very much upon the state of the system a considerable length of time ago, when the initial conditions were established to start up the process.

The changes in the parameter values also do not change the overall form of the trajectories, which are still looping around the non-trivial equilibria. However, they do move the loops on the phase plane (Figure 5.7).

Stella is unlikely to get the loops using any other method of integration than fourth-order Runge–Kutta. The Euler method quickly results in increasing oscillations

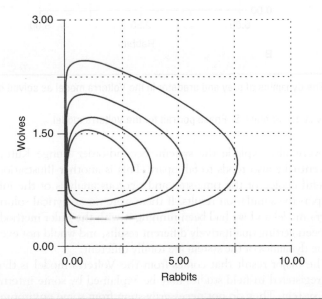

Figure 5.5   The Volterra model solved with the Euler method.

The trajectory unwinds further away from the equilibrium in the center, until the system crashes.

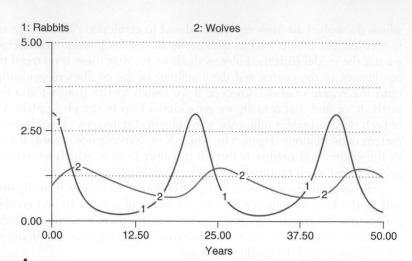

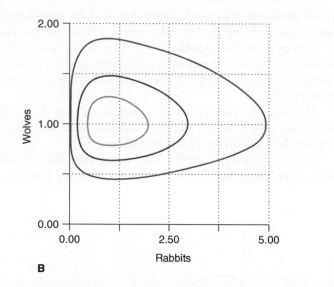

Figure 5.6   The dynamics of prey and predator in the Volterra model as solved by the Runge–Kutta fourth-order method.
A. Graphs for the Wolves and Rabbits. B. Phase portrait for the Volterra model.

that eventually explode the system. Second-order Runge–Kutta persists for longer, but eventually also tends to fall apart. This is another illustration of the importance of careful choice of the time-step and rigorous analysis of the influence of the time-step upon the simulation results. If there were no analytical solution available for the Volterra model and we had been running it with the Euler method in Stella, we would have been getting qualitatively different results, and would not even be suspecting that the true dynamics of the system are totally different.

The major result that comes from the Volterra model is that population cycles often registered in field studies may be explained by some internal dynamic features of the system. They do not necessarily stem from some environmental forcings, such as the seasonal variations in climatic factors. Cycles may occur simply as a result of interaction between the two species.

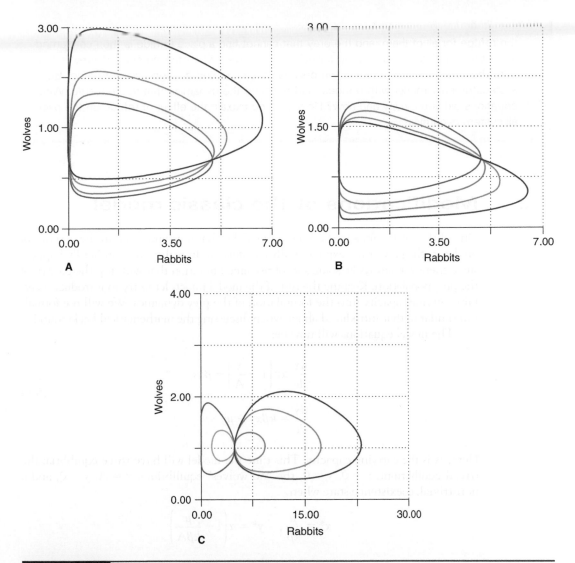

Figure 5.7   Sensitivity analysis in the predator–prey system.
A. Variations in α, the prey birth rate, quite unexpectedly affect the predator more than the prey. B. By increasing the predation rate β, we increase the maximal size of the prey population! C. By changing μ, we have flip-flopped the loop over the equilibrium point: $\mu = 0.1, 0.3, 0.5, 0.7, 1, 1.2$. When $\mu = 0.5$, the loop shrinks to a point.

Exercise 5.1

1. Can you think of any examples of other systems that do not have predators and preys but can be described by similar equations and demonstrate the kind of behavior that is found in the predator–prey model?

2. In some predator–prey systems the prey can find a refuge to hide from predators and avoid being consumed. Usually there are only a certain fixed number of individuals that the refuge can house. When the population of prey is large there is not enough space in

the refuge for all of them, and the prey that cannot find a place to hide is then consumed as in the standard predator–prey formalism. However, when the number of prey is limited, their consumption slows down because they can find enough places to hide. Build a predator–prey model with refuges, and describe the dynamics that you observe. What equations did you modify and how? How can you explain the effect of refuge on the overall system dynamics?

5.2   Modifications of the classic model

One of the weak points in the Volterra model is that it assumes that in the absence of predators the prey grows exponentially *ad infinitum*. In reality this can hardly happen, since there will always be some sort of resource limitation that will stop the growth of the prey population. Keeping the rest of the model intact, let us try to introduce a certain carrying capacity into the formulation of the prey dynamics. We will use formalism similar to that introduced above when discussing the mathematical background.

The model equations will now be:

$$\frac{dx}{dt} = \alpha x \left(1 - \frac{x}{A}\right) - \beta xy$$

$$\frac{dy}{dt} = k\beta xy - \mu y$$

(5.5)

Here, A is the carrying capacity. This time the model will have three equilibria: the trivial equilibrium, $x = 0$, $y = 0$; the "no wolves" equilibrium, $x = A$, $y = 0$; and a non-trivial, coexistence state when

$$x^* = \frac{\mu}{k\beta}, \quad y^* = \alpha\left(1 - \frac{\mu}{k\beta A}\right)$$

It always makes sense to test the equilibria for feasibility – in our case, to make sure that both x and y are positive. Contrary to the classic Volterra model, here there is a condition for y to be positive (feasible):

$$A \geq \frac{\mu}{k\beta}$$

The feasibility condition has a certain ecological meaning. Basically, it requires that the prey-carrying capacity A should be high enough that, at carrying capacity, the prey can also support the feeding of the predator. Let us use a Stella implementation of the model to see if the equilibria are stable or not. It should be easy to modify the Volterra model to include the carrying-capacity term.

Now the oscillations in population dynamics gradually dampen out, converging to the non-trivial steady state, as long as it exists (Figure 5.8). Both other steady states seem to be unstable – that is, small displacements from them result in trajectories that again tend to the non-trivial state, when both populations coexist. Also note that now

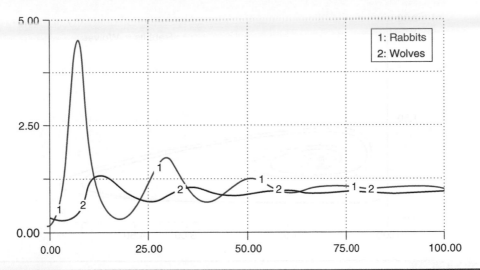

Figure 5.8 Dynamics of Rabbits and Wolves with carrying capacity introduced for Rabbits.

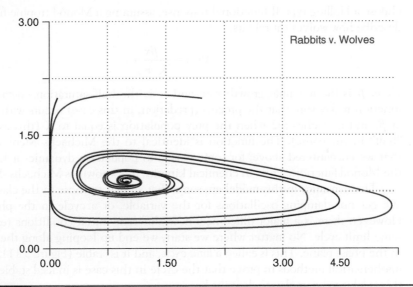

Figure 5.9 Phase portrait for the Volterra model with prey saturation run with different initial conditions.

the system dynamics does not depend upon the initial conditions. The coexistence state appears to be stable, and the oscillatory behavior is only transient (Figure 5.9).

As might be expected, the model also becomes more robust with respect to the numerical method for its solution. We can safely run the model with Euler method and much larger time-steps, yet still arrive at the same steady state (Figure 5.10).

Let us consider some further adjustments for the Volterra model. As noted above, another simplification in the model, that was hardly realistic, was the assumption regarding the linear trophic function. The wolves remained equally hungry, no matter how many rabbits they had already eaten. This seems unlikely. Let us now

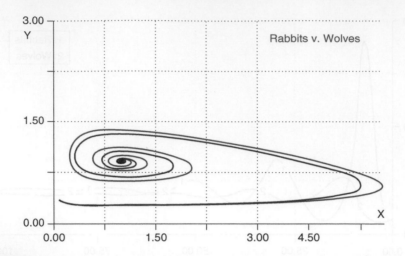

Figure 5.10   Phase portrait for the Volterra model with prey saturation run using Runge–Kutta (blue) and Euler (red) methods.

choose a Holling type II functional response, assuming a Monod trophic function, to describe how wolves eat rabbits:

$$V(x) = \frac{\beta x}{K + x}$$

Here, β is the maximal growth rate and K is the half-saturation coefficient. The function makes sure that the process (predation, in this case) occurs with saturation at β, and it reaches $\beta/2$ when the prey population is equal to K (this explains the "half" in the name). The function is identical to the Michaelis–Menten function that we encountered above: for some reason in population dynamics it is known as the Monod function, while in chemical kinetics it is known as Michaelis–Menten.

The dynamics in this model are somewhat similar to those in the classic model. We get non-damping oscillations for the variable, or a cycle in the phase plane. However, there is a major difference: now, different initial conditions result in the same limit cycle. No matter where we start, we end up looping along the same trail in the phase plane. This is called a *limit cycle*, and it is stable (Figure 5.11). There are mathematical methods to prove that the cycle in this case is indeed stable; however, this is a bit too complex to describe here.

As in the previous case, when prey growth was stabilized by carrying capacity, here again the model can be solved by the Euler method as well as by Runge–Kutta. Whenever you have a "stable" situation that attracts the trajectories, Euler works too. The cycle it generates will be slightly different from that which the Runge–Kutta method derives, but qualitatively the behavior of the system will be identical.

* * *

Kolmogorov (1936) considered a very general system that covers all the cases studied above. He analyzed a system of two ordinary differential equations:

$$\frac{dx}{dt} = \alpha(x)x - V(x)y$$

$$\frac{dy}{dt} = K(x)y$$

(5.6)

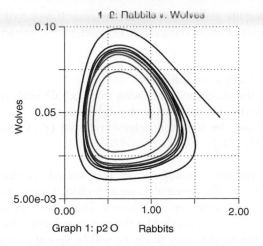

Graph 1: p2 O   Rabbits

Figure 5.11   Phase portrait for the Volterra model with prey saturation and type-2 trophic function for predation. Note that different initial conditions result in the same limit cycle.

We can see that Volterra's system is a special case of this system; however, there are many other systems that can be also described by these equations – the Volterra system is just one of them. The functions $\alpha(x)$, $V(x)$ and $K(x)$ can be any, although as long as we are describing population dynamics they have to comply with certain obvious restrictions:

1. $d\alpha/dx < 0$; $\alpha(0) > 0 > \alpha(\infty)$ – this is to say that the prey birth rate is decreasing as the prey population grows (the derivative of over x is less than 0), going from positive to negative values. This is something we were getting with the carrying capacity function in (5.5), which is quite a natural assumption for populations with intraspecific competition and a limited resource. With this assumption, even with no predator to control it the prey population grows, but it is then stabilized at a certain value given by the equation $\alpha(x^\circ) = 0$.
2. $dK/dx > 0$; $K(0) < 0 < K(\infty)$ – this is to make sure that the predator birth rate increases with the prey population. It starts with a negative value, when there is no food available, and then increases to positive values.
3. $V(x) > 0$ for $x > 0$; and $V(0) = 0$ – this is to make sure that the trophic function is positive for all positive values of the prey population. It also equals zero when there are no prey.

Under these conditions, system (5.6) has either two or three positive equilibria:

1. The trivial equilibria $x = 0$; $y = 0$
2. $x = x^\circ$ (where x° is the solution to $\alpha(x) = 0$; $y = 0$
3. Point (x^*, y^*), which is the solution to

$$\alpha(x^*)x^* - V(x^*)y^* = 0$$
$$K(x^*) = 0$$

at $\alpha(x^*) > 0$, that is when $x^* < x^\circ$.

For this quite general system, Kolmogorov has proved that there may be one of the three types of phase trajectories, no matter what particular form of the functions we choose:

1. $x^* > x^o$, $(x^o, 0)$ is a stable knot
2. $x^* < x^o$, the trajectory originating from $(x^o, 0)$ leads to (x^*, y^*), and the point (x^*, y^*) is either a stable focus or a stable knot
3. $x^* < x^o$, and the trajectory originating from $(x^o, 0)$ winds around a limit cycle around (x^*, y^*).

The important conclusion from this analysis is that the systems considered above are structurally quite stable: we may change some of the functions and yet will be still getting the same type of behavior. The crucial condition is that there is some kind of self-limitation provided for the prey population. As long as there is a carrying capacity in the model, most likely we will be seeing a "controlled" behavior – that is, trajectories will remain in the positive quadrant and will not tend to infinity. In some cases, they will equilibrate to a certain point. It is also possible that they will keep oscillating, winding around a stable limit cycle.

Exercise 5.2

Build the predator–prey model with limited growth and predation based on these equations:

Rabbits(t) = Rabbits (t − dt) + (R_births − Predation)*dt
INIT Rabbits = 1
R_births = alpha*Rabbits*(1−Rabbits/A)
Predation = beta*Wolves*Rabbits/(Rabbits + SS)

Wolves(t) = Wolves(t − dt) + (Uptake − W_mortality)*dt
INIT Wolves = 0.05
Uptake = k*Predation
W_mortality = mu*Wolves
A = 2
alpha = 0.25
beta = 3.6
k = 0.1
mu = 0.2
SS = 0.5

You can also download it from the book website. Run sensitivity analysis for this model with respect to the half-saturation parameter SS. What do you observe when SS varies in the [0.5, 1.5] interval? How can you explain that?

5.3   Trophic chains

Assuming the predator–prey type of interaction between species, we may add more variables to our model to build a trophic chain, where a predator of a lower level becomes a prey for a higher-level predator. There are numerous examples of trophic chains around us, and in almost every ecosystem we can find interactions like this.

Certainly, in most real systems the predator–prey trophic chain is "contaminated" by other kinds of processes and interactions. There may be temperature dependencies, or there may be other limiting factors involved. There may be competition between several predators for the same prey, etc. However, it would be interesting to look at the "bare bones" of this kind of a system. Let us assume that all other processes are negligible, and investigate the dynamics of a simplest trophic chain "as is" (Figure 5.12).

Processes in trophic chains may cause some fairly unpredictable outcomes. Consider the situation with some of the Southern salt marshes. These marshes stretch from Chesapeake Bay to the central Florida coasts, and are some of the most productive grasslands in the world. The marshes mitigate coastal flooding, filter mainland runoff and act as nurseries for commercially important fish and other species. The marshes also protect barrier islands, which buffer shorelines from erosion. The marshes' ecology and wellbeing is very much dependent on the plant population. Cordgrass dominates the marsh, binding the sand particles and providing its animals with habitat. Without the plants to hold the sediment, the salt marsh ecosystem collapses. The marsh grasses have a predator – the periwinkle snail, which can eat a lot of grass. Luckily, there is also a predator for the snail: the blue crab. However, there is also a predator for the blue crab: the humans, who unfortunately like this delicacy and can harvest the crabs very efficiently. If the crabs are harvested, the snails are no longer under pressure and become abundant. As a result, they start eating a lot of grass. In fact, overgrazing by periwinkle snails will convert a southern salt march into a barren mudflat within 8 months. By overharvesting the blue crab, humans damage the landscape! That would be quite hard to predict.

Consider a chain length of five, with five trophic levels. Let us suppose that there is a constant flow of resource that goes into the first trophic level. Further on, the trophic interactions are according to the simplest linear trophic function, $R_i = u_i T_i T_{i+1}$, with similar parameters describing the rate of uptake for all the trophic levels: $u_i = 0.1, i = 1,...,5$.

We want to see how the system dynamics are affected by the value of the resource that enters the system. It is easy to assemble this model in Stella and do a sensitivity test for N; however, it is even more exciting to watch the dynamics in Madonna, using the slider window. As we know, Madonna runs many times faster than Stella, which allows us to see on the fly how the model trajectories are changed by modifications in certain parameters.

We observe quite unexpected behavior (Figure 5.13): when there is no input of external resources, the trophic levels 1, 3 and 5 are eliminated; this shuts down the throughflow for the system and allows the remaining two trophic levels, 2 and 4, to equilibrate at a non-zero value. As we start increasing the inflow of energy into the system (N is gradually growing), there is enough of a throughflow to keep all five trophic levels from extinction. The equilibrium values gradually increase, but still levels 1, 3 and 5 have lower biomasses than levels 2 and 4. However, as N continues to grow, the difference between the equilibrium values for the odd and even trophic levels starts to decrease, and when N = 0.1 all the species equilibrate at the same value $T_i = 1$.

If we further increase the flow into the system, the equilibrium values split once again, but now the equilibrium biomasses in the odd trophic levels are higher than

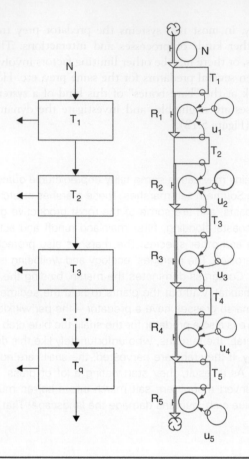

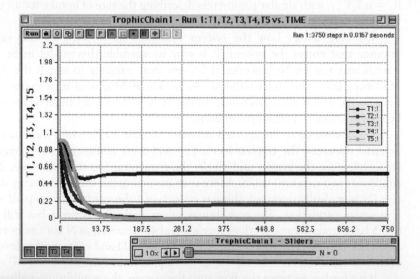

Figure 5.12   A general diagram of a trophic chain of length q and a Stella model that describes it. Here, a species preying on another species is in turn prey to another predator. N is the external resource flowing into the system. T_i are the biomasses or numbers of organisms in the trophic levels.

Figure 5.13   Dynamics of five trophic levels in a trophic chain. Odd and even trophic levels behave differently.

those ones in the even levels. By further increasing the inflow into the system we do not change the values in the even levels, whereas the odd levels gradually continue to increase their equilibrium biomass.

To check whether this is just a coincidence that might go away if parameter values are modified, or whether it is something real regarding the system dynamics, we may take a look at the equations and figure out the equilibria. In the most general form, the equations for the model are:

$$T_1 = N - u_1 T_1 T_2$$
$$T_2 = u_1 T_1 T_2 - u_2 T_2 T_3$$
$$...$$
$$T_i = u_{i-1} T_{i-1} T_i - u_i T_i T_{i+1}$$
$$...$$
$$T_k = u_{k-1} T_{k-1} T_k - u_k T_k$$

The last equation yields an equilibrium at:

$$T_{k-1} = \frac{u_k}{u_{k-1}}$$

which means that this equilibrium is independent of the flow of material into the system.

Also:

$$T_{i-1} = u_i \frac{T_{i+1}}{u_{i-1}}$$

which allows us to calculate back, starting from T_{k-1}, all the equilibria for odd (even) trophic levels if k is even (odd). Note that all of them are constant and independent of N. From the first equation, we have either $T_1 = N/(u_1 T_2)$, or $T_2 = N/(u_1 T_1)$. Therefore, if we know all the equilibria for odd trophic levels, we can calculate the value for T_2, and then use $T_{i+1} = u_{i-1} T_{i-1}/u_i$, to calculate all the remaining equilibria. Similarly, if k is odd and we know all the even equilibria, we can calculate T_1, and then build up the equilibrium values for all the remaining even trophic levels.

What is important is that we get every other trophic level constant and independent of the amount of flow into the system, whereas material accumulates only on the remaining trophic levels. We have an alternating pattern of equilibria, where every other trophic level simply passes material through to the next trophic level. The analytic treatment confirms some of the assumptions that we made from watching the dynamics of the system in Madonna. Moreover, it confirms that this is really the way the system behaves beyond the simulation period and parameter values chosen.

The overall dynamics look quite similar to the second case discussed above, when we introduced carrying capacity for the prey population (equation 5.5). This might well be expected, if we realize that at carrying capacity we have a constant flow of external resources into the system, which is exactly the formulation we are considering now: N = const. So the fact that the system equilibrates and the equilibrium appears to be stable is quite consistent with what we observed in the simple two-species system. What is somewhat surprising is the distinctly different behavior observed in the odd and even trophic levels.

* * *

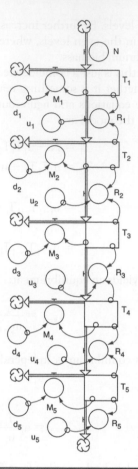

Figure 5.14   A Stella model of a five-level trophic chain with mortality.

In the model above we assumed that natural mortality is negligible compared with the predator uptake. Suppose this is not so. Let us consider a trophic chain, which has a certain fraction of biomass removed from each trophic level due to mortality (Figure 5.14), and see how the model dynamics is influenced by changes in the amount of resources N provided to the system.

The apparently subtle change in the model formulation results in quite substantial differences in the system dynamics. Once again, we can easily put the model together in Stella or, even better, in Madonna. If we look at how the system reacts to changes in the flow of the external resource N, we may see that now, for substantially high flow into the system, all the five trophic levels can coexist and equilibrate at certain values that appear to be stable. If we start to decrease the external flow N, the species equilibrate to lower and lower values, until the last, fifth, trophic level becomes extinct. The fourth level then follows and so on, until all species become extinct when there are no external sources of energy or material (N = 0) (Figure 5.15).

This result may have an interesting ecological interpretation. The more resources flow into a trophic chain, the longer the trophic chain that can be sustained. Not only do the equilibrium values increase; also, entirely new trophic levels spring up.

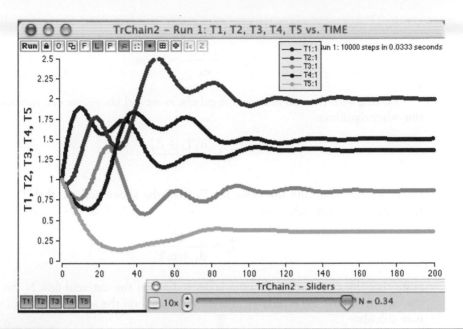

Figure 5.15   Dynamics of five trophic levels in a trophic chain with mortality.
The length of the trophic chain is defined by the amount of resource flowing into the system.

This kind of phenomenon has been observed in real-life systems. In agriculture, it has been noticed that when larger amounts of fertilizers are applied new pests appear, which effectively extends the existing trophic chain, adding a new level to it.

At this time, however, we still can make qualitative conclusions only about the system we have analyzed, and only for the parameter values that we have used. With respect to parameter values, the system seems to be quite robust. We may start modifying the coefficients in a fairly wide range (as long as they stay ecologically feasible – that is, positive and perhaps less than 1 for most of the rate coefficients, like mortality). The system behavior seems to be the same. However, if we want to consider a trophic chain with more species involved, we may need to put together another model and repeat the analysis. It is most likely that, qualitatively, the dynamics will be the same, but still we can never be 100 percent sure unless we perform some analytical treatment.

A full analytical solution to this problem can be found in Svirezhev and Logofet (1983). Here, let us take a quick glance at what the equilibria can look like, and what makes species fall out of the system. The system of algebraic equations that defines the equilibria in this model is quite simple:

$$N - d_1 T_1 - u_1 T_1 T_2 = 0$$
$$u_1 T_1 - u_2 T_3 - d_2 = 0$$
$$\dots$$
$$u_{i-1} T_{i-1} - u_i T_{i+1} - d_i = 0$$
$$\dots$$
$$u_4 T_4 - u_5 - d_5 = 0$$

Always check to see if there is something you can get out of an analytical study. Even if a full solution is impossible, see if you can get some constraints on parameters or try to figure out equilibrium conditions.

From the last equation, we immediately get:

$$T_4 = \frac{u_5 + d_5}{u_4} = \text{const.}$$

Playing with the odd and even numbers, as we did above, we can now calculate the other equilibria:

$$T_2 = \frac{u_3 T_4 + d_3}{u_2}$$

where we can substitute the value for T_4 from the above and see that $T_2 = \text{const.}$ Knowing T_2, we can calculate

$$T_1 = \frac{N}{d_1 + u_1 T_2}$$

Note that this time the equilibrium is dependent on the external flow N. So far, all the equilibria have been positive at any time. Based on the second equation, we can now calculate

$$T_3 = \frac{u_1 T_1 - d_2}{u_2}$$

For this equilibrium we need to make sure that $T_1 > d_2/u_1$, otherwise the equilibrium is negative and makes no ecological sense. This condition translates immediately into a requirement for N: the flow of external resource has to be larger than a certain value. Similarly, for T_5 to be non-negative we need $T_3 > d_4/u_3$ or, substituting for T_3,

$$T_1 > \frac{d_2}{u_1} + \frac{u_2 d_4}{u_1 u_3}$$

This explains why, with decreasing N, the equilibria for T_1, T_3 and T_5 are getting smaller and smaller, and eventually the species ceases to exist as the equilibria become negative. However, this does not explain the fate of the other two trophic levels, T_2 and T_4, which are supposedly constant and independent of N. So what is going on?

Let us take a closer look at the model dynamics in the animation above. Note that actually at first, when we start cutting the input of N, the equilibria for T_2 and T_4 are indeed fixed and do not change. It is the other three equilibria that show a downward trend. It is only after T_5 hits zero that T_2 and T_4 start to change. But note: when T_5 becomes extinct, we no longer have the same five-level trophic chain. Instead we have only four trophic levels, and the equations that we are to solve now change. Now, for four trophic levels, we have T_1 and T_3 constant and independent of N, whereas T_2 and T_4 are defined by N and decrease with N. Indeed, this is what we see in the animation. Now T_1 and T_3 stay fixed until T_4 hits zero, when once again the system and the equations are redefined. Again the system has an odd number of levels, and now T_2 becomes fixed while T_1 and T_3 start to fall.

Now that we have figured out what goes on in the system, we can with far greater confidence describe the system behavior with an arbitrary number of trophic levels. There is strong evidence that the equilibria are stable, and we have understood how the odd and even trophic levels are alternating their behavior as the flow of resource into the system changes. We also know that the parameters of the model define the intervals in the N continuum that correspond to the particular numbers of trophic levels in the system. Let us look at how the system evolves in the other direction, when we start with N = 0, and then start increasing N. Once N > 0, there is a resource that can support one species. As N increases, the population in this trophic level keeps growing until N passes a threshold, after which another species in the next trophic level appears. At this point the first trophic level stabilizes, and from now on all the resource is transmitted to the new trophic level, the population of which starts to grow. Next, after N passes another threshold, another, third trophic level appears. Now the second trophic level freezes, while the first and the third (odd) trophic levels start to grow. Then, at some point, as N passes another threshold, a fourth (even) trophic level becomes established. From now on, odd levels become frozen, and even levels start to grow biomass. And so on.

* * *

In both the trophic chains considered above, we had the input of external resource independent of the biomass in the first trophic level. We assumed that it was the resource that was always limiting growth, and there were as many organisms in that trophic level as were needed to uptake all the resource that was made available. This is different from what we had in the classic model. What will the trophic chain look like if the resource is not limiting? This may appear to be a fairly subtle change in the system; ho wever, the dynamics will be quite different.

Let us put together a simplified version with only three trophic levels:

$$
\begin{aligned}
&T_1(t) = T_1(t - dt) + (N - R_1)*dt \\
&\text{INIT } T_1 = 1 \\
&N = u_0*T_1 \\
&R_1 = u_1*T_1*T_2 \\
&T_2(t) = T_2(t - dt) + (R_1 - R_2)*dt \\
&\text{INIT } T_2 = 2 \\
&R_1 = u_1*T_1*T_2 \\
&R_2 = u_2*T_2*T_3 \\
&T_3(t) = T_3(t - dt) + (R_2 - R_3)*dt \\
&\text{INIT } T_3 = 1 \\
&R_2 = u_2*T_2*T_3 \\
&R_3 = u_3*T_3 \\
&u_0 = 0.1 \\
&u_1 = 0.1 \\
&u_2 = 0.1 \\
&u_3 = 0.1
\end{aligned}
$$

Note that in this model N is not constant; instead, it is a linear function of T_1. Now the model looks exactly the same as the "classic" model but with one additional trophic level. We can import theses equations into Madonna, or quickly assemble the model in Stella or one of the other packages to do some preliminary qualitative analysis. With the model "as is," we get the familiar oscillations (Figure 5.16). However, if we change the coefficients u_i even slightly, we get a dramatically different picture: either the species become extinct, or they start to grow exponentially (Figure 5.17).

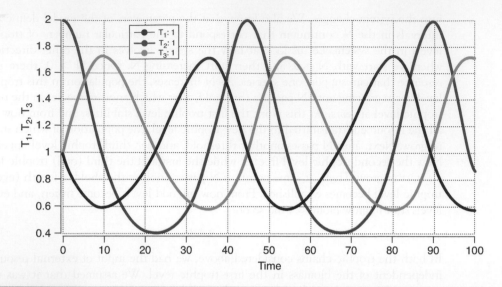

Figure 5.16   Dynamics in a three trophic level model with no resource limitation.

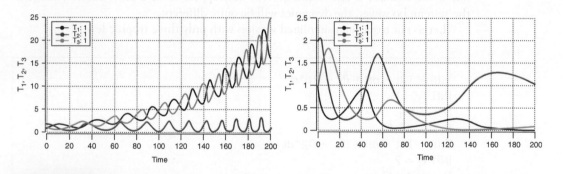

Figure 5.17   Dynamics in a three trophic level model with no resource limitation with unequal rate coefficient. The system either dies off or species produce infinite growth.

If u_0 or u_2 are even slightly increased, trophic levels T_1 and T_3 grow exponentially while T_2 keeps oscillating approaching a positive equilibrium. A similar trend is produced when u_1 or u_3 are decreased. If u_0 or u_2 are even slightly decreased, trophic levels T_1 and T_3 go extinct while T_2 keeps oscillating approaching a positive equilibrium. A similar trend is produced when u_1 or u_3 are increased.

A quick analytical look at the equilibria gives us only a very general idea about the underpinnings of these trends. First, we find that there are two equations for equilibrium in the second trophic level: $T_2 = u_0/u_1$, and $T_2 = u_3/u_2$. Second, we see that for the equilibria in the first and third trophic levels we have $u_1T_1 = u_2T_3$. The equilibrium in the second trophic level is therefore feasible only if $u_1u_3 = u_0u_2$.

These calculations explain some of the qualitative dynamics we observed above. If $u_1u_3 = u_0u_2$, we get stable oscillations; if $u_1u_3 > u_0u_2$, we have the downward trend that leads to species extinctions. Otherwise, we have oscillations following an exponential growth trend. We could have been expecting this from what we saw in the model; however, it might have been hard to guess the exact relationship between the parameters that defines the course of the trajectories. We also see that there is a relationship between T_1 and T_3, which makes them behave in a similar way – something we also observed from the model output.

However, this is probably all we can say about the system, based on this primitive analysis. We do not know what makes T_1 and T_3 grow to infinity or vanish from the system, when the parameters are chosen in some specific way. Unlike the "classic" model, which produced the loop in the phase plane for any combination of parameters, now a loop is possible only for specific values. Moreover, it would be hard to imagine in real life an exact equality of the kind $u_1 u_3 = u_0 u_2$. Therefore, we may conclude that a three- or more trophic level system of the predator–prey type is unstable and unlikely to exist in reality.

What will happen if, instead of three, we have four trophic levels? Will the results be the same? The answer is a definite NO. To our surprise, the system always persists, even though it goes through some dramatic oscillations which in many cases appear to resemble chaos. Once again, it is strongly recommended that you reproduce the model in one of the modeling packages. Below are the equations that you can simply paste into Madonna and enjoy the model performance yourself:

$$T_1(t) = T_1(t-dt) + (N - R_1)*dt$$
$$\text{INIT } T_1 = 1$$
$$N = u0*T_1$$
$$R_1 = u_1*T_1*T_2$$
$$T_2(t) = T_2(t - dt) + (R_1 - R_2)*dt$$
$$\text{INIT } T_2 = 2$$
$$R_1 = u_1*T_1*T_2$$
$$R_2 = u_2*T_2*T_3$$
$$T_3(t) = T_3(t - dt) + (R_2 - R_3)*dt$$
$$\text{INIT } T_3 = 1$$
$$R_2 = u_2*T_2*T_3$$
$$R_3 = u_3*T_3*T_4$$
$$T_4(t) = T_4(t - dt) + (R_3 - R_4)*dt$$
$$\text{INIT } T_4 = 1$$
$$R_3 = u_3*T_3*T_4$$
$$R_4 = u_4*T_4$$
$$u_0 = 0.1$$
$$u_1 = 0.1$$
$$u_2 = 0.1$$
$$u_3 = 0.1$$
$$u_4 - 0.1$$

The variety of designs that the trajectories produce when we start modifying the parameters is truly remarkable. A few examples appear in Figure 5.18. In the left-hand column we are looking at the regular graphs of state variables vs time; in the right-hand column we have the scatter graphs, where T_1 and T_2 are displayed as functions of T_3.

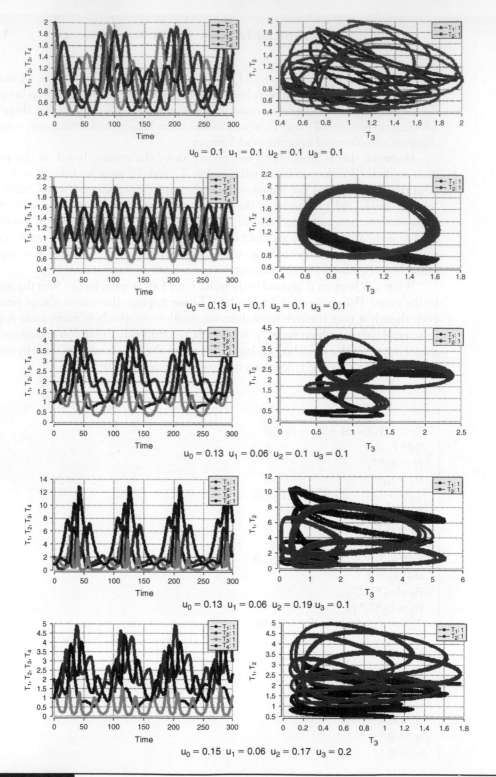

Figure 5.18 Adding another trophic level (fourth) stabilizes the system and makes it persist, even though some of the oscillations seem to be chaotic.

The left-hand column shows the dynamics of the four populations; the right-hand column graphs are phase dynamics of populations of the first two trophic levels as functions of the third trophic level population. These show how irregular the oscillations may become.

It is not clear what the trajectories are that we are getting. Is there a period in the oscillations that we observe? Or is it really a chaotic regime that we are generating? Or is what we see an artifact of the computational method that we chose? In this case, the system has a non-trivial equilibrium:

$$T_1 = \frac{\dfrac{u_2 u_4}{u_1}}{u_3}; \quad T_2 = \frac{u_0}{u_1}; \quad T_3 = \frac{u_4}{u_3}; \quad T_4 = \frac{\dfrac{u_0 u_2}{u_1}}{u_3}$$

The equilibrium is unstable: take a small step away from the equilibrium, and the system plunges into the oscillation mode that we see. However, apparently the system stays within the bounds of reasonable change: no species become extinct and no species continue to infinity. But this is probably all that we can say based on this simple analysis. To answer the questions above, we need far more sophisticated mathematical analysis – which is beyond this course.

It is interesting to note that once again we have a distinctly different system behavior for odd and even lengths of trophic chains. Trophic chains with odd number of levels seem to be impossible to realize: either they become extinct, or they shoot to infinity. If another trophic level is added, the system stays within certain bounds, though the behavior is very irregular – perhaps chaotic. It remains unclear what the overall behavior of the system is with multiple trophic levels. We have considered several special cases, but there is no general picture emerging. The system is clearly structurally unstable: small changes in the system formalization alter the overall performance quite dramatically. One general conclusion that appears is that odd and even length trophic chains behave differently. It would be interesting to see if we can find something like this in real ecological systems.

A somewhat related situation is observed in the Yellowstone National Park in the USA. One of the trophic chains in that ecosystem is the following

Aspen ⇒ Elks ⇒ Wolves ⇒ Humans

Humans served as the top predator, as hunters went in and killed wolves. At some point it was decided that the wolves were a nuisance in the park; besides, there were complaints from local farmers. Therefore, in the 1920s the US government eliminated the gray wolf from Yellowstone. Immediately, the elk population peaked. By examining tree rings, it was shown that the park's aspens stopped regenerating soon after that.

Next, whole river valleys started to erode. The annual river floods washed away soils that had been accumulating for thousands of years. There was hardly any substantial vegetation to hold the soil in place. It turns out that the wolves affected a whole river system! Very much like our models predict, with a low wolf population we get high elk numbers and low aspen biomass.

In 1995, it was decided to reintroduce wolves into the ecosystem. The trees had not regenerated in the park for more than half a century, but they are now returning in some areas. Their recovery, according to researchers, is not simply because the wolves are hunting the elk; it is also because the wolves have reintroduced the fear factor, making the elk too nervous to linger in an aspen grove and eat. The return of the park's key predator, *Canis lupus*, has created a more biologically diverse and healthier ecosystem.

Now we are tending towards a system with high biomass of aspen, low elk population, high wolf numbers, and low human hunting pressure. This is again similar to what our theoretical model showed.

5.4   Spatial model of a predator–prey system

The models we have looked at so far have been local – that is, spatially they had no resolution, assuming that the whole area that we were modeling was uniform, and that the same populations with the same parameters of growth and death were distributed across the area. We did not know or care about any spatial differences. But what if that is not the case?

Suppose we do care about spatial differences. Suppose that the populations have different numbers across the landscape. How can we model the system in this case?

First, let us decide on how to represent space. In Chapter 2, we saw several ways to make space discrete so that we can put the spatial dimension into a model. We need to decide on the form and size of the spatial segments that we wish to use. In doing that, as always in modeling, we will be looking at the goal of the study and the spatial resolution of the data that are available. Then we will select modeling software for these spatial simulations.

Stella may not be the best tool for this. Theoretically, we could replicate our model several times and have several stocks for prey and several stocks for predator, representing their numbers in different spatial locations. We could also add some rules of transition between these stocks, representing spatial movement between different places. The Stella model would look like Figure 5.19 (see page 165). In this case, we assume that organisms migrate to the compartment where the existing population size is lower.

This could probably work for two, three or four locations – maybe even ten – but then the Stella model would become almost incomprehensible. We could use the array functionality in Stella, which would make it a little bit easier to handle. If you are unfamiliar with arrays in Stella, read the pages of the Help File. It does a really good job of explaining how to set up arrays in Stella. For example, the model above on a 3 × 3 grid of 9 cells can be presented with a diagram that looks quite simple (Figure 5.20; see page 165); however, the equations are not simple at all:

Rabbits[col1,row1](t) = Rabbits[col1,row1](t−dt) + (R_births[col1,row1]−Predation [col1,row1]−R_migration[col1,row1]) * dt
INIT Rabbits[col1,row1] = 1
Rabbits[col1,row2](t) = Rabbits[col1,row2](t−dt) + (R_births[col1,row2]−Predation[col1,row2]−R_migration[col1,row2]) * dt
INIT Rabbits[col1,row2] = 2
Rabbits[col1,row3](t) = Rabbits[col1,row3](t−dt) + (R_births[col1,row3]−Predation[col1,row3]−R_migration[col1,row3]) * dt
INIT Rabbits[col1,row3] = 3
Rabbits[col2,row1](t) = Rabbits[col2,row1](t−dt) + (R_births[col2,row1]−Predation[col2,row1]−R_migration[col2,row1]) * dt
INIT Rabbits[col2,row1] = 3
Rabbits[col2,row2](t) = Rabbits[col2,row2](t−dt) + (R_births[col2,row2]−Predation[col2,row2]−R_migration[col2,row2]) * dt
INIT Rabbits[col2,row2] = 2
Rabbits[col2,row3](t) = Rabbits[col2,row3](t−dt) + (R_births[col2,row3]−Predation[col2,row3]−R_migration[col2,row3]) * dt
INIT Rabbits[col2,row3] = 1

Rabbits[col3,row1](t) = Rabbits[col3,row1](t−dt) + (R_births[col3,row1]−Predation[col3,row1]−R_migration[col3,row1]) * dt
INIT Rabbits[col3,row1] = 1
Rabbits[col3,row2](t) = Rabbits[col3,row2](t−dt) + (R_births[col3,row2]−Predation[col3,row2]−R_migration[col3,row2]) * dt
INIT Rabbits[col3,row2] = 2
Rabbits[col3,row3](t) = Rabbits[col3,row3](t−dt) + (R_births[col3,row3]−Predation[col3,row3]−R_migration[col3,row3]) * dt
INIT Rabbits[col3,row3] = 3

INFLOWS:
R_births[column,row] = alpha*Rabbits[column,row]
OUTFLOWS:
Predation[column,row] = beta*Rabbits[column,row]*Wolves[column,row]
R_migration[col1,row1] = gamma*(Rabbits[col1,row1]−Rabbits[col2,row1]) + gamma*(Rabbits[col1,row1]−Rabbits[col1,row2])
R_migration[col1,row2] = gamma*((Rabbits[col1,row2]−Rabbits[col1,row1]) + (Rabbits[col1,row2]−Rabbits[col2,row2]) + (Rabbits[col1,row2]−Rabbits[col1,row3]))
R_migration[col1,row3] = gamma*((Rabbits[col1,row3]−Rabbits[col1,row2]) + (Rabbits[col1,row2]−Rabbits[col2,row3]))
R_migration[col2,row1] = gamma*((Rabbits[col2,row1]−Rabbits[col1,row1]) + (Rabbits[col2,row1]−Rabbits[col2,row2]) + (Rabbits[col2,row1]−Rabbits[col3,row1]))
R_migration[col2,row2] = gamma*((Rabbits[col2,row2]−Rabbits[col2,row1]) + (Rabbits[col2,row2]−Rabbits[col2,row3]) + (Rabbits[col2,row2]−Rabbits[col1,row2]) + (Rabbits[col2,row2]−Rabbits[col3, row2]))
R_migration[col2,row3] = gamma*((Rabbits[col2,row3]−Rabbits[col1,row3]) + (Rabbits[col2,row3]−Rabbits[col2,row2]) + (Rabbits[col2,row3]−Rabbits[col3,row3]))
R_migration[col3,row1] = gamma*((Rabbits[col3,row1]−Rabbits[col2,row1]) + (Rabbits[col3,row1]−Rabbits[col3,row2]))
R_migration[col3,row2] = gamma*((Rabbits[col3,row2]−Rabbits[col3,row1]) + (Rabbits[col3,row2]−Rabbits[col2,row2]) + (Rabbits[col3,row2]−Rabbits[col3,row3]))
R_migration[col3,row3] = gamma*((Rabbits[col3,row3]−Rabbits[col3,row2]) + (Rabbits[col3,row3]−Rabbits[col2,row3]))

Wolves[col1,row1](t) = Wolves[col1,row1](t−dt) + (Uptake[col1,row1]−W_mortality[col1,row1]−W_migration[col1,row1]) * dt
INIT Wolves[col1,row1] = 1
Wolves[col1,row2](t) = Wolves[col1,row2](t−dt) + (Uptake[col1,row2]−W_mortality[col1,row2]−W_migration[col1,row2]) * dt
INIT Wolves[col1,row2] = 2
Wolves[col1,row3](t) = Wolves[col1,row3](t−dt) + (Uptake[col1,row3]−W_mortality[col1,row3]−W_migration[col1,row3]) * dt
INIT Wolves[col1,row3] = 3
Wolves[col2,row1](t) = Wolves[col2,row1](t−dt) + (Uptake[col2,row1]−W_mortality[col2,row1]−W_migration[col2,row1]) * dt
INIT Wolves[col2,row1] = 3
Wolves[col2,row2](t) = Wolves[col2,row2](t−dt) + (Uptake[col2,row2]−W_mortality[col2,row2]−W_migration[col2,row2]) * dt
INIT Wolves[col2,row2] = 2

Wolves[col2,row3](t) = Wolves[col2,row3](t−dt) + (Uptake[col2,row3]−
W_mortality[col2,row3]−W_migration[col2,row3]) * dt
INIT Wolves[col2,row3] = 1
Wolves[col3,row1](t) = Wolves[col3,row1](t−dt) + (Uptake[col3,row1]−
W_mortality[col3,row1]−W_migration[col3,row1]) * dt
INIT Wolves[col3,row1] = 1
Wolves[col3,row2](t) = Wolves[col3,row2](t−dt) + (Uptake[col3,row2]−
W_mortality[col3,row2]−W_migration[col3,row2]) * dt
INIT Wolves[col3,row2] = 2
Wolves[col3,row3](t) = Wolves[col3,row3](t−dt) + (Uptake[col3,row3]−
W_mortality[col3,row3]−W_migration[col3,row3]) * dt
INIT Wolves[col3,row3] = 3

INFLOWS:
Uptake[column,row] = k*Predation[column,row]
OUTFLOWS:
W_mortality[column,row] = mu*Wolves[column,row]
W_migration[col1,row1] = delta*((Wolves[col1,row1]−Wolves[col2,row1]) + (Wolves[col1,
row1]−Wolves[col1,row2]))
W_migration[col1,row2] = delta*((Wolves[col1,row2]−Wolves[col1,row1]) + (Wolves[col1,
row2]−Wolves[col2,row2]) + (Wolves[col1,row2]−Wolves[col1,row3]))
W_migration[col1,row3] = delta*((Wolves[col1,row3]−Wolves[col1,row2]) + (Wolves[col1,
row2]−Wolves[col2,row3]))
W_migration[col2,row1] = delta*((Wolves[col2,row1]−Wolves[col1,row1]) + (Wolves[col2,
row1]−Wolves[col2,row2]) + (Wolves[col2,row1]−Wolves[col3,row1]))
W_migration[col2,row2] = delta*((Wolves[col2,row2]−Wolves[col2,row1]) + (Wolves[col2,
row2]−Wolves[col2,row3]) + (Wolves[col2,row2]−Wolves[col1,row2]) + (Wolves[col2,
row2]−Wolves [col3,row2]))
W_migration[col2,row3] = delta*((Wolves[col2,row3]−Wolves[col1,row3]) + (Wolves[col2
row3]−Wolves[col2,row2]) + (Wolves[col2,row3]−Wolves[col3,row3]))
W_migration[col3,row1] = delta*((Wolves[col3,row1]−Wolves[col2,row1]) + (Wolves[col3,
row1]−Wolves[col3,row2]))
W_migration[col3,row2] = delta*((Wolves[col3,row2]−Wolves[col3,row1]) + (Wolves[col3,
row2]−Wolves[col2,row2]) + (Wolves[col3,row2]−Wolves[col3,row3]))
W_migration[col3,row3] = delta*((Wolves[col3,row3]−Wolves[col3,row2]) + (Wolves[col3,
row3]−Wolves[col2,row3]))
alpha = 1
beta = 1
delta = 0.02
gamma = 0.01
k = 0.1
mu = 0.1

In particular, it is a real headache to define the equations of movement, migration. We assume that our cells are arranged as in Figure 5.21, and both wolves and
rabbits can move to the next cell if the population size there is lower than in the
current cell. There will be lots of clicking on the Stella diagram to define all the
connections. As the number of spatial cells grows, the model description quickly

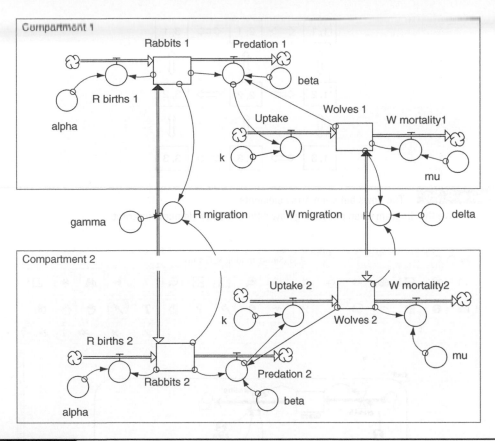

Figure 5.19 A two-compartment Stella model of a predator–prey system.

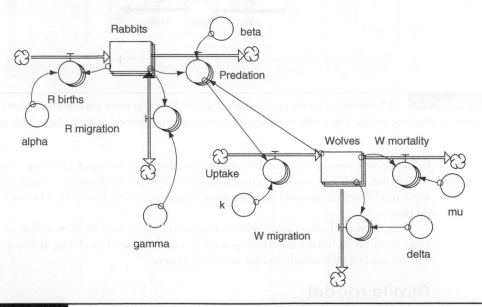

Figure 5.20 Several compartments can be modeled using the array functionality in Stella. The diagram is tidier, but it is still quite cumbersome to describe the intercompartmental flows.

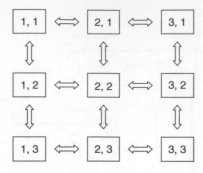

Figure 5.21 The flows between array elements.

It helps to have the diagram when describing how different array elements interact.

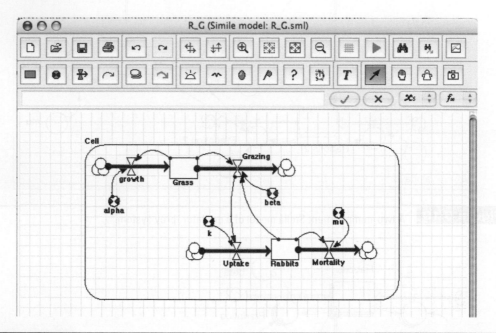

Figure 5.22 A Simile model for the predator–prey system. Note that there are many more icons to use when constructing models. The whole model can be described as a submodel (called a Cell, in this case).

becomes very cumbersome; it becomes especially hard to input the data, visualize the output, or define various scenarios that involve spatial dynamics. Imagine defining a model with a hundred or more array elements! There has to be a better way to do this.

Let us take a look at some other software tools that may be more suited to these tasks than Stella. One potentially powerful tool for spatial modeling is Simile, and we will explore an example in that modeling system.

Simile model

The predator–prey system itself is very simple to put together, especially if we already know Stella conventions. The basic interface in Simile is almost identical

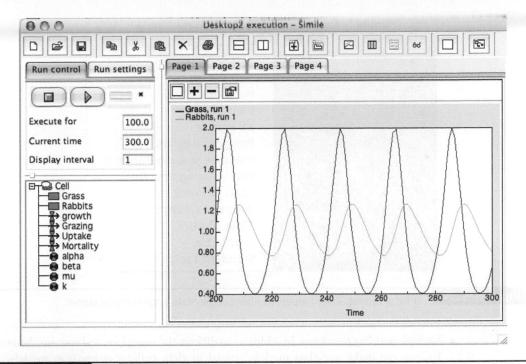

Figure 5.23   Output from the Simile predator–prey model using the Plotter helper to create a time-dependent graph.
This is identical to what we were generating in Stella.

(Figure 5.22). Here, we slightly modified the model, describing Grass as prey and Rabbits as predator. That would be one trophic level below what we were considering above, but there is really no need for much change in how we formulate the model. Whereas in its systems dynamics Simile follows Stella's formalism quite closely, it also goes way beyond Stella's functionality in a lot of ways. As you may notice, in Figure 5.22, there are quite a few more icons or building blocks in Simile. We will not go into much detail describing all of them – that can always be done by downloading the free trial version of the package and exploring the different examples and contributed models. The Help file and the Tutorial for Simile is nowhere nearly as foolproof as in Stella, so be prepared to spend quite some time if you decide to explore the more advanced features of the software.

Among these features let us mention the following.

- *Modularity*. In Simile, you can create a "submodel" that can be then used in other models. This is handy for disaggregation of models, for creating spatial models or for substituting one model component for another.
- *C++ code*. Simile generates C++ code, which can be used within the framework of other systems, interfaces or environments. It can be ported to different compilers producing optimized computer code.
- *Extendable interfaces*. All input/output is handled by Tcl/Tk programs called "Helpers." Users can create their own Helpers to suit the needs of a particular application, and port these programs into the software. For example, the output for the model in Figure 5.22 shown in Figure 5.23 comes from a particular Helper designed to plot model results.

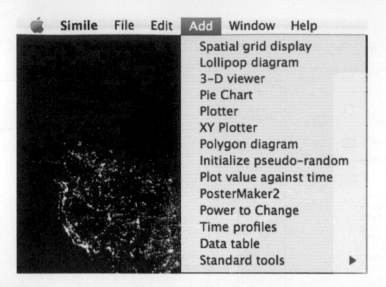

Figure 5.24   A list of various Simile helpers available to generate graphic model output.

- *Multiple modeling paradigms.* In addition to differential equations and systems dynamics, we can make age-class models, spatial models, cellular automaton models, a object-oriented model or an agent-based model. Moreover, we can integrate these paradigms, making, for instance, an individual-based model run on top of cellular automata.

As you can see, Simile is ready for a much more advanced user to build quite sophisticated models. Let us see how we can make the predator–prey model run spatially using this modeling package.

Probably the most straightforward way of doing this is to replicate the submodel called "Cell" in our diagram, making, say, 100 copies of it. Now this submodel is repeated 100 times. We can already assemble these replicas into a grid and generate some spatial output. Suppose we initialize Rabbits as a random variable with values between 1 and 2, using the built-in function: rand_const(1,2). From the list of Helpers available for data output (Figure 5.24) we can choose "Spatial Grid Display" and see the results as color-coded 2D array or cells (Figure 5.25). The intensity of blue shows the random distribution of Rabbits across our area. There was an important but poorly documented step in generating this graph. The 1D array of cells had to be converted into a 2D array. In order to do that, the user has to figure out the sequence of rows and columns and distribute the elements of the 1D array into these rows and columns. This can be done by introducing two new variables:

```
col = fmod(index(1)−1, size) + 1
row = floor((index(1)−1)/size) + 1,
```

where index(1) is the way to present the 1D array of indices of cells, size is the number of cells per row, and the two built-in functions fmod and floor help handle lists: floor(A) is the largest integer less than A, so floor(12.4) = 12; and fmod(A,B) is the remainder of A/B – say, fmod(78,10) = 8. By using these two functions we split the 1D index into the column and row parts, putting the array element into a certain position on the 2D grid.

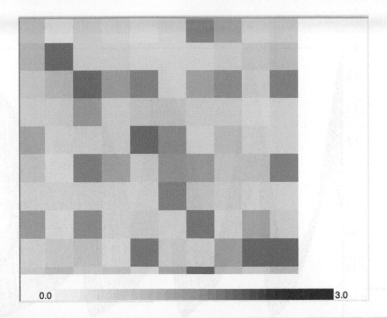

0.0   3.0

Figure 5.25   A two-dimensional graphic visualization that the Spatial Grid display helper generates. The intensity of the color corresponds to the population numbers of Rabbits in different cells.

There is actually an easier (but also not very well-documented) way to do this if you define the array as being 2D. You do this by double clicking on the background of your stack of cells, which opens a dialogue box:

Here we can input the dimensions of the array, making it two-dimensional. Let us specify the dimensions 10,10. Now the array will be treated as rows and columns, and we will not need to worry about the conversion of a linear array into a matrix.

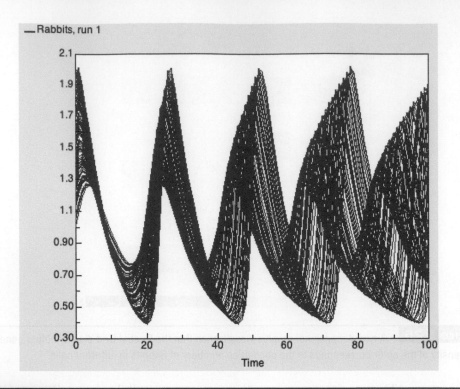

— Rabbits, run 1

Figure 5.26 Using the Plotter helper to output an ensemble of trajectories for all the cells in the model. The initial conditions are generated in random in the [1,2] interval, and each cell then develops on its own.

To view the results of our spatial runs, we can choose the helper called "grid display." When defining the grid display we will be requested to "click on the variable containing the positions of IDs of the columns" – click on the "col" variable. Then we will be asked to choose the variable to display, and will click on Rabbits.

If we now run the model, we can observe how Rabbit populations vary in all the cells (Figure 5.26). Note that in this case the graphic display produces an ensemble of 100 curves, which originate somewhere in the interval [1,2] and then oscillate like in the predator–prey model considered before.

So far, the cells have been working independently. There has been no interaction between variables in different cells. That is not particularly interesting. Let us now make the Rabbits move horizontally. Suppose that, as in the Stella model we considered above, we want to make Rabbits move from cells with higher density to cells where there are less Rabbits. This is similar to the diffusion process. For each cell we add the migration flow (Figure 5.27), which calculates the movement of Rabbits in each of the four directions: front, back, left and right. First, we define an array of Rabbits in all cells – R_A. Then

Migration = delta * ((if col > 1 then Rabbits-element([R_A], (row − 1) * size + col − 1) else 0) + (if col < size then Rabbits-element([R_A],(row − 1) *size + col + 1) else 0) + (if row>1 then Rabbits-element([R_A],(row − 2) * size + col) else 0) + (if row < size then Rabbits-element([R_A], row * size + col) else 0))

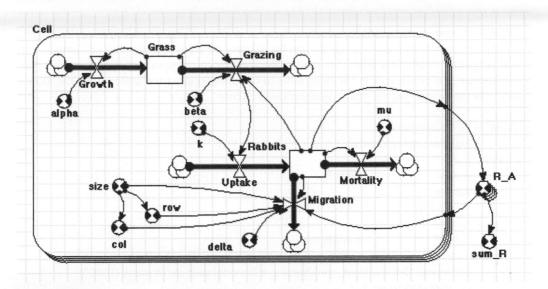

Figure 5.27   Spatial predator–prey model in Simile with migration added for Rabbits. The R_A variable stores the values for Rabbits in all cells as an array. The decision for migration is based on the number of Rabbits in adjacent cells. Rabbits jump to the neighboring cell if the population there is less than in the current cell.

This was pretty clumsy, but straightforward. For each cell, we compare the number of Rabbits with the numbers in the four adjacent cells. If the difference is positive, we get a positive flow from the cell to the neighboring cell. If it is negative, we get a flow from the neighboring cell into the center cell. Here, we used the element built-in function element([A],i), which returns the ith element of array A. Note that here we are translating the 2D definition in terms of (row,cell) back into the 1D definition.

To test how this works, we will initialize the model differently. Let us make the Rabbits biomass equal, say, three only in one cell (e.g. $i = 25$), and make the biomass equal one in all other cells. Let us also switch off all the ecological predator–prey dynamics by setting the growth, death and predation rates to zero. If testing a particular process, horizontal dispersion in this case, it is important to ensure that nothing is interfering with it. If we run the model, we will see how rabbits gradually disperse across the area (Figure 5.28). Note that we have also added a variable, sum_R, to the diagram. This variable is equal to sum([R_A]), another built-in function which returns the sum of elements of an array. This is useful to check that we are not losing or gaining rabbits; it works as a mass conservation check. As long as sum_R does not change, we are OK.

What is also nice about Simile is that we can change the size of the area and the number of cells just by changing the "size" variable and the number of instances of the "Cell" array. This can be done by double clicking on the Cell submodel and then specifying the dimensions. For example, we can switch from the 10×10 grid that we were exploring above to a 100×100 grid in just a moment, and start generating similar dispersion patterns on a much finer grid of cells (Figure 5.29). Imagine building a similar model on a 100×100 grid in Stella!

Simile can also save equations; however, here it is done using a programming language, Prolog, which makes it a bit harder to read for somebody unfamiliar with the conventions of that language – especially when the model becomes more complex. For simple models like the one we are studying, it is still quite easy to understand what the statements are about. Below is the Grass–Rabbits model as described in Simile:

Model R_G_array10000
      Enumerated types: null
Variable R_A
      R_A =   [Rabbits]
      Where:
          [Rabbits] = Cell/Rabbits
Variable sum_R
      sum_R =   sum([R_A])
Submodel Cell
      Submodel Cell is a fixed_membership submodel with dimensions [10000].
      Enumerated types: [
Compartment   Grass
      Initial value = 2
      Rate of change = + Growth − Grazing
Compartment   Rabbits
      Initial value = if (index(1) == 2550) then 300 else 1
      Rate of change = + Uptake − Mortality − Migration
      Comments:
          For random initialization rand_const(1,2)
Flow Grazing
      Grazing =   beta*Grass*Rabbits
Flow Growth
      Growth =   alpha*Grass
Flow Migration
      Migration =   delta*((if col > 1 then Rabbits-element([R_A],(row − 1)*
size + col−1) else 0) + (if col<size then Rabbits-element([R_A],(row−1)*size + col + 1) else
0) + (if row > 1 then Rabbits-element([R_A],(row−2)*size + col) else 0) + (if row<size then
Rabbits-element([R_A],row*size + col) else 0))
      Where:
          [R_A] = ../R_A
Flow Mortality
      Mortality =   mu*Rabbits
Flow Uptake
      Uptake =   k*Grazing
Variable alpha
      alpha =   1
Variable beta
      beta =   1
Variable col
      col =   fmod(index(1) − 1,size) + 1
Variable delta
      delta =   0.1

Variable k
      k =   0.1
Variable mu
      mu =   0.1
Variable row
      row =   floor((index(1) − 1)/size) + 1
Variable size
      size =   100

1.0 ▭▭▭▭▭▭▭▭▭▭▭▭▭ 3.0

Figure 5.28   Spatial output for the model with migration.
First we use a simplified initial condition to make sure that we can generate a pattern of dispersion, as we might expect to see in a model that is similar to the diffusion process.

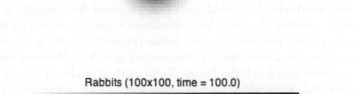

Rabbits (100x100, time = 100.0)
1 ▭▭▭▭▭▭▭▭▭▭ 3

Figure 5.29   The same model but with 10,000 cells active.
Switching from one model dimension to another is easy; it requires only changing one parameter and the definition of the array size.

Now that we are confident about how rabbits move horizontally, we can switch the ecological processes back on and see how the system performs in space. Once again initializing Rabbits and Grass randomly over the landscape, we can see how, due to dispersion, the patches become blurred; every now and then, when Grass is

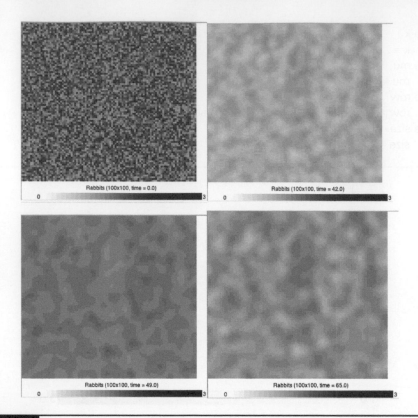

Figure 5.30   Spatial output for the model with randomly generated initial conditions. The diffusion creates blurred patterns of distribution of Rabbits.

depleted, the overall population falls to a low then, following general predator–prey dynamics, Rabbits reappear (Figure 5.30). We can also output the results as time graphics for each cell. Figure 5.31 presents ensembles of 10,000 curves for Rabbits and Grass in each of the 10,000 cells. This graphic and the quantity of computations that stand behind it should really be appreciated. Interestingly, in spite of all this spatial variability, the totals for Rabbits and Grass follow exactly the classic predator–prey pattern that we have seen before (Figure 5.32). Well, almost exactly, as we can see from the scatter-plot XY diagram in Figure 5.33. Whereas previously for just two variables in one cell the Runge–Kutta method produced an exact ellipsoid, winding over and over itself again and again, with 10,000 instances of the same model the behavior becomes quite different. There is certainly far more reason to expect that it is the error that accumulates and takes us slowly off track. Let us check: is it the error that causes this, or something else?

The first remedy to decrease computation error is to switch to higher-order numerical methods or to decrease the time-step. There is nothing better in Simile than Runge–Kutta, so higher-order methods are not an option. However, we can easily decrease the time-step. Above, we had DT = 0.1. Let us make it DT = 0.01. Now it will take us almost 10 times longer to run the model, yet unfortunately we are not getting any different output. Still the trajectory keeps winding towards the center. So what else could be causing it?

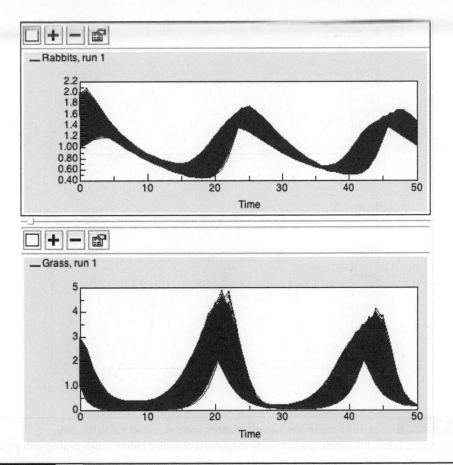

Figure 5.31   Using the plotter, we can view dynamics for all the 10,000 cells.

Let us go back to the original model. In order to get there, we will remove the horizontal fluxes (Migration = 0 if delta = 0), and initialize all the cells the same. Now we are simply running a bunch of predator–prey models simultaneously. To make the model run faster, we can also make the spatial dimensions smaller: let us set the size equal to 2, and the dimension of cells equal to 4. If we now run the model, we will finally get the expected ellipse (Figure 5.34A). Next, let us initialize the four cells that we have randomly selected. The result is somewhat unexpected (Figure 5.34B), and answers our dilemma: it is the random numbers in the initial conditions that make the total population dynamics so different. If we increase the number of cells (size = 10), the populations tend to be less chaotic and tend towards a limit cycle (Figure 5.34C). The graphic in Figure 5.34D is produced by the same 10,000 cells with horizontal migration switched on (delta = 0.1), as we had in Figure 5.33, but after some 1,500 time-steps. We see that here also there is a clear trend to the center, where the population almost equilibrates.

This is quite remarkable, since, as you may recall, one of the major critiques of the classic Lotka–Volterra model was that it depends so much upon the initial

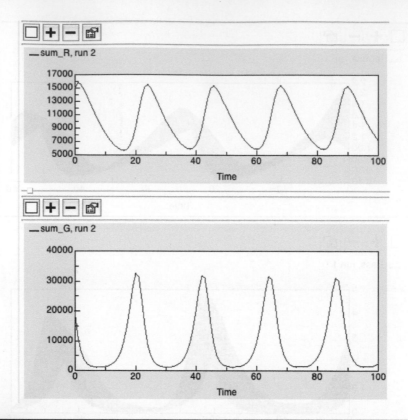

Figure 5.32   Output for the total numbers of Rabbits and Grass in all the 10,000 cells.
The totals seem to follow the classic predator–prey oscillations observed before, when dealing with a spatially aggregated model.

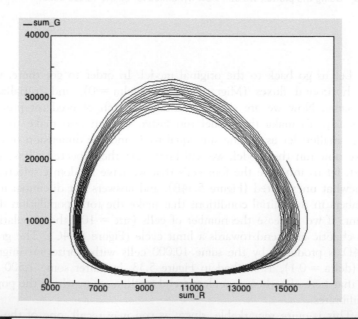

Figure 5.33   A scatter graph (or XY graph) where the numbers for Grass are displayed as a function of the number for Rabbits. It shows that the oscillations are damping.

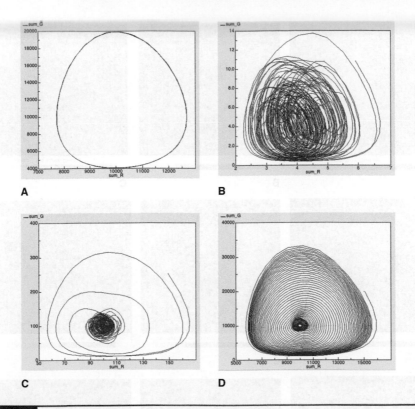

Figure 5.34 Resolving the mystery of dampened oscillations.
A. When we have no spatial heterogeneity, the population is spatially uniform, and we have an ideal predator–prey ellipse as in the classic model. B. With the population randomly initialized in just four cells we get a chaotic behavior that fills the whole interior of the ellipsoid. C. With 100 cells randomly initialized, the area of chaotic dynamics shrinks to a smaller domain. D. With 10,000 cells there is no more chaos and the trajectories tend to a small limited cycle, around which they keep oscillating. This behavior no longer depends upon the initial conditions, as long as the cells are initialized with different values.

conditions. The classic model describes a population over a certain area, where spatial heterogeneities are ignored and all the organisms are lumped into one number representing the total population. However, in reality they are certainly unevenly distributed over space. If we split the space into just a few regions and present the dynamics in this spatial context, we get results that are significantly different from the classic model. Actually, it turns out that the stable oscillations are an artifact of the averaging over space. With several spatial entities we have a converging dynamic, which also no longer depends upon the initial conditions.

If we take a closer look at the spatial distributions that correspond to this quasi-equilibrium state, we may find some weird spatial patterns (Figure 5.35). Starting from the randomly distributed initial conditions (Figure 5.35A), after some 1,000 iterations, as the trajectory on the phase plane converges toward the center of the ellipsoid a spatial pattern emerges that, while changing to a degree, still persists, as can be seen from the series of snapshots taken approximately every 50 iterations

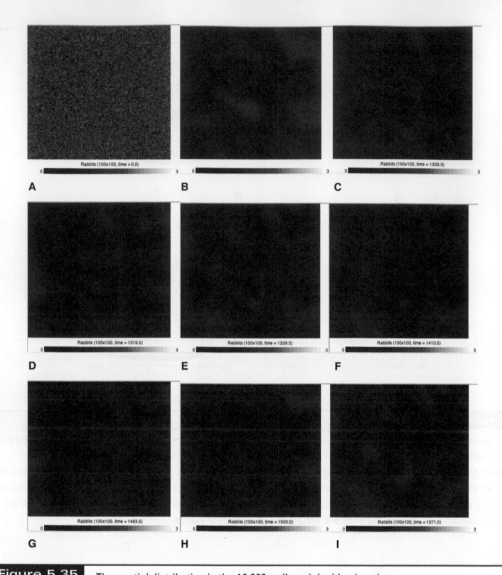

Figure 5.35   The spatial distribution in the 10,000-cell model with migration.

Starting with random initial conditions (A), after some 1,000 iterations a pattern is formed, which then persists (B–I). Thus there is a pattern that emerges both in time and space.

(Figure 5.35B–I). It is not clear how and why this pattern emerges, but it is interesting to register that emergent patterns can result from this kind of non-linear dynamics.

Using the so-called association submodel concept in Simile we could put together much more elegant solutions for this model; however, these models also become far more difficult to build and comprehend.

Let us put together an association submodel called NextToCell. It will be defined by two relationships: "self" and "neighbor." These are cell attributes that are provided by the stack of cells with the submodel in each of them. The existence of NextToCell submodel is defined by the condition cond1.

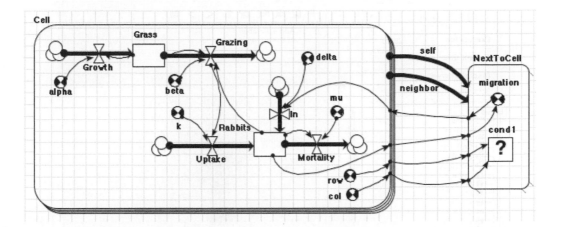

cond1 = ! (col_self == col_neighbor and row_self == row_neighbor) and abs(col_self-col_neighbor) < 1.5 and abs(row_self-row_neighbor) < 1.5

This condition is true only if the coordinates (col, row) of the two cells are adjacent to each other – that is, the difference between the col and row coordinates is less than 1.5 and the cell is not itself. In this way we can describe all eight cells in the vicinity of a given cell. For each of these neighbor cells we define a variable called

migration = Rabbits_neighbor − Rabbits_self

This is the difference between the number of Rabbits in the cell and the neighboring cell. This value is then fed back into the model and is used to define the flow called

In = delta*sum({migration_self}).

Here we are summing all the migrations for the eight neighboring cells and, with the diffusion rate of delta, using this sum to update the number of Rabbits in the current cell. Note that when

Rabbits_neighbor > Rabbits_self

the flow is positive, and it is negative otherwise. This should be sufficient to describe the diffusion process of Rabbits in our system. Indeed, if we run the model we get some very plausible distribution that looks very similar to what we have been generating above – but, we have to agree, this formulation is way more elegant.

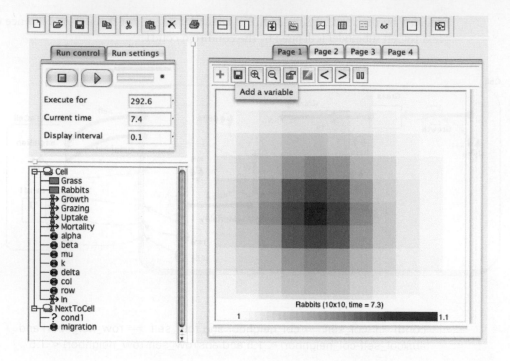

SME model

Let us explore yet another way to build and run spatial models. The Spatial Modeling Environment (SME) is not quite a modeling system, since it does not require a language or formalism of its own. It can take the equations from your Stella model and translate them into an intermediate Modular Modeling Language (MML), which is then translated into C++ code. At the same time, SME will link your model to spatial data if needed.

Let us first put the same Grass–Rabbits model into Stella and make sure that it runs properly. As a result, we will end up with the following system of Stella equations:

Grass(t) = Grass(t − dt) + (G_growth − Grazing)*dt
INIT Grass = 2
INFLOWS:
G_growth = alpha*Grass
OUTFLOWS:
Grazing = beta*Grass*Rabbits

Rabbits(t) = Rabbits(t − dt) + (Uptake − R_mortality)*dt
INIT Rabbits = 1
INFLOWS:
Uptake = k*Grazing
OUTFLOWS:
R_mortality = mu*Rabbits
alpha = 1
beta = 1
k = 0.1
mu = 0.1

For these Stella equations, we do <Edit -> Select All> and then <Edit -> Copy>.

Next, we open a Text Editor on our computer (on a Macintosh it will be BBEdit, or TextEdit; in Windows it is probably the NotePad) and paste the equations into the file, then save the file using the .eqns extension and naming it R_G1.eqns.

We now need to get SME running. SME is open source and is available for download from Source Forge, the main repository of open-source projects. The URL is http://sourceforge.net/projects/smodenv. SME is available for Linux and Mac OSX operating systems; there is no Windows version so far. Once we have downloaded and installed SME, we need to set up the SME project.

Having chosen a name for our project – let us say R_G, representing Rabbits&Grass – we open the Terminal window and enter the command:

>SME project R_G

If the installation has been done properly, this sets up the project directory. Now we can put the equations file that we created in Stella into the directory Models. We will call the model R_G1 and perform the SME command:

>SME model R_G1

Now we get:

Current project directory is /Documents/SME/Projects/
Current project is R_G
Current model is xxx
Current scenario is xxx
Current model set to R_G1
Current project set to R_G

It is not important at this time, but let us also choose a scenario name. We will see what that is later on. Using the command

>SME scenario S1

we get:

Current project directory is /Documents/SME/Projects/
Current project is R_G
Current model is R_G1
Current scenario is S1

Now we can import and configure the model:

>SME import

This will take the equation file and translate it into the MML (modular modeling language) specification. There will probably never be any need to see the result, but for the sake of curiosity it is possible to look at the file Models/R_G1.MML for the MML specification and then look at Models/R_G1/R_G1_module.xml, which is the same file in an intermediate XML specification.

At the same time the first config file has been generated in Config/R_G1.MML. config. This file still contains just a list of all variables and parameters of the model.

Let us do the build command now:

>SME build

Something is processed, there are some messages, and at the end it can be seen that some C++ code has already been compiled. This is not important at this time, since we will probably still need to do some more configuring before we get something meaningful. What is important is that a couple of more config files are generated. See what is in the Config directory now:

R_G1.biflows, R_G1.conf, R_G1.S1, and R_G1.S1.conf.out

The most important file is R_G1.conf. This will be the config file that we will be working with most of the time. At this time it has the following list of parameters:

```
# global  DS(1.0,0) n(1) s(4332) ngl(0) op(0) OT(1,0,20) d(0) UTM(0,0.0,0.0) UTM(1,1.0,1.0)
$ R_G1_module
* ALPHA          pm(1)
* BETA           pm(1)
* GRASS          s(1) sC(C)
* GRAZING        ft(u)
* G_GROWTH       ft(u)
* K              pm(0.100000)
* MU             pm(0.100000)
* RABBITS        s(1) sC(C)
* R_MORTALITY    ft(u)
* TIME
* UPTAKE         ft(u)
```

If we compare this file with the Stella equations above, we see that it contains information about all the parameters that we had there. In the equations:

```
alpha = 1
beta = 1
k = 0.1
mu = 0.1
```

we find the same values in the R_G1.conf file.

What we have lost are the initial conditions. That is because in Stella we defined the initial conditions in the state variables boxes, rather than as parameters. SME does not like that. Let us quickly go back to Stella and fix it by defining initial conditions in terms of some auxiliary parameters:

```
INIT Grass = G_init
INIT Rabbits = R_init
G_init = 2
R_init = 1
```

Note the tiny difference between this set of equations and what we had above. We will now have to do another >SME import and >SME build. Keep in mind that whenever we alter the equations, we need to do a re-import and a rebuild. We do not need to re-import and rebuild if we only modify the parameters in the config file. However, if any of the parameters are redefined as spatial, a rebuild is needed. We will get back to this later.

So another SME import modifies the R_G1.MML.config file – but when we run SME build the R_G1.conf file will not be changed. This is a level of protection to make sure that the config file with all the valuable spatial information is not inadvertently overwritten, by re-importing and rerunning the Stella equations that do not contain this data. This might be a little confusing; however, it is important to protect the spatial version of the config file.

The output from the last rebuild can be found in R_G1.S1.conf.out, and if this is really what you want to do, you can delete your R_G1.conf file and rename the R_G1.S1.conf.out into R_G1.conf. This is what we will do now to get the following as the config file for the model.

```
# global DS(1.0,48) n(1) s(4332) ngl(0) op(0) OT(1.0,0.0,20.0) d( 0) UTM(0,0.0,0.0)
UTM(1,1.0,1.0)
$ R_G1_module
* ALPHA            pm(1)
* BETA             pm(1)
* GRASS            s(1) sC(C)
* GRAZING          ft(u)
* G_GROWTH         ft(u)
* G_INIT           pm(2)
* K                pm(0.100000)
* MU               pm(0.100000)
* RABBITS          s(1) sC(C)
* R_INIT           pm(1)
* R_MORTALITY      ft(u)
* TIME
* UPTAKE           ft(u)
```

Note that the initial conditions are now properly defined in this file. We are ready to run the model in SME. However, first let us take another look at the config file. We have already guessed that pm( ) is a parameter in Stella. Whatever value the parameter had in Stella, it was automatically transferred into the config file. Also, the state variables (GRASS and RABBITS in this case) are described by two commands, s( ) and sC(C). What are they? The best available documentation for SME is on the web at http://www.uvm.edu/giee/SME3/ftp/Docs/UsersGuide.html. Most of the commands are described there, though not in the most foolproof way. For the state variables, we learn that s(1) means that we will be using the first-order precision numeric method. We might also learn that the two commands that were generated by the SME build command are actually not quite consistent with the latest documentation: the sC(C) command could be erased and instead the s command should be s(C1C). However, SME will still run with the sC(C) command. "C" means that the variable should be clamped – that is, it will not be allowed to become negative. It is not unusual to find these kinds of glitches in open-source code; after all, these guys are not paid to write the fancy tutorials and documents to make their software useful! We have to either bear with them (after all, the software is free) or, even better, help them. We can always contribute our bug reports and pieces of documentation that we put together while exploring the program.

Next we need to configure the output. So far it is undefined; we do not know what the program will output and where will it go. Let us use the P(0,0) command to see how the state variables change. The lines for GRASS and RABBITS will now be:

```
* GRASS     P(0,0) s(C1C)
* RABBITS   P(0,0) s(C1C)
```

Note that we have also got rid of the outdated sC(C) command. Just one more thing before we run the model. Take a look at the first line in the config file, the one that starts with #global. This is a set of general configuration commands that are placed there by default by the translator. The two important ones that we may want to change right away are the OT() and the d() commands. Check out the SME documentation to learn more about them. The d() command sets up the debug level – that is, the amount if information that will be provided into the command line interface. When we have d(0), that is the minimal amount. If we want to see what equations are solved in which order and what actually happens during our model run, we probably need to bump up the debug level, making it d(1) or d(2).

The OT command defines the time-step, the start and the end of the simulation. Right now we have OT(1.0,0.0,20.0), which means that we will run the model with a time-step of 1, starting from day 0 and finishing on day 20. This will not allow us to go beyond 20. If we wish to have a longer simulation time, we need to change it to, say, OT(1.0,0.0,100.0). Now we can make up to 100 steps.

Finally we can run the model, using

>SME run

See what happens. In the command line interface we get:

```
[AV-Computer:SME/Projects/R_G] voinov% SME run
*** Spatial Modeling Environment, Copyright (C) 1995 (TXU-707-542), Tom Maxwell
*** SME comes with ABSOLUTELY NO WARRANTY
*** This is free software, and you are welcome to redistribute it
*** under the terms of the GNU General Public License.
Current project directory is /Documents/SME/Projects/
Current project is R_G
Current model is R_G1
Current scenario is xxx
Running SME model R_G1 in serial mode, cmd:
/Documents/SME/Projects//R_G/Driver/R_G1
-ppath /Documents/SME/Projects/ -p R_G -m R_G1
-ci /Documents/SME/Projects//R_G/Config/R_G1.conf -pause 0 -scen xxx
info: Setting Project Name to R_G
info:
Allocating module R_G1_module; ignorable: 0
info: Reading Config Files
info: Opening config file: /Documents/SME/Projects/R_G/Config/R_G1.conf:
info: Reading config file
warning: this program uses gets(), which is unsafe.
SME>
```

Here, the driver stops and waits for us to tell it what to do next. It looks like gibberish, but may actually contain some important information – especially if we run into errors. To run the model for 5 days, we use

SME>r 5

If we have the debug level set at d(1), we will probably get:

```
info: Setup Events
info: CreateEventLists
info: ProcessTemporalDependencies
info: ProcessSpatialDependencies
info: CreateEventLists
```

```
info: FillInitializationList
info: Split & Sort Lists
info: Setup Variables
info: Setting Up Frames & Schedules
info: Allocating Memory
info: Posting Events
info:  Opened  xml  File:  /Documents/SME/Projects/R_G/Models/R_G1/xxx/R_G1_
module.xml.
info:  ******************** Executing  Event  R_G1_module:StateVarInit  at  time
0.000000
...
info:  ******************** Executing  Event  R_G1_module:FinalUpdate__S__  at
time 5.000000
TCL> 5
SME>
```

The model now stops again, and another r command is required to continue. Let us run it till day 100:

```
SME > r 100
```

Now it stops and waits again. To quit, we do

```
SME > X
```

It is important to ensure that Enter is pressed after each of these commands.

This is it. Now where are the results? Go to Projects/R_G1/DriverOutput. Here, we might notice that two more files have been generated:

```
GRASS.PTS.P_p_0
RABBITS.PTS.P_p_1
```

These files cannot be seen until we have quit the model run; they appear only after the X command has been issued. Now that we have exited SME, the files should be there. These are simple timeseries, with output for GRASS and RABBITS respectively. The first column is the time, the second column is the value of the state variable. One way to look at these results is to simply copy and paste the files into Excel or another spreadsheet program. We can draw the graph and see that, after a couple of oscillations, the GRASS population crashes followed by the slow dying off of the RABBITS. This is not exactly what we would expect from a standard predator–prey model. Where are those nice population numbers, going up and down indefinitely?

Of course, we were running the model with the first-order Euler method. That is a pretty rough approximation. Let us switch to a more accurate numeric method. We open the config file and change to the fourth-order method:

```
* GRASS        P(0,0) ε(C1C)
* RABBITS      P(0,0) s(C4C)
```

Note that previously we had s(C1C), now we have s(C4C). This does it. If we rerun the model (SME run, then r 100), exit (X), go to the DriverOutput directory and paste the output files into Excel, then we will get what we were expecting – nice lasting oscillations of both variables.

However, where are the spatial dynamics? We could get all this in Stella without the trouble of setting up the model in SME. But how can we expect anything spatial if we have not defined anything spatial in our model? So far, we have simply replicated the Stella model. Now let us go spatial. First of all we will need some maps to describe the areas for grass and rabbits. Suppose we choose the area shown in Figure 5.36.

These days, the simplest way to generate these maps is to use ArcInfo or ArcGIS, the monopolist on the GIS market. However, if we run GRASS, an open-source GIS (do not confuse with one of the variables in this model) that will also work. Anyway, what we need to do is generate a simple ascii file that will first of all describe the study area in our model. This will have 1s inside the study area and 0s everywhere else. It may look like this in one of the formats that SME takes, i.e. the MapII format:

```
FILETYPE=INTERCHANGE
ROWS=62
COLUMNS=67
CELLSIZE=200.000000
FORMAT=DEC
INFO="hunt.wsh"
DATA=0 0 0 0 0 0 0 0 0 0 0 0 0 0 0 0 0 0 0 0 0 0 0 0 0 0 0 0 0 0 0 0 0 0 0 0 0 0 0 0 0 0 0 0 0 0 0 0 0 0 0 0 0 0 0 0 0 0 0 0 0 0 0 0 0 0 0
00000000000000000000000000000000000000000000000000000000000000000000
00000000000000000000000000000000000001100000000000000000000000000000
00000000000000000000000000000000000011000000000000000001100000000000
00000000000000000000000000000000011111100000000000111110000000000
00000000000000000000000000000111111111110001111111111110000000
00000000000000000000000000011111111111111111111111111111000000
00000000000000000000000000111111111111111111111111111111110000000
00000000000000000000000001111111111111111111111111111111100000000
00000000000000000000000011111111111111111111111111111110110000000
00000000000000000000000111111111111111111111111111111110000110000000
00000000000000000000011111111111111111111111111111111110000000000
00000000000000000000011111111111111111111111111111111110000000000
00000000000000000000111111111111111111111111111111111100000000000
00000000000000000001111111111111111111111111111111111100000000000
00000000000000000001111111111111111111111111111111111100000000000
00000000000000000001111111111111111111111111111111111100000000000
00000000000000000011111111111111111111111111111111111100000000000
00000000000000000111111111111111111111111111111111111000000000000
00000000000000011111111111111111111111111111111111110000000000000
00000000000000111111111111111111111111111111111111110000000000000
00000000000011111111111111111111111111111111111111111100000000000
00000000000011011111111111111111111111111111111111111110000000000
00000000000111111111111111111111111111111111111111111110010000000
00000000011111111111111111111111111111111111111111111111100000000
000000000111111111111111111111111111111111111111111111111111000000
00000000111111111111111111111111111111111111111111111111111000000
0000000111111111111111111111111111111111111111111111111111110000
0000000111111111111111111111111111111111111111111111111111110000
0000000111111111111111111111111111111111111111111111111111111000
0000001111111111111111111111111111111111111111111111111111111000
0000001111111111111111111111111111111111111111111111111111111000
0000001111111111111111111111111111111111111111111111111111111000
0000001111111111111111111111111111111111111111111111111111110000
0000001111111111111111111111111111111111111111111111111111110000
0000001111111111111111111111111111111111111111111111111111100000
0000001111111111111111111111111111111111111111111111111111000000
0000001111111111111111111111111111111111111111111111111000000000
000000111111111111111111111111111111111111111111111000011100000
0000001111111111111111111111111111111111111111111110000000000000
00000011111111001111111111111111111111111110000011111100000000
000000111111100001111111111111111111111111110000000010000000000000
00000011111100000011111111111111111111111111000000000000000000
0000001111110000000001111111111111111111111100000000000000000000
00000011111000000000011111111111111111111111110000000000000000
00000011110000000000011111111111111111111111100000000000000000
000000110000000000000011111111111111111111111000000000000000
00000010000000000000011111111111111111111111100000000000000000
0000000000000000000000011111111111111111111100000000000000000
000000000000000000000011111111111111111111100000000000000000
000000000000000000000111111111111111000000000000000000000
000000000000000000000011111111100110000000000000000000000
0000000000000000000000011111000000000000000000000000
00000000000000000000000111100000000000000000000000000000
0000000000000000000000001110000000000000000000000000000
000000000000000000000000011000000000000000000000000000
0000000000000000000000000000000000000000000000000000000000000000
LABELS
```

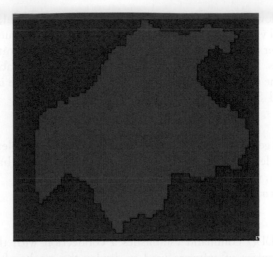

Figure 5.36   A map that defines the study area in SME.

This is very similar to the ArcInfo ascii format. Everything will be the same except for the first header lines:

```
ncols      62
nrows      67
xllcorner    443396.06231037
yllcorner    241834.36232137
cellsize    30
NODATA_value -9999
00000000000000000000000000000000000000000000000000000000000000
00000000000000000000000000000000000000000000000000000000000000
00000000000000000000000000000000000110000000000000000000000000
00000000000000000000000000000001110000000000000110000000000000
00000000000000000000000000001111110000000000001111100000000000
...
```

The bottom line is that it does not matter what software is used, as long as we can get our map files into one of these ascii formats. One thing to watch out for, especially when preparing data in Windows, is the EOL – end of line – symbol. It is different in Windows and Unix. Since SME is using Unix conventions, it is likely to choke when it has to read files with the "wrong" EOL symbol. Try the dos2unix command to convert the files – usually it comes with the standard distribution of Unix or Linux, although it does not come with Darwin, the Mac OS-X Unix kernel. Luckily, on a Macintosh there are other ways to handle this. For example, BBEdit allows a choice of format for the text files.

Now that we have the map, let us get it into the model. As we remember, it is mostly the config file that SME uses to link the model with the data. We define the model as spatial using the g() command:

```
$ R_G1_module   g(A,/Documents/SME/Projects/R_G/Data/Maps/Area.arc,default,
                /Documents/SME/Projects/R_G/Data/Maps/Area.arc) AL(0,0)
```

The long path /Documents/SME/Projects/R_G/Data/Maps/Area.arc points to the Area.arc map that is the ArcInfo ascii format for the study area that we have chosen. Now we have initialized the model as being spatial, but have not identified any spatial variables: all of them are still treated as single numbers. To do that we can use another SME command – oi(). It is called "override initialization," and has two parameters. If the first parameter is positive, then the variable is assumed to be constant. If the second parameter is positive, then the variable is assumed to be spatially distributed. So if we configure, say, GRASS as

* GRASS   s(C4C) oi(0,1)

we should get what we want – a spatially distributed variable.

Next let us deal with the graphic output. This is handled by the so-called Viewserver, which we now need to start up.

Let us add yet another command to the previous line:

* GRASS   DD() s(C4C) oi(0,1)

The DD() command establishes a connection with the Viewserver – a very important piece of software used to display the results of spatial simulation. The Viewserver should be started using the command startup_viewserver. It is better to do it from a separate terminal window, since the Viewserver generates a long command line output that will clog the terminal that is being used to run SME.

Let us also generate spatial output for the other model variable, RABBITS:

* RABBITS   DD() s(C4C)

Note that in this case we do not even need to declare the variable as spatial. In the model it is dependent upon an already spatial variable (GRASS), so it will become spatial automatically.

Once we have started the Viewserver and done another SME run, we can see that the Viewserver receives output from the running model and a new data set is added to the list on the left panel of the Viewserver. If we highlight one of the data sets and then choose a 2D animation viewer and click the "Create" button, we will get an image of the map that is now dynamically changing as the variables change their values across the whole area. You can watch how the Grass and Rabbits alternate their biomasses, changing from minimal (blue) to maximal (red) numbers (Figure 5.37).

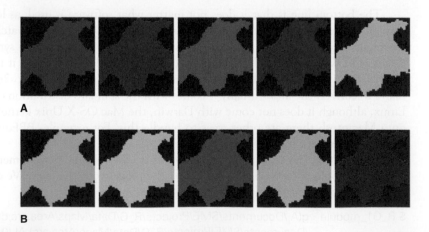

Figure 5.37   Simple spatial dynamics when all cells are the same.
A. Grass (max. 2.043, min. 0.3279), B. Rabbits (max. 1.285, min. 0.7642).

Let us now make the model spatially heterogeneous. Suppose we have a spatially heterogeneous initial condition for the GRASS biomass, and that grass is not uniformly distributed but has different biomass in different locations. We will initialize the GRASS variable with a map that has different values in different cells. Let us use the map in Figure 5.38.

Note that the spatial extend of this map is different from that of the map above. That is OK. SME will crop this map to match it to the area defined above by the study area map. How do we input this new map? Back to the config file. This time instead of defining the initial condition for GRASS as a constant parameter pm(2), we will use a map:

```
* G_INIT      d(A,/Documents/SME/Projects/R_G/Data/Maps/Biomass.arc,
              /Documents/SME/Projects/R_G/Data/Maps/Area.arc)
```

Again, we have to provide the full path to the map file that we want to use. There is actually a better way to do it using the Environment file. This file should reside in the Data directory, and it contains all the paths that we may wish to use in the configuration files. For this model, we will put the following two lines into the Projects/R_G/Data/Environment file:

```
MAPS = /Documents/SME/Projects/R_G/Data/Maps
RMAP = /Documents/SME/Projects/R_G/Data/Maps/Area.asc
```

The first line defines the Maps directory, which we seem to be constantly referring to. The other line is the full name of the reference map, or the study area map, which is used to crop all the other maps in the project.

Now some of the lines in the configuration file can be much shorter:

```
$ R_G1_module g(A,${RMAP},default,${RMAP}) AL(0,0)
...
* G_INIT       d(A,${MAPS}/Biomass.arc,${RMAP})
...
```

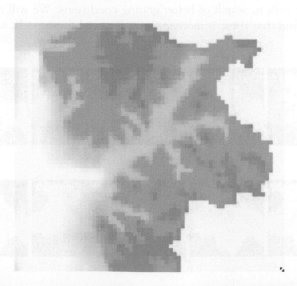

Figure 5.38   Another map used to define spatially heterogeneous initial conditions for Grass.

Moreover, the biomass map that we used to initialize the model has values between 0 and 61. The initial condition that we used before was 2. It would be nice if we could scale the map to some values that would be closer to those we had originally, and we can use the S() command to do that. The syntax of this command is S(a,b), which means that if x is the input value then the result of this command is y = a*x + b. So finally if we use the command

* G_INIT d(A,${MAPS}/Biomass.arc,${RMAP}) S(.01e + 00,1.0)

this means that we will input the map from the Biomass.arc file, then each value will be multiplied by 0.01 and added to 1. That will be the result used in the simulations. Also note that we no longer need the oi(0,1) command, since we now have the initial condition that initialized the variable as a spatial one, which ensures that all the rest of the variables connected to the spatial one will also be spatial.

This is a little more interesting: now there are some spatial variations, and there are some differences in how various cells evolve (Figure 5.39). However, there is still no interaction between cells, and the real spatial context is not present. We simply have a whole bunch of models running in sync, but they do not interact with each other.

Making cells "talk" to each other is a little more complex than anything we have done so far. Whereas until now we have simply used some predefined commands, and the model we built in Stella, from now on if we are to define some meaningful spatial interaction we will need to do some programming.

There are some modules that we can use in the Library of Hydro-Ecological Modules (LHEM – http://giee.uvm.edu/LHEM); however, there are not too many things we can do with those pre-designed modules. If we really want to be able to build complex spatial models, we will probably need to be capable of some level of C++ programming. SME supports so-called User Code and offers full access to its classes and methods, which can significantly help us in designing our own code for spatial dynamics.

Suppose for the Rabbits & Grass model we wish to allow rabbits to move between cells in search of better grazing conditions. We will assume that whenever rabbits find that there is more grass in the neighboring cell, a certain proportion of

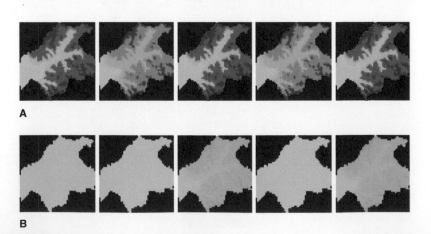

Figure 5.39   Spatial dynamics with no migration.
A. Grass (max. 1.490, min. 0.5), B. Rabbits (max. 1.142, min. 0.86).

rabbits from the current cell will move to the cell with more grass. Let us write the code that will describe this behavior of the predator:

```
/**********************************************************/
#include "Rabbit.h"
/**********************************************************/

void MoveRabbits( CVariable& Rabbits, CVariable& Grass, CVariable& Rate )
// moves rabbits toward more grass, if there are less rabbits there
// arguments come from MML.config file, first arg is always variable being configured.
{
        Grid_Direction il;
        float fr, R_moved = 0.;

        DistributedGrid& grid = Rabbits.Grid();
        grid.SetPointOrdering(0);
        // sets grid ordering to default ordering (row-col) (ordering #0)

        Rabbits.LinkEdges();
        Grass.LinkEdges();
        static CVariable* R_Flux = NULL;
        if(R_Flux = = NULL )
                R_Flux = Grass.GetSimilarVariable("R_Flux");
        // intermediate increment to Rabbits
        R_Flux- > Set(0.0);

        for( Pix p = grid.first(); p; grid.next(p) )
        {
                const OrderedPoint& pt = grid.GetPoint(p);
                // sets currentPoint

                if( !grid.onGrid(pt) ) continue;
                // (onGrid = = False) - > Ghost Point

                float g_max = Grass(pt);
                Pix p_max = p;

                // for each point calculate where is the max Grass in the vicinity
                for( il = firstGD(); moreGD(il); incrGD(il) )
                {
                // enum Grid_Direction {NE = 2, EE, SE, SS, SW, WW, NW, NN};

                        Pix rp = grid.NeighborPix( p, il );
                        // relative to pt, takes enum Grid_Direction as arg

                        if( rp )
                        {
                                const OrderedPoint& rpt = grid.GetPoint(rp);
                                if ( Grass(rpt) > g_max )
                                {      g_max = Grass(rpt);
                                        p_max = rp;
                                }
                        }
                }
        }
```

```
    const OrderedPoint& pt_max = grid.GetPoint(p_max);
    // sets currentPoint

    // if there is a cell in the vicinity where there is more Grass, then a
    // portion of Rabbits moves to that cell
    if ( g_max>Grass(pt) )
       fr = ( Rabbits(pt)>Rabbits(pt_max) ) ?
           (Rabbits(pt)−Rabbits(pt_max)) * Rate(pt) : 0;

    (*R_Flux)(pt_max) + = fr;
    (*R_Flux)(pt)− = fr;
    R_moved + = fr;

  }// end area loop

  Rabbits.AddData(*R_Flux);
  printf ("\ninfo: Rabbits moved = %f," R_moved);
}
/*************************************************************/
```

So here we have only Rabbits moving horizontally from one cell to another in search of a better life. How do we tell SME that there is something new that the model wants to take into account?

First, we go all the way back to the MML.config file that we can find in the Config directory. In this file we add a command for Rabbits:

```
* RABBITS   UF( Rabbit,MoveRabbits,GRASS,RATE)
```

Here, Rabbit is the name of the file that contains the above C++ code. Actually its name is Rabbit.cc, and it resides in the UserCode directory. MoveRabbits is the name of the function in this file that we use. GRASS and RATE are two variables that are passed to this function. While GRASS has always been there, RATE is new. The way we get it into the config file is by modifying the Stella model and adding another variable. Once again, we have to export the equations and then do the "SME import" command. Alternatively, we can modify the equation file that we created earlier from Stella equations. We simply need to add one line:

```
rate = 0.5
```

and then we can also do this by hand in the R_G1.MML.config file. Note, however, that this is somewhat risky, since it is very easy to forget about some of these small modifications of the equation file, and there is no way we can import these modifications from the equations to the Stella model. As a result, once we have finally decided that we wish to modify the Stella model for some other reason later on, most likely we will forget about these modifications. When taking the equations from Stella and creating a new equation file, we will lose all these previous changes. The model will suddenly perform quite contrary to expectations, and it will take a while to figure out why and to redo all the little updates. So while every now and then it seems very simple to modify just the equation file, actually it is much better if all the modifications are done directly to the Stella model.

As we remember, whenever the equations or the MML.config file is changed we need to do the SME import command. Then we can do the SME build command, and update the config file to add the RATE parameter to it as well. Remember − either it has to be done by hand, or the file can be renamed to use the R_G1.S1.conf.out instead. As a result, we get:

```
* RATE              pm(0.5)
```

Are we ready to run? Almost, but there is still one glitch to fix. The variables that we have been passing to the newly designed function to move Rabbits are all assumed to be spatial.

```
MoveRabbits( CVariable& Rabbits, CVariable& Grass, CVariable& Rate )
```

However, the RATE parameter as we defined it above is a scalar. There is an easy fix. Just add the override command

```
* RATE              pm(0.5) oi(0,1)
```

and you will be back in the game. Alternatively, you could also define this parameter as a map:

```
* RATE              d(A,${RMAP},${RMAP}) S(.5e + 00,0.0)
```

Here we used the study area map to initialize this parameter, which, with this scaling factor, is identical to what we did above. However, this could be any map, which would probably be the only reasonable way to define this parameter if we wanted it to be spatially heterogeneous.

Alternatively, if we do not want this parameter to be spatial, we must not refer to it as if it were spatial in the code. Replace Rate(pt) for Rate.Value(). Rate.Value() is a scalar, it will not need to be initialized by a map or a spatial variable. It will take pm(0.5).

Finally, we are ready to hit the "SME run" command and watch something moving across the landscape – rabbits hoping from one place to another, grass dying and regrowing back when the predators leave, and so on (Figure 5.40).

Certainly, this was not as easy as putting together a model in Stella, or even Simile. However, for somebody comfortable with C++ it would not be a big deal and actually may turn out to be simpler than learning the new formalism required for Simile. Once we are in the programming language mode, we have all the power we

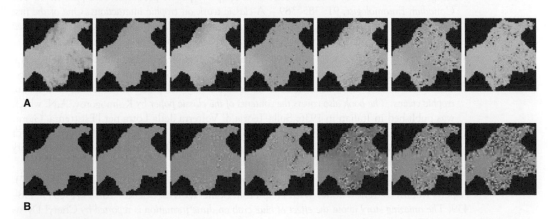

Figure 5.40 Spatial dynamics with migration towards the cells with higher density of Grass. Clusters of high density are formed when Rabbits from several cells jump into a cell with higher Grass abundance. A. Grass (max. 1.490, min. 0.5), B. Rabbits (max. 1.142, min. 0.86).

need to create any complex model. So in a way, SME may be treated as a nice interface between Stella and C++ power modeling.

5.5   Conclusions

A very simple model can produce an amazingly diverse collection of behavior patterns. The fact that the predator–prey model contains non-linearity makes it a very exciting system to explore. After many generations of mathematicians and modelers studying the system, it still every now and then produces some interesting results, especially if we add some detail in either the structural or the spatial interpretation. There are probably hundreds if not thousands of papers about the dynamics in such or similar two-species systems.

What is always most intriguing about models is when we find some emergent properties that were not at all expected when we first looked at the system. For example, the fact that pure species interactions may produce persistent oscillations in population numbers could be hardly expected. With everything constant in the system, with no external forcings, no climatic or environmental conditions involved, we still get variability in species populations.

Systems with linear functional response are usually more predictable. It is when we find feedbacks that have a non-linear effect in the system that we should expect surprises. These systems need especially careful analysis. They are also hardest to analyze analytically.

Providing for spatial heterogeneity only adds to the list of surprises. Why would the spatial distribution make the predator–prey oscillation converge? Why does it stabilize the system? How general can these conclusions be? Does this mean that more diversity in the system also means more stability? How far can we go in this sort of generalization?

These are all exciting questions that beg further research.

Further reading

Holling, C.S. (1959). Some characteristics of simple types of predation and parasitism. *The Canadian Entomologist*, 91: 385–389 – *A classic work on trophic interactions. One of the first to give a qualitative description of functional responces in population ecology. Almost always cited when trophic functions are considered and used, especially for modeling.*

Svirezhev Yu, M. and Logofet, D.O. (1983). *Stability of Biological Communities*. Mir Publishers – *This is an excellent example of what analytical studies can do in research of population dynamics. Two-species communities, including Volterra model are very well presented, as well as the theory of trophic chains. The book also covers the contents of the classic paper by Kolmogorov, A.N. which was published in Italian in 1936: Sulla Teoria di Volterra della Lotta per l'Esisttenza. Giorn. Instituto Ital. Attuari, 7, 84–80. Unfortunately the book is also quite hard to get. Some of the ideas have been further developed in Logofet D.O. (1993). Matrices and Graphs: Stability Problems in Mathematical Ecology, CRC Press.*

To read more details about the wolves in the Yellowstone see the article by Virginia Morell (2007). Aspens Return to Yellowstone, With Help From Some Wolves. *Science*, Vol. 317(5837): 438–439. *The amazing story about the effect of blue crab on dune formation is reported by* Cheryl Dybas *on an NSF web site at* http://www.nsf.gov/od/lpa/news/02/tip020916.htm.

Simile can be found at the Simulistics Inc. web site at http://www.simulistics.com/. *You can download a trial version that will let you run the models but will not allow saving your changes. It is a good way to explore the software and the models that we have in this book.*

The Spatial Modeling Environment, SME, is an open source project on SourceForge. See http://sourceforge.net/projects/smodenv. *Some example projects and latest developments related to the SME can be found at* http://www.uvm.edu/giee/IDEAS/.

Some ideas about the role of spatial interactions in adding stability to the system can be found in Maynard-Smith, J. (1978). *Models in Ecology.* Cambridge University Press. *Later on these effects were studied for so-called metapopulations, which are collections of interacting populations of the same species. There is even special software packages developed to study such populations. RAMAS is one of those (see* http://www.ramas.com/mpmodels.htm). *To learn more about metapopulations see for example,* Hanski, I., Gaggiotti, O. eds. (2004). *Ecology, genetics, and evolution of metapopulations.* Elsevier Academic Press.

6. Water

6.1 Modeling as a hydrology primer
6.2 Unit model
6.3 Spatial model
6.4 Conclusions

SUMMARY

There are critical natural resources that are essential for human survival, and water is certainly one of them. The dynamics of water, its quantity and quality mirror what is happening at the watershed, and can serve as an indicator of overall environmental quality. We first consider various parts of the hydrologic cycle, and some of the different processes that move water and that define its quality and quantity in different storages. We then put these processes together into a unit model that can describe dynamics of water in a small, confined and spatially homogeneous plot or cell. A variety of temporal, spatial and structural scales and resolutions may be considered, as dictated by the goal of the modeling effort. We then present several ways in which water can be described over spatially heterogeneous area. The lumped modeling approach uses relatively large spatial compartments or hydrologic units, which are then connected over a stream network. In the grid-cell approach, local dynamics are replicated across an array of grid cells that are driven by raster maps for variables and parameters. If time is not important, it is better to focus on spatial aspects using a GIS approach.

Keywords

Excludable and rival resources, scoping model, rainfall, snow/ice, surface water, groundwater, unsaturated zone, infiltration, precipitation, Julian day, evaporation, National Climatic Data Center, photoactive radiation, bi-flow, porosity, transpiration, percolation, field capacity, soil moisture, hydraulic conductivity, soil types, *Melaleuca*, Delay function, TR-55, retention, curve number, surface roughness, horizontal water transport, vertical water transport, lumped models, hydrologic units, HSPF, SWAT, grid-based models, SME, GIS-based models, stormwater, rain barrel, retention pond, rain garden, LIDAR, ArcGIS, watershed management.

* * *

Water, energy and land are the three most crucial limiting resources on this planet. This makes it especially important to understand how the systems related to these resources operate, the most efficient ways to control the depletion of these resources, and how the resources can be restored if damaged. In this chapter, we start with water.

Water is essential for life on this planet. The water content of a human body is about 60 percent. Humans can survive for more than 3 weeks without food, but for only

3 days without water. There are some reports of longer survival times, up to as many as 7–8 days; however, irreversible damage to the organism is most likely to occur earlier than that, and in any case it will be thirst rather than hunger that will kill first.

Water is also required for other organisms and plants to persist. It is an important transport mechanism that delivers nutrients to the plants. At the same time, it provides a mechanism for pollution reduction through dilution. While most ecologists will tell you that "pollution dilution is not a solution," until recently it was probably the main – if not the only – way to remove toxins and waste from our environment. Or rather to make them less toxic, since dilution certainly does not *remove* them. In 2000, *Fortune* magazine predicted that water "will be to the 21st century what oil was to the 20th."

Note that as long as we rely upon purely renewable water (as well as energy), it is non-rival and non-excludable. That is, solar energy and rainfall are available, more or less uniformly, over vast territories. Whoever is there has access to that water and energy. We cannot prevent our neighbor from having equal access to sunshine or rainfall, or collecting it in some way. We cannot exclude someone from using it, and since there is no rivalry it makes no sense to attempt to do so. Certainly there may be geographical differences. We know that there is very much more water in the Pacific North West than in the Sahara, but these are regional distinctions. Locally, everybody in the Pacific North West still has equal access to rainfall and sunshine, just as everybody in the Sahara has equal access to the rainfall and sunshine there. However, as soon as we need to dip into reserves, into fossil water or energy, or even into the temporary reserves (lakes, reservoirs, or forest and crop biomass), immediately the resources become excludable and rival (Daly and Farley, 2004). We can put a fence around a reservoir, privatize a forest, or outlaw pumping water from underground – like Israel did in Palestine. This changes the whole political landscape, and requires different types of management. As resources become scarcer and we dip into stocks, we are creating potential for conflict situations (water and energy wars).

Let us consider some simple models related to the water cycle, and figure out how they can be used to increase our understanding of what is happening with water.

6.1 Modeling as a hydrology primer

As in other models, we should first decide on the spatial and temporal scales that are to be used in our hydrologic model. At varying temporal scales processes look fairly different. Consider a major rainfall event when, say, during a thunderstorm there is a downpour that brings 10 cm of rain in 1 hour, then the storm moves away and there is no more rain over the next 23 hours.

If we assume a 1-minute time-step in our model, we will need to take into account the accumulation of water on the surface, its gradual infiltration into the soil, and the removal of water by overland flow. If we look more deeply into the unsaturated layer, we can see how the front of moisture produced by the infiltrating water will be moving downwards through the layer of soil, eventually reaching the saturated layer. After the rain stops, in a while all the surface water will be removed, either by overland flows or by infiltration. A new equilibrium will be reached in the unsaturated layer, with some of the water accumulating on top of the saturated layer and effectively causing its level to rise somewhat, and the rest of the water staying in the unsaturated layer, increasing the moisture content of soil.

Now suppose that the model time-step is 1 day. The picture will be totally different. In 1 day we will see no surface water at all, except in rivers or streams. In other parts of the landscape, the water will already have either got into the soil or run downhill to a nearby stream or pond. The unsaturated layer will not show any water-front propagation; it will have already equilibrated at the new state of moisture content and groundwater level. The processes look quite different in the model. And we probably already needed to know something about the hydrologic processes in our system to figure all this out.

Similarly, the spatial resolution is important. If all the variables are averages over a certain area, then within this area we do not distinguish any variability, and the amounts of surface water, snow/ice, unsaturated and saturated water are considered to be the same. If we are looking at a 1-m^2 cell this does not cause any problem, and it is easy to imagine how to measure and track these variables. However, if we are considering a much larger area – say 1 km^2 – then within a single cell we may find hills, depressions, rivers and ravines. The geology and soils may be also quite different, and need to be averaged across the landscape. We may be able to track many more processes, but the model cost will increase accordingly as we will need far more data and greater computer power to deal with these spatially detailed models.

For the first iteration of our modeling process, let us assume that the area of interest is a small watershed with quite uniform geo-morphological conditions, with more or less homogeneous soils, and let us suppose that we wish to figure out the amount of water that drains off this watershed into the river downstream. With this goal in mind, we can probably consider the system using a daily time-step – at least as a first iteration. A simplified conceptual model of hydrologic processes for this system is presented in Figure 6.1. This diagram is only the tip of the iceberg, with a lot of fairly complex processes that may be further described in much more detail. At this point, it is important to decide on the most important features of the system that need be considered.

We chose the following four variables for this general model:

1. SURFACE WATER – water on the surface of the land (in most cases it is in rivers, creeks, ponds and depressions).
2. SNOW/ICE – at freezing temperatures surface water becomes ice, which then melts as temperature rises above 0°C.
3. UNSATURATED WATER – the amount of water in the unsaturated layer of ground. Imagine the ground as a sponge; when we pour water onto it the sponge will hold a certain amount before it starts dripping. All the time water can still be poured onto and held by the sponge, it is in the unsaturated condition.
4. SATURATED WATER – the amount of water in the saturated ground. Once the sponge can no longer hold additional water, it becomes saturated. As with surface water, if we add water to the saturated zone, its level increases.

These variables are connected by a variety of processes that we also need to understand in order to build a meaningful model. When working on complex models, it helps considerably if we split the whole system into components, or modules, and develop some simplified models for these modules. It is very likely that some modifications will be needed when pulling all the modules together again; however, as previously discussed, it is so much easier to deal with a simplified model than to get lost in the jungle of a spaghetti diagram of a complex model with numerous processes and interactions, and no clear understanding of what affects what.

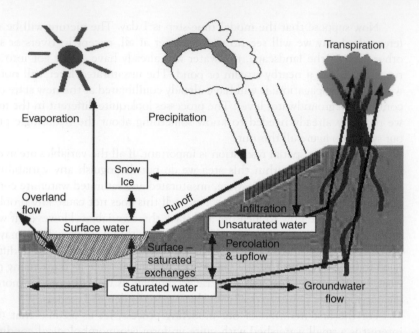

Evaporation   Precipitation

Transpiration

Snow
Ice

Overland
flow

Runoff

Infiltration

Surface water

Unsaturated water

Surface –
saturated
exchanges

Percolation
& upflow

Saturated water

Groundwater
flow

Figure 6.1   Conceptual model of unit hydrology.

Note that this diagram describes certain processes as if they were spatially distributed with a horizontal dimension present (runoff "moves" water from rainfall to a pond, saturated water also moves). In fact, when we run the model we assume that all these variables are uniformly distributed over the whole area and are represented by "point" quantities or concentrations.

Modeling is truly an iterative process. As stated many times before, we want to know the spatial and temporal scales *before* we start building the model. But how do we figure them out if we have only a vague idea about the system? What are the processes involved? At what times are they important, and do we want to include them at all? Or perhaps there are some other important processes that we are simply unaware of.

Indeed, there is no prescribed sequence of events. Perhaps you want to start with a so-called "scoping model" – a model that would put together whatever you already know about the system in a rather qualitative format, omitting all the details that are not clear, outlining the system in general and the processes that we think are important. This you can start discussing with colleagues and with potential future users of the model. These users are the ones who formulated the initial goal of the study, so they are most likely to know something about the system. Start talking to them or, even better, engage them in a *participatory modeling process* – something we will be discussing in a lot more detail in Chapter 9.

In any case, do not think that there is anything final in your decisions about the scales and processes. There will always be a reason and a chance to come back and make improvements. That is the beauty of computer models: they exist in virtual reality, to build them you do not have to have something cut, ploughed, extracted or destroyed, and you can easily modify or refocus them if necessary.

Water on the surface

The *surface water* variable is used to model water on the surface of the land. If we are looking at an area with no steep gradients and fairly high potential rainfall (for example, the Florida Everglades or other wetlands), then surface water can accumulate in significant amounts before it is absorbed by the soil. In this case it is necessary to consider the process that connects the accumulated surface water and the underlying unsaturated layer. This process is known as *infiltration*. In most terrestrial areas with steeper slopes, most of the surface water will drain off into rivers, creeks, ponds and depressions in which it will accumulate over a layer of saturated water. Therefore, there will be no infiltration. Instead, there will be an exchange process between the surface water and the saturated layer.

It is hard to isolate a unit of surface water without connecting it with the surrounding neighborhood. Much of the surface-water transport is due to horizontal fluxes, and therefore a box-model approach will be only approximate when modeling surface-water dynamics. However, with appropriate spatial and temporal scaling we can think of an aggregated unit model to represent surface water in a homogeneous unit cell, assuming that we are modeling the total amount of water over a large enough area and one that can somehow be isolated from the other territories. This can be a small watershed, or an agricultural field, for which we can monitor the inflows and outflows. A simple conceptual model can be described as in Figure 6.2. There are two major processes involved: *precipitation* and *infiltration*.

Precipitation is probably the process that is intuitively most obvious. We deal with precipitation in our everyday lives when we decide whether we might need an umbrella on going out for the day. The amount of precipitation is what we are concerned with when building a hydrologic model. It is also important to know in what form (liquid or solid – rain or snow) the precipitation will arrive. Precipitation is recorded, by most of the meteorological stations, in millimeters or inches per day. A sample data sheet for precipitation registered at Baltimore Washington Airport, MD in 1996 is shown in Figure 6.3.

In Figure 6.3 0.0T stands for traces, which means that the precipitation was recorded at levels below measurement accuracy. In many cases it is possible to find meteorological data for a specific area at the National Climatic Data Center (NCDC: http://www.ncdc.noaa.gov/). For example, on entering this site and choosing Maryland, then the station at Baltimore Washington Airport, the relevant data can be found. A graphic can also be generated for a table such as that reproduced here. The data can be downloaded in numeric format to use in a model. Temperature is important for us to decide whether the precipitation is rain or snow. The Snow/Ice model below describes this process.

Infiltration is the process by which water from the surface is taken into the ground by means of gravitational and capillary forces. The rate of infiltration defines how much water will be left on the surface to contribute to the rapid runoff, and how

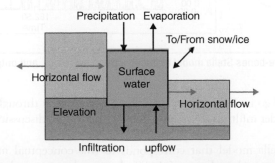

Figure 6.2   A simplified conceptual model for local surface hydrology.

DAILY

Station BALTIMORE W ARPT Parameter Prcp   % Coverage   100
PO Code MD   Latitude   N39:11:00 Begin M/Yr 08/1948
Stn ID   465   Longitude W076:40:00 End M/Yr   12/1996
County ANNE ARUNDEL   Elevation(m.) 45.1   # Record Years 49

1996	Jan	Feb	Mar	Apr	May	Jun	Jul	Aug	Sep	Oct	Nov	Dec	Annual
1	0.01	0.0T	0	0.65	0.0T	0	0.04	0.0T	0	0.0T	0.07	1.17	
2	1.02	0.34	0.21	0.0T	0.0T	0	0.01	0	0	0.36	0	0.41	
3	0.09	0.38	0.0T	0	0.04	0	0.09	0.08	0.01	0	0	0	
4	0	0	0	0.59	0.0T	0	0	0.21	0	0	0		
5	0	0	0.02	0.0T	0.65	0.0T	0	0	0	0	0.57		
6	0.0T	0	0.18	0.04	0	0	0	0.48	0	0.0T	0.32		
7	2.51	0.03	0.53	0.06	0.3	0	0	0	0	0.0T	0.51		
8	0.82	0.19	0.18	0.0T	0.33	0	0.15	0	0.08	1.33	2.85	0.03	
9	0.3	0	0	0.43	0.34	0.07	0.01	0.41	0.01	0.22	0.02	0	
10	0.0T	0	0	0.0T	0	0.02	0	0	0.0T	0.0T	0		
11	0	0	0	0	0.94	0	0	0	1.35	0	0	0.23	
12	0.78	0.0T	0	0	0	0.17	1.12	1.46	0.2	0	0	0.07	
13	0	0	0	0	0	0	2.28	0.94	0.08	0	0	2.73	
14	0	0.03	0	0	0.0T	0.11	0	0	0	0.05	0.05		
15	0	0.14	0.17	1.38	0.08	0	0.11	0	0	0	0	0.01	
16	0.0T	0.58	0	0.3	0.45	0	0	0.06	0.72	0	0	0.01	
17	0.03	0	0.0T	0	0.04	1	0	0	0.9	0	0	0.02	
18	0.03	0	0	0	0	0.35	0.17	0	0.01	1.88	0.0T	0.08	
19	0.54	0	0.73	0.0T	0	1.38	0.32	0	0	0.36	0.02	0.2	
20	0	0.16	0.0T	0	0	0.13	0	0	0	0.07	0	0	
21	0	0.18	0.0T	0.0T	0.55	0	0	0.0T	0	0.01	0	0	
22	0	0.14	0	0	0	0	0.17	0	0.92	0.0T	0	0.0T	
23	0.0T	0.12	0	0.12	0	0	0	0.0T	0	0.02	0	0.0T	
24	0.21	0.03	0	0	0.0T	0.36	0	0	0.0T	0	0	0.22	
25	0	0	0.0T	0	0.0T	0	1.03	0	0	0	0.04	0	
26	0.0T	0.04	0	0.15	0.06	0	0.14	0	0	0.0T	0.73	0	
27	0.46	0.0T	0	0.01	0.56	0	0	0.93	0.03	0.0T	0.0T	0.05	
28	0	0.0T	1.3	0	0.16	0	0.0T	0.29	0.85	0.07	0	0.0T	
29	0.0T	0	0.25	0.06	0.59	0.41	0.32	0	0.0T	0	0	0.1	
30	0	---	0	0.56	0	0.19	1.04	0	0	0	0.19	0	
31	0.0T	---	0.0T	---	0	---	0.27	0	---	0	---	0.0T	
Total	6.8	2.36	3.57	3.76	5.88	4.08	7.38	4.17	5.85	4.32	3.77	6.77	58.31
Extr	2.51	0.58	1.3	1.38	0.94	1.38	2.28	1.46	1.35	1.88	2.85	2.73	2.73

Figure 6.3   Precipitation data at Baltimore Airport in Maryland (USA).
Notice the treacherous inches/day used as a unit in this data set.

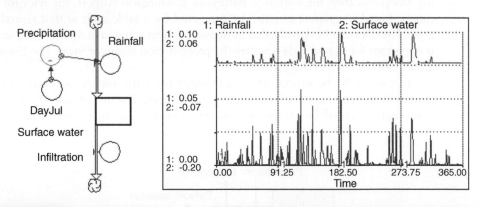

Figure 6.4   A bare-bones Stella model for local surface hydrology, and output from this model.

much will go into the ground and then travel slowly through the porous media. We will consider infiltration in more detail below, when discussing the unsaturated water storage.

A Stella model that corresponds to this conceptual model of surface hydrology is presented in Figure 6.4. We have only one stock and two flows, and no feedbacks. In this case we assume that the surface water is delivered by rain and then gradually infiltrates into the ground. The rainfall is fast, whereas infiltration is slow. However, rainfall occurs only sometimes, whereas infiltration is continuous. The equations are:

Surface_Water(t) = Surface_Water(t – dt) + (Rainfall – Infiltration) * dt
INIT Surface_Water = 0.01
DOCUMENT: The surface water is assumed to be a function of two processes. Rapid rainfall provides surface water, which then gradually infiltrates into the ground.

Rainfall = Precipitation*0.0254
DOCUMENT: Converting rainfall in inches/day to m/day

Infiltration = 0.01
DOCUMENT: Infiltration rate (m/day). In reality this rate depends upon soil characteristics, habitat type, slope, pattern of rainfall.

DayJul = mod(time-1,365) + 1
DOCUMENT: Julian day, 1 thru 365. This is a counter that resets the day to zero after 365 iterations. Needed to use the same graph function for several years of model runs.

Precipitation = GRAPH (DayJul)
(1.00, 0.02), (2.00, 0.34), (3.00, 0.00), (4.00, 0.07), (5.00, 0.00), (6.00, 0.18), (7.00, 0.46),
...
(354, 0.00), (355, 0.00), (356, 0.15), (357, 0.00), (358, 0.00), (359, 0.00), (360, 0.02), (361, 0.00), (362, 0.00), (363, 0.00), (364, 0.00), (365, 0.00)
DOCUMENT: Rainfall from Beltsville, MD 1969. {in/d}

Note a few interesting features here, which may be helpful in other models. First, notice the units. We have put together the model in meters and days, as would normally be the case in science. However, the data came from a US meteorological station where they still use inches for measurements. Therefore, we need the converter

$$Rainfall = Precipitation * 0.0254$$

where we use the conversion factor 1 inch = 0.0254 m. It is extremely important to make sure that all units are consistent throughout the model. While Stella offers some background functionality to help track the units, it is really in your best interest to make sure that you are always aware of the units in each parameter and process and ascertain that the units match, both in time and space. The more involved you are in the model structure and formulation and the less you rely on some of the built-in automatic features, the more you will learn about the system and the better you understand it.

Mind the units. They can help test your model for consistency. Do not rely on the automatic unit checks offered by some software packages; you will understand your system better if you track the units yourself.

Another trick is the introduction of the DayJul variable, which is the Julian day calculator. The data we have from the station are for only 365 days. In Stella, once the data in a Graphic function are exhausted, the very last value is taken and repeated for all the further times. So if we tried to describe Precipitation as GRAPH (TIME) for TIME > 365, we would be getting Precipitation = 0.00, since GRAPH (365) = 0.0 in this case. If we wish to run the model over longer periods of time and might want to use the same forcing data, we need to reset the time counter to 1 when TIME is later than 365. That is exactly what the DayJul variable does:

$$DayJul = MOD(TIME - 1, 365) + 1$$

The Stella built-in MOD function returns the remainder from division of the first argument by the second. Unfortunately, the most straightforward expression, MOD(TIME,365), which might be expected to resolve the problem, does not work because it returns a zero every now and then. Therefore, we need the formulation above. The nice thing is that we can now run the model for as long as we want using the same pattern of climatic data given by the graphic function.

Running the model, we may observe a certain lag in the dynamics between precipitation and surface water accumulation (Figure 6.4). Surface water trails the amount of rainfall. Some rainfall events produce no surface water at all, which means that these rainfalls are entirely infiltrated as they arrive and nothing remains on the surface. If the area is flat with low runoff, the presence of standing surface water can also be interpreted as a flooding event. This is what may be observed in wetland areas. In steeper areas with high rates of overland flow, the amount of surface water generated by the model may be tracked by the amount of runoff.

Evaporation is yet another process that is important when analyzing the dynamics of surface water. We explore this in more detail below.

Exercise 6.1

1. In Stella, you can clamp your state variables to make sure that they never become negative. For example, in this model the Surface_Water is non-negative. Note that the "non-negative" option is checked in the variable definition box:

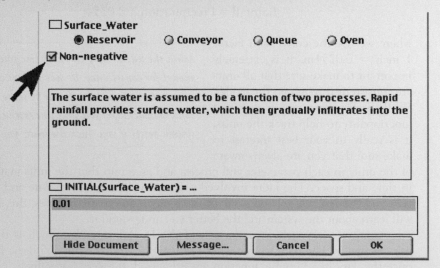

By default, the variables are clamped. This may be somewhat confusing and hide some of the errors, when the variable is actually negative but you do not see it. It is good practice to make sure that your processes (flows) are described properly, and do not deplete state variables beyond the levels that are intended. Uncheck the non-negativity in this model, and see what really happens to the Surface_Water. Redefine the flows in the model to make sure that Surface_Water does not go negative.

2. Let us suppose that the whole area is paved. How do we describe this in the model? What happens to Surface_Water? Does the result look plausible? Are there any other processes that we may be missing?

Certainly there is also surface water runoff, which removes the excess water horizontally. However, strictly speaking this flow is not a process to consider within the framework of a unit model. It is driven by an elevation gradient, whereas elevation in this case, like other spatial characteristics, is assumed to be uniform over the whole modeled area. Therefore we will look at this process in more detail when describing the spatial implementation of the model. For the unit model, it is enough to provide some sort of a function that will remove either all of the available surface water (works well for a highland, terrestrial area) or take away a certain proportion of the available water (wetlands, water bodies – areas where we want surface water to be constantly present). This can be done by the *evaporation* flow.

Evaporation is the process that converts water into water vapor, which is then diffused into the atmosphere. In this way water is lost from the surface, no matter whether it is from open water (lakes, rivers) or from soil. It is important that evaporation moves water directly from the surface to the atmosphere; this is different from the transpiration process, which moves water into the atmosphere from the soil through plants.

There are a number of climatic factors that affect evaporation (Figure 6.5), most of which are quite obvious from everyday experience. We know that after getting wet we will dry faster if we stay in the sun, and even faster if it is windy. When it is humid outside, things do not dry as quickly as when it is dry. So the evaporation is higher when it is sunny, hot, windy and dry. Humidity, wind and temperature data are usually available from standard meteorological records. In many cases the necessary meteorological data for a specific area can be found at the National Climatic Data Center mentioned above (NCDC: http://www.ncdc.noaa.gov/) by choosing a state and then

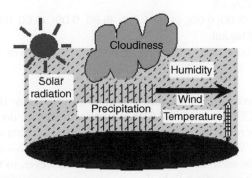

Figure 6.5   Major factors affecting evaporation.

finding a station that is located close to the site being modeled. It is most problematic to obtain data on solar radiation (also known as photoactive radiation – PAR). For some reason it is not one of the standard observations, and direct measurements are rare. Therefore, in our model we will estimate solar radiation based on the latitude of a site and the generally available information about precipitation. All these factors are put together to estimate evaporation in a simple model in Figure 6.6. The corresponding equations are as follows:

```
A = 720.52-6.68*LatDeg
Air_temp_degC = ((Air_temp_degF-32)*5/9 + Air_temp_minC)/2.
Air_temp_minC = (Air_temp_minF-32)*5/9
B = 105.94*(LatDeg-17.48)^0.27
C = 175-3.6*LatDeg
cloudy = if Precipitation > 0 then max(0,10-1.155*(vap_press/(Precipitation*25.4*30))^0.5) else 0
Cl_factor = 0.15
DayJul = mod(time-DT,365) + 1
Evap_M = EvapBelt*0.0254
Hyd_evap_calc = •Hyd_evap_rc*SolRadGr/585*pan_CW*pan_CT*pan_CH
LatDeg = 39.0
pan_CH = 1.035 + 0.240*(Humidity/60)^2-0.275*(Humidity/60)^3
pan_CT = 0.463 + 0.425*(Air_temp_degC/20) + 0.112*(Air_temp_degC/20)^2
pan_CW = 0.672 + 0.406*(Wind/6.7) + 0.078*(Wind/6.7)^2
SolRad = A + B*COS(T) + C*SIN(T)^2
SolRadGr = max(0,SolRad*(1-Cl_factor*cloudy))
T = 2/365*PI*(DayJul-173)
vap_press = Humidity*6.1078*EXP(17.269*Air_temp_degC/(Air_temp_degC + 237.3))
Wind = Wind_speed*1.852/24
•Hyd_evap_rc = 0.0028
Air_temp_degF = GRAPH (DayJul)
(1.00, 44.0), (2.00, 42.0), (3.00, 51.0), (4.00, 42.0), (5.00, 38.0), (6.00, 43.0), (7.00, 44.0),...
Air_temp_minF = GRAPH (DayJul)
(1.00, 19.0), (2.00, 21.0), (3.00, 22.0), (4.00, 26.0), (5.00, 19.0), (6.00, 21.0), (7.00, 32.0),...
EvapBelt = GRAPH (DayJul)
(0.00, 0.00), (1.00, 0.00), (2.00, 0.00), (3.00, 0.00), (4.00, 0.00), (5.00, 0.00), (6.00, 0.00),...
Humidity = GRAPH (DayJul)
(1.00, 67.0), (2.00, 71.0), (3.00, 69.0), (4.00, 50.0), (5.00, 65.0), (6.00, 88.0), (7.00, 90.0),...
Precipitation = GRAPH (DayJul)
(1.00, 0.00), (2.00, 0.00), (3.00, 0.00), (4.00, 0.00), (5.00, 0.00), (6.00, 0.05), (7.00, 0.41),...
Wind_speed = GRAPH (DayJul)
(1.00, 129), (2.00, 113), (3.00, 148), (4.00, 160), (5.00, 102), (6.00, 66.0), (7.00, 179),...
```

The climatic data are entered as graphs to represent the time series downloaded from the NCDC website. Note that in this model we do not have any state variables; we only reproduce some empirical relationships that correlate evaporation with known data. We do not really need to use Stella; all this could be done in a spreadsheet program such as Excel or Open Office. However, in this case Stella is useful to describe the cause–effect links that are important to estimate evaporation. The model is based on an empirical relationship by Christiansen (see Saxton and McGuinness,

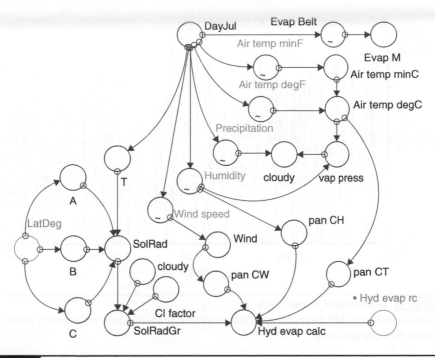

Figure 6.6 Stella model for evaporation. With no state variables this could easily be a spreadsheet in Excel.

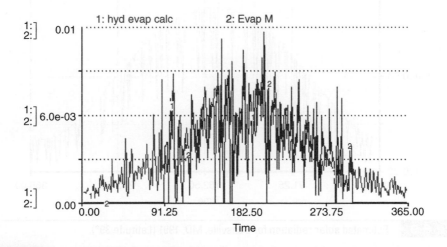

Figure 6.7 Goodness of fit for evaporation.
Comparison of model results with available data for evaporation for Beltsville, MD, 1991.

1982). The solar radiation is estimated by a simplified version of an algorithm developed by Nikolov and Zeller (1992). We can compare the results of this analysis with existing measurements of evaporation to see how well the model works (Figure 6.7).

There is a lot of variability in evaporation caused by the differences in climatic data. In the model we have managed to obtain a good estimate of the general trend, but have failed to reproduce all the changes in evaporation. The data for wind speed, precipitation

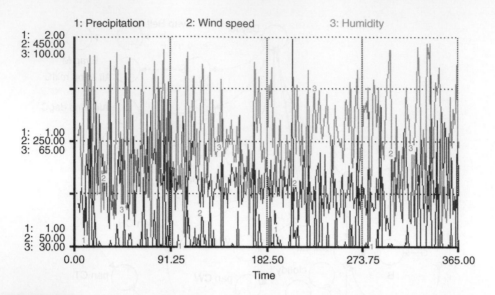

Figure 6.8   Variability of climatic data.

Data measured at Beltsville, MD, meteorological station in 1991. There is hardly any seasonal pattern in the data for rainfall, humidity and wind.

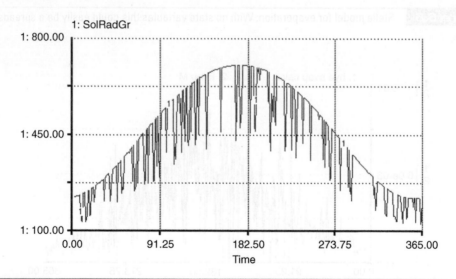

Figure 6.9   Estimated solar radiation for Beltsville, MD, 1991 (Latitude 39°).

and humidity show significant variability (Figure 6.8). The model of solar radiation also shows significant variability caused by the cloudiness effect (Figure 6.9). The basic bell-shaped trend for radiation that is defined by the latitude of the site is smooth. Added to it is the stochastic pattern of climate that generates the cloudiness in our model.

Also note that this model can be formulated as a pre-processor that is run to generate the missing time series to run the full model. There are no feedbacks that would point into this module from anywhere else. The only purpose is to generate the missing time series for PAR based on the existing climatic time series and the latitude/longitude of the site we are modeling. We may want to run this model only once, generate the missing time series and then save some processing time by simply feeding that data into our model along with the other climatic data. This may not be of great value in the case of a simple model like this one; however, as the models become more complex and computer-intensive, it certainly makes sense to tease out all the pieces that are separate from the overall dynamics and do not need to be rerun every time we run the full model.

> *separate the parts of the model that need to be run only once for setup to prepare the data. Generate the time series that these parts are designed for, and then feed that into the model. There is no need to recalculate the same things many times.*

Exercise 6.2

1. As we have seen, our model of evaporation seems to produce less variability than the data (compare Evap_M and Hyd_evap_calc). Try to tweak the model parameters to increase the variability in model output.
2. Check out the sensitivity of the model to changes in the parameter that describes the effect of cloudiness. What happens if the effect of clouds is increased quite dramatically (say, tripled)? How can we avoid some of the unrealistic behavior in the model that we observe in this case?
3. What is more important for the rate of evaporation: the latitude of the site, or the climatic conditions? What changes in climate can compensate the effect of the latitude, and vice versa?
4. Include the evaporation flow in the model of surface hydrology and see what difference it makes under different climatic conditions.

Snow and ice

Water on the surface may fall out of the cycle under certain climatic conditions. That is why we add this other variable: SNOW/ICE. At temperatures below freezing the surface water becomes ice, which then melts as the temperature rises above 0°C. Snow and ice can appear directly during a precipitation event if the temperature is below freezing, or may occur because of freezing of the already available surface water.

A simple conceptual model of snow formation is presented in Figure 6.10. It is closely associated with the model of surface-water dynamics, and is actually just a

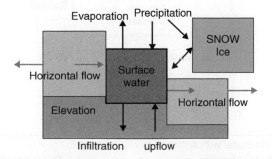

Figure 6.10   A simplified conceptual model for local surface hydrology, with snow/ice included.

part of that model. The accompanying Stella model is shown in Figure 6.11. We will want to supplement the equations above with the following:

Snowice(t) = Snowice(t − dt) + (Freeze + Snowfall) * dt
INIT Snowice = 0
DOCUMENT: The amount of snow and ice on the surface (m)
Freeze = if Temperature < 0 then Surface_Water/DT else -Melt
DOCUMENT: Freezing/melting of water/snow. Formulated as a biflow. When temperature is above 0°C snow (if available) is melting at a constant rate. Otherwise water is freezing. All available water is assumed to freeze immediately.

Snowfall = if Temperature< = 0 then Precipitation*0.0254 else 0
DOCUMENT: Snow accumulation from precipitation; use 0.0254 to transfer inches into m.

Melt = 0.01
DOCUMENT: How much snow can melt per day (m/d)

Temperature = 25*SIN(DayJul*PI/180/2)^2-5 + RANDOM(-3,3)
DOCUMENT: Temperature (°C) is modeled by a combination of the SIN function and the RANDOM function. The amplitude of the SIN is increased to 25. Power 2 is used to make it always positive. The DayJul*PI/180/2 conversion is used to switch to radians and stretch the SIN period over the whole year. −5 is the lowest temperature generated. All temperatures are modified by a random value between −3 and 3.

Notice here that we are using a so-called bi-flow to describe the conversion of water into ice and back – the "Freeze" flow. Stella allows only positive flows. Whenever a flow becomes negative, it is clamped to zero. Sometimes this is a useful feature, but it can cause a lot of confusion if it is forgotten. If it is clear that the

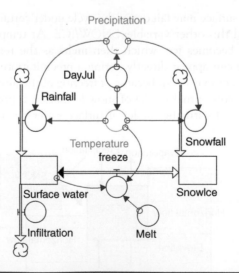

Precipitation

DayJul

Rainfall

Temperature
freeze

Snowfall

Surface water

Snowice

Infiltration   Melt

Figure 6.11   A Stella model with snow/ice formation added.

One important process, called sublimation, is missing from this model. This is not important in warm climates, when snow does not stay on the ground for long periods of time.

flow is supposed to be negative sometimes, it is important to ensure that it is described as a bi-flow by clicking on the radio button at the top of the flow dialogue box.

Another feature to note here is instantaneous conversion of all available surface water into snow or ice whenever the temperatures fall below zero. Remember why we divide Surface_Water/DT? Also note the effort made to provide proper documentation directly in the body of the model. This can save a great deal of trouble later on, when we return to your model after a period of time and are trying to figure out once again what an equation was for and why a particular parameter looks so weird.

Also notice that temperature is described as a formula in a similar way to that described in Chapter 2. While the formula is somewhat different, the result is quite the same: cycles of warm and cold temperature over a 365-day period with some random noise imposed on top of them. Which of the two formulas is better? It is really hard to say.

> *There is no such thing as too much documentation. Never economize on commenting and describing your model and what you did with it. This will make you very proud of yourself, and happy when you need it later on!*

The model results are shown in Figure 6.12. We can see that surface water is delivered by rain and then gradually infiltrates into the ground. It will freeze into snow/ice when the temperature is below 0°C, and under freezing conditions precipitation also arrives as snow. We observe a rapid accumulation of snow during the early, cold months of the year. Later on snow/ice disappears, and the dynamics are similar to those generated by the surface water dynamics model. Towards the end of the year there are again freezing temperatures, and thus some snow/ice is produced.

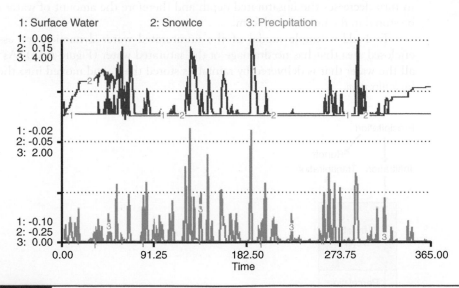

Figure 6.12   Output from the snow/ice model.
Snow/ice is present only during the first few cold months and then quickly disappears. More snow appears at the end of the year when the temperature drops below zero.

The model is good enough for southern regions, where snow rarely stays on the ground for too long. In the north, where there is persistent snow cover throughout the winter, we will need to incorporate another important process – snow sublimation, or evaporation of snow. Once again, we can see that it is the particular project goals and site-specific features that drive the temporal, spatial and, in this case, structural resolution of our models.

Exercise 6.3

1. Why do we model snowmelt as a constant rate rather than a proportion of the available snow? Make this modification to the model and analyze the difference in model performance. Which model seems to be more realistic?
2. Temperature in this model is presented as a random fluctuation over the sine function of time. What alternative methods could be used to generate temperature for this model?

Water in the unsaturated layer

Unsaturated water is the amount of water in the unsaturated layer of the ground. Remember the sponge analogy? We pour water on a sponge and it is absorbed and held until the sponge is full of water, at which stage it starts dripping from the sponge. While water can be still added to and held by the sponge, it is in the unsaturated condition. In the unsaturated layer there is almost no horizontal movement of moisture; all flows are in the vertical dimension.

Modeling the unsaturated layer is somewhat more complicated, because it is very closely connected to the dynamics of the saturated layer. As water seeps down from the unsaturated layer, it adds to the saturated layer and the water table goes up. This in turn decreases the unsaturated depth and therefore the amount of water that can be stored in the unsaturated form.

To build a simple model of the unsaturated water dynamics, we assume an enclosed area that has no drainage of the saturated water (Figure 6.13). As a result, all the water that is delivered by rainfall is stored there, and moved into the ground by *infiltration*.

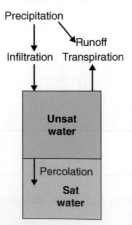

Figure 6.13   A simplified conceptual model for water in the unsaturated layer.

The factors that influence infiltration may be grouped into three categories (Figure 6.14):

1. Those related to climatic conditions. The amount of water infiltrated depends upon the duration and intensity of rainfall. A 24-hour drizzle can be entirely accommodated by the soil, whereas the same amount of water received during a 20-minute downpour will most probably end up in the surface-water runoff. Temperature also matters. When the ground is frozen, the intensity of infiltration is reduced.
2. Those related to surface characteristics. Landuse and land cover translate into the imperviousness of the surface. A parking lot will leave little water to infiltrate, whereas a forest may capture the entire amount of water arriving. On the other hand, forests can intercept the incoming rainfall with leaves and trees in such a way that a certain portion of the incoming water never reaches the ground. This moisture is only exposed to evaporation. Slope also matters. In a flat area there is more time for water to enter the ground, while on a hill it starts traveling downwards along the surface as soon as it hits the ground.
3. Those related to soil characteristics. Sand is an excellent medium for infiltration. On the contrary, clay can block almost all infiltration. Moreover, if the soil is already saturated with water (the soil moisture content is high) there will be little space left in the pores for additional water to infiltrate.

A typical infiltration event evolves in both space and time (Figure 6.15). As the rainfall starts, some water begins to seep into the ground, gradually increasing the soil water content (curves 1–3) at the top of the soil layer. As more water comes with the rain it keeps entering the soil pores. The gravitation removes some water from the top layers and makes it travel further deeper into the ground. If this vertical movement is fast enough to free up space on the top for the additional incoming water, then all the rain is absorbed. If the soil characteristics do not allow water to travel fast enough through the soil, then the pores on the top are all filled (curve 4–6) and the additional water will be left on the surface to travel with overland flows. This is when ponding may occur. The wave of saturated water propagates downwards through the soil. Once the rain stops, the pores at the top start to dry out and get ready to accommodate a new rainfall event.

Loss of water from the unsaturated layer occurs by transpiration (upwards) and percolation (downwards). *Transpiration* is a process that removes water from the soil and transfers it as water vapor into the atmosphere – just as in evaporation. The major difference is that in transpiration plants are responsible for water transport. They suck moisture from the soil with their roots, move it up into the canopy and

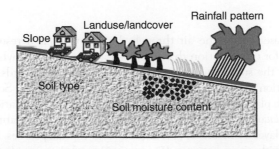

Figure 6.14 Major factors affecting infiltration.

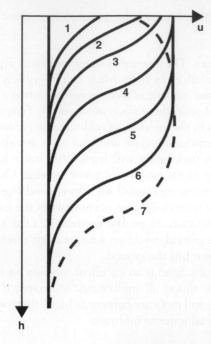

Figure 6.15 Propagation of a water front through the unsaturated layer during a rainfall event. Note that this simple model does not describe the spatial dynamics in the vertical. See how the amount of water in unsaturated storage (u) changes as a function of depth (h). Curve 1, start of a rainfall event; curves 2–4, increase of unsaturated moisture until saturation is reached; curves 5–6, propagation of the saturation front downwards; curve 7, end of rainfall event, dryout from top.

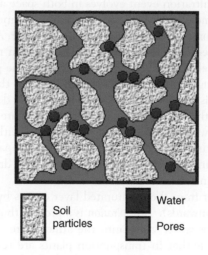

Soil particles

Water

Pores

Figure 6.16 How water travels through porous media. When all the pores are filled with water, soil is saturated.

then release it into the air through their leaves. As a result, there is more water available for transpiration than there is for evaporation – which only picks up moisture from the surface and the very few centimeters towards the surface of the soil. Transpiration can access water as deep as the roots extend. So transpiration is a function of the plant biomass which can change over the simulation period.

Percolation is the process by which water from the unsaturated storage enters the saturated layer by means of gravitational and capillary forces. Soil consists of material particles with air in between (Figure 6.16), and these voids or pores can potentially be filled by water. When all the pores are filled the soil is referred to as saturated, and vertical movement of water is very much slowed down. While the pores are not

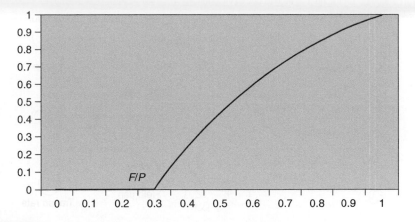

Figure 6.17   Percolation rate as function of soil moisture content.

filled, the force of gravitation and the capillary effects pull down the water until it reaches the saturated layer. In this case, the vertical motion is dominant.

Soil porosity is calculated as a fraction of pore space in a unit volume of soil:

$$P = \frac{V_{pores}}{V_{total}}$$

Some of the water becomes attached to soil particles and stays in the soil. The proportion of water that cannot be removed by gravitation is called field capacity:

$$F = \frac{V_{min\ water}}{V_{total}}$$

In this case, as shown in Figure 6.17, some water remains attached to the soil particles even after most of it has been drawn downwards by gravitation. Porosity and field capacity are important characteristics of soil. They define the maximum amount of water that soil can hold, and the minimum amount that stays in the soil after a rainfall event. Additional amounts of water can be sucked out by the action of roots, in the transpiration process discussed above.

Similarly, the soil moisture content is defined as:

$$U = \frac{V_{water}}{V_{total}}$$

Soil moisture content is the amount of moisture related to the total volume of the unsaturated layer. In reality, the unsaturated layer is not homogeneous and therefore U varies with depth. Since in our model we assume the whole unsaturated zone to be uniform, we will assume that V_{total} is the total volume of the unsaturated layer and V_{water} is the total amount of water in this layer. In saturated soils, $U = P$ – that is, all the pores are filled with water so $V_{water} = V_{pores}$; in completely dry soils, $U = 0$. Note, however, that due to percolation U can get no lower than F, the field capacity.

Water moves at different speeds through different soil types. As in the infiltration process, the rate of percolation (also called hydraulic conductivity, h_c) can vary

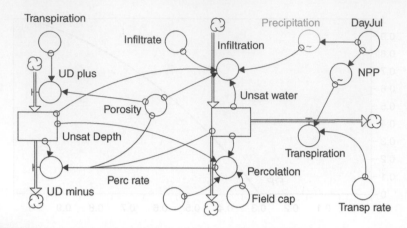

Figure 6.18 A simple Stella model for water in unsaturated layer.

from 1 m/d in sandy soils to 1 mm/d in clay. The percolation rate is also affected by the soil moisture content: $p = f(U)h_c$. The more moisture there is in the soil, the higher the rate of percolation. If $U < F/P$, the moisture is at field capacity and percolation is 0. It tends to 1 when U approaches 1.

As the water percolates downwards, it adds to the amount of water already present in the unsaturated storage. It takes only $P - U$ water to fill in the unsaturated storage so that it becomes saturated.

The Stella model for water in unsaturated layer is presented in Figure 6.18. The corresponding equations are as follows:

Unsat_Depth(t) = Unsat_Depth(t − dt) + (UD_plus − UD_minus) * dt
INIT Unsat_Depth = 1.2
UD_plus = Transpiration/Porosity
DOCUMENT: Unsaturated depth is increased by the effect of transpiration, which removes water from the saturated layer and can make it unsaturated (m/day) NB. Note how porosity comes into play. Why do we do that?

UD_minus = if (Unsat_Water> = Unsat_Depth*Porosity) then Unsat_Depth/DT else Percolation/Porosity
DOCUMENT: Unsaturated depth is decreased due to the percolation (m/d) of water from the unsaturated zone to the saturated, which raises the water table. If the amount of unsaturated water exceeds the potential unsaturated capacity (Unsat_Water> = Unsat_Depth*Porosity), this means that no unsaturated layer can remain, all soil becomes saturated, unsaturated depth becomes zero.

Unsat_Water(t) = Unsat_Water(t − dt) + (Infiltration − Percolation − Transpiration) * dt
INIT Unsat_Water = 0.11
DOCUMENT: Amount of water in the unsaturated layer measured as height of water column if "squeezed" from the soil (m).

Infiltration = min(Infilt_rate,Precipitation*0.0254,Porosity*Unsat_Depth-Unsat_Water)
DOCUMENT: The amount of water infiltrated is the minimum of infiltration rate, the amount of precipitation available (0.0254 converts inches to m), and the unsaturated capacity (m/d). The

unsaturated capacity is the potential capacity (the volume of pores in the soil) minus Unsat_
Water (the space already occupied).

Percolation = if Unsat_Depth = 0 then Unsat_Water/DT
else if Unsat_Water< = Field_cap*Unsat_Depth then 0
else Perc_rate
DOCUMENT: Percolation flow (m/d). The amount of water removed by gravity from the unsaturated layer. This process can remove only water in excess of field capacity.

Transpiration = NPP*Transp_rate
DOCUMENT: The transpiration flow (m/day)

DayJul = mod(time-1,365) + 1
DOCUMENT: Julian day, 1 thru 365
This is a counter that resets the day to zero after 365 iterations. Needed to use the same graph
function for several years of model runs.

Field_cap = 0.13
DOCUMENT: The amount of moisture in soil that is in equilibrium with gravitational forces.
(dimless)

Infilt_rate = 0.5
DOCUMENT: Rate of infiltration – the amount of water (m) that can be moved into the unsaturated layer from the surface

Perc_rate = 0.01
DOCUMENT: Rate of water removal by gravitation (m/day). Depends upon soil characteristics.

Porosity = 0.35
DOCUMENT: Proportion of pores in the soil. They can be potentially filled with water (dimless)

Transp_rate = 0.005
DOCUMENT: The amount of water that plants can remove from soil by the sucking action of
their roots (m of water/kg biomass*m²/d)

NPP = GRAPH (DayJul)
(0.00, 0.00), (33.2, 0.00), (66.4, 0.00), (99.5, 0.04), (133, 0.4), (166, 0.925), (199, 0.975), (232,
0.995), (265, 0.985), (299, 0.855), (332, 0.105), (365, 0.00)
DOCUMENT: An estimate of plant growth over the year (kg/m²)

Precipitation = GRAPH (DayJul)
(1.00, 0.02), (2.00, 0.34), (3.00, 0.00), (4.00, 0.07), (5.00, 0.00), (6.00, 0.18), (7.00, 0.46),
...
(354, 0.00), (355, 0.00), (356, 0.00), (357, 0.00), (358, 0.00), (359, 0.00), (360, 0.02), (361, 0.00),
(362, 0.00), (363, 0.00), (364, 0.00), (365, 0.00)
DOCUMENT: Rainfall from Beltsville, MD 1969. {in/d}

In this model we reproduce the dynamics that may be observed in a wetland
that gets flooded during the wet season and dries out during the dry period. The vegetation that is removing significant amounts of water by transpiration controls the
state of the wetland. The resulting dynamics of unsaturated water and unsaturated
depth are shown in Figure 6.19. When the transpiration rate is 0.005 m/kg · m²/d,
the plants can remove almost all the water and keep the area dry for most of the
year. When the transpiration rate declines to 0.003, there is a succession of wet and
dry periods. Certain species are known to be more effective in sucking the water out
of the soil (e.g. *Melaleuca quinquenervia* – the Australian cajeput, which is sometimes

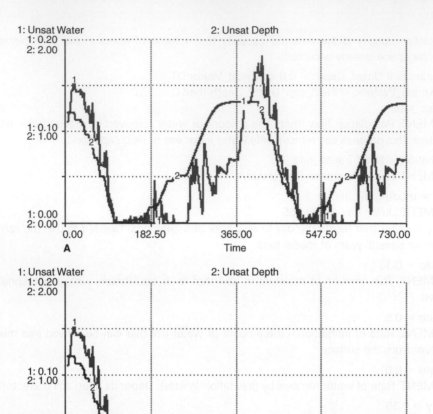

Figure 6.19   Dynamics of unsaturated water and unsaturated depth. A. High transpiration; B. Low transpiration.

used for artificial drainage of wetlands). If for some reason the transpirative efficiency declines, the dry periods can become shorter and the area gets entirely flooded (Unsat_Depth = 0) for considerable periods of time.

Exercise 6.4

1. What other processes are important to define the flooding regime in the area? What parameters need to be changed to restore the drainage of the area and make sure that it does not get flooded? What ecological processes correspond to these changes in parameters?

2. In fact when the Unsat_Depth decreases, certain amounts of unsaturated water get lost because they are no longer in the unsaturated zone. Imagine the water table rising and therefore the saturated water occupying more of the unsaturated zone. The unsaturated water in that marginal loss of the unsaturated depth will be lost from the Unsat_Water stock. Add this process to the Stella model. Is there a big difference in the results?

Water in the saturated layer

Saturated water is the amount of water in the saturated ground. Once a sponge can no longer hold additional water, it becomes saturated. As with surface water, if we add water to the saturated zone, its level increases. Also as the surface water, the saturated layer is involved in both horizontal and vertical transport. However, saturated water travels only through pores and cracks, and therefore its rate of motion is considerably slower. A simple conceptual model is shown in Figure 6.20. Some of these processes, such as percolation, have already been considered above. In addition to those shown in the figure, there may also be the exchange between surface water and saturated water. The saturated/surface water interface accounts for a series of processes that can move water from the surface directly into the saturated storage and back.

For example, a direct flux of water from the surface into the saturated storage is assumed to take care of the situation when all the soil is saturated but some water is drained by horizontal flows or by transpiration. In this case, if there is surface water present, it will immediately replenish the saturated tank. Once some pores in the saturated storage become vacant, water from the surface will make every attempt to refill them. Only after the surface water has been removed will the unsaturated layer start to appear.

The reverse flow from saturated storage to the surface occurs every time the elevation drops below the water table. This is the flow that feeds the rivers and streams at baseflow (Figure 6.21). Whenever the saturated layer of groundwater hits the surface, moisture starts to seep out. This can be often seen on steep riverbanks, cliffs or ravines (Figure 6.22).

A simple Stella model for vertical saturated water dynamics is presented in Figure 6.23. In this model we present saturated water together with the unsaturated water flows. In fact it is very difficult to separate the two, because the boundary between the

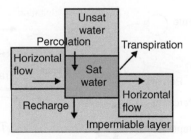

Figure 6.20   A simplified conceptual model for water in the saturated layer.

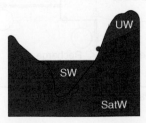

Figure 6.21   How water can re-emerge from the saturated layer and return to the surface water flow.

Figure 6.22   Calvert Cliffs in Maryland.

The layer of clay underlies the unsaturated layer. Clay has very low permeability, and water travels horizontally on top until it reaches the shore of Chesapeake Bay. Note the dry unsaturated layers on top of the wet saturated layers below.

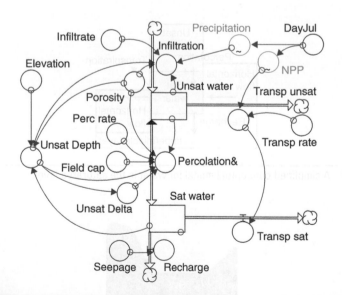

Figure 6.23   Stella model for water in saturated layer.

saturated and unsaturated zones moves and the two storages become closely related. What was previously the saturated zone may turn to be unsaturated, and *vice versa*.

The model equations are as follows:

Sat_Water(t) = Sat_Water(t − dt) + (Percolation& − Recharge − Transp_Sat) * dt
INIT Sat_Water = 4
DOCUMENT: Amount of water in the saturated layer, measured in m from some base datum

Percolation& = if Unsat_Depth > 0 then
(if Unsat_Water< = Field_cap*Unsat_Depth then 0 else Perc_rate)+
(if Unsat_Delta > 0 then Unsat_Water*Unsat_Delta/Unsat_Depth/DT
else Unsat_Delta*Porosity/DT)
else Unsat_Water/DT
DOCUMENT: Percolation flow (m/d) + the compensation for the change in the water table height. First term is percolation, the amount of water removed by gravity from the unsaturated layer. This process can remove only water in excess of field capacity. Second term tells how much water was added to (or removed from − hence the biflow) the unsaturated zone when water table went down (up).

Recharge = Seepage*Sat_Water
DOCUMENT: Loss of saturated water to deeper aquifers (m/d)

Transp_Sat = Transp_Unsat
DOCUMENT: Assuming that transpiration from the saturated layer occurs at a rate equal to that from the unsaturated layer

Unsat_Water(t) = Unsat_Water(t − dt) + (Infiltration − Percolation& − Transp_Unsat) * dt
INIT Unsat_Water = 3
DOCUMENT: Amount of water in the unsaturated layer measured as height of water column if "squeezed" from the soil (m).

Infiltration = min(Infilt_rate,Precipitation*0.0254,Porosity*Unsat_Depth-Unsat_Water/DT)
DOCUMENT: The amount of water infiltrated is the minimum of infiltration rate, the amount of precipitation available (0.0254 converts inches to m), and the unsaturated capacity (m/d).
The unsaturated capacity is the potential capacity (the volume of pores in the soil) minus Unsat_Water (the space already occupied).

Percolation& = if Unsat_Depth>0 then
(if Unsat_Water< = Field_cap*Unsat_Depth then 0 else Perc_rate) +
(if Unsat_Delta > 0 then Unsat_Water*Unsat_Delta/Unsat_Depth/DT
else Unsat_Delta*Porosity/DT)
else Unsat_Water/DT
DOCUMENT: Percolation flow (m/d) + the compensation for the change in the water table height. First term is percolation, the amount of water removed by gravity from the unsaturated layer. This process can remove only water in excess of field capacity. Second term tells how much water was added to (or removed from – hence the biflow) the unsaturated zone when water table went down (up).

Transp_Unsat = NPP*Transp_rate
DOCUMENT: The transpiration flow (m/d)

DayJul = mod(time-1,365) + 1
DOCUMENT: Julian day, 1 thru 365
This is a counter that resets the day to zero after 365 iterations. Needed to use the same graph function for several years of model runs.

Elevation = 30
DOCUMENT: Elevation of surface from base datum (m)

Field_cap = 0.13
DOCUMENT: The amount of moisture in soil that is in equilibrium with gravitational forces. (dimless)

Infilt_rate = 0.05
DOCUMENT: Rate of infiltration—the amount of water (m/d) that can be moved into the unsaturated layer from the surface

Perc_rate = 0.005
DOCUMENT: Rate of water removal by gravitation (m/d). Depends upon soil characteristics.

Porosity = 0.35
DOCUMENT: Proportion of pores in the soil. They can be potentially filled with water (dimless)

Seepage = 0.0001
DOCUMENT: Rate of loss of saturated water to deep aquifers (l/d)

Transp_rate = 0.005
DOCUMENT: The amount of water that plants can remove from soil by the sucking action of their roots (m of water/kg biomass*m^2/day)

Unsat_Delta = DELAY(Unsat_Depth,DT)-Unsat_Depth
DOCUMENT: Increment in water table height (m) over one DT.

Unsat_Depth = Elevation-Sat_Water/Porosity
DOCUMENT: Depth of unsaturated zone (m), defined as Elevation – amount of saturated water * porosity. Note that sat. water is the water "squeezed" out of the ground, by multiplying it by porosity we get the actual height of saturated layer

NPP = GRAPH (DayJul)
(0.00, 0.00), (33.2, 0.00), (66.4, 0.00), (99.5, 0.04), (133, 0.4), (166, 0.925), (199, 0.975), (232, 0.995), (265, 0.985), (299, 0.855), (332, 0.105), (365, 0.00)
DOCUMENT: An estimate of plant growth over the year (kg/m^2)

Precipitation = GRAPH (DayJul)
(1.00, 0.02), (2.00, 0.34), (3.00, 0.00), (4.00, 0.07), (5.00, 0.00), (6.00, 0.18), (7.00, 0.46), (8.00, 0.22), (9.00, 0.08), (10.0, 0.00), (11.0, 0.00), (12.0, 0.38), (13.0, 0.1), (14.0, 0.00),
...,
(354, 0.00), (355, 0.00), (356, 0.00), (357, 0.00), (358, 0.00), (359, 0.00), (360, 0.02), (361, 0.00), (362, 0.00), (363, 0.00), (364, 0.00), (365, 0.00)
DOCUMENT: Rainfall from Beltsville, MD 1969. {in/d}

We consider the amount of water in the saturated zone, as if it were squeezed out of the ground. The actual height of the saturated layer will then be Sat_Water/ Porosity, where porosity is the proportion of pores in the ground. The depth of the unsaturated layer Unsat_Depth is now calculated as the difference between the elevation and the height of the saturated layer.

Notice the use of the DELAY function in this model. The Unsat_Delta is calculated as the difference between the unsaturated depth before and the depth now. If Unsat_Delta is positive, it means that there was a deeper unsaturated layer before than there is now. This can only be the case if the water table is rising, so we need to move some water that previously was in the unsaturated storage into the saturated

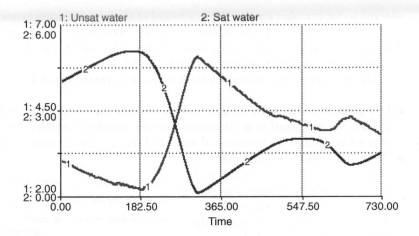

Figure 6.24   Dynamics of saturated and unsaturated water.
As the water in the saturated layer drops, the water table goes down, leaving more space for unsaturated water. As a result, there is then more water in the unsaturated layer.

one. Otherwise, if the Unsat_Delta is negative, the unsaturated layer is increasing its depth, which means that whatever water was contained in the pores of the saturated layer is now added to the unsaturated storage.

The resulting dynamics of water in saturated and unsaturated storages are shown in Figure 6.24. The model variables tend to an equilibrium height that slightly changes due to the transpiration of plants.

As with surface-water transport, strictly speaking groundwater flow is not a process to consider within the framework of a unit model. It is driven by spatial gradients, whereas in the unit model all spatial characteristics are assumed to be uniform over the whole modeled area. Therefore we will look at this process in more detail when describing the spatial implementation of the model. At this point we can simply provide a function that will remove a certain proportion of the available groundwater to keep it close to a steady state.

Exercise 6.5

1. Why do the dynamics of unsaturated water look so much smoother in this model when compared to the model for unsaturated water only?
2. Can you make all the ground saturated and the unsaturated layer disappear from the system? What parameters need to be changed? What is their ecological meaning?

6.2   Unit model

Now that we have considered some bits and pieces of the hydrologic cycle, we may revisit our goals and scales. Probably, for the goals that we had previously decided we will need to include most of the processes described above. At any stage of model building, it is always important to be clear about all the assumptions that have been adopted so far. It makes sense to keep a record of those, since many times a modeler can get carried away with the process and forget about some of the simplifications that were made at one of the earlier stages. It also adds credibility to the model if you can always explain all the assumptions to the model users.

The major processes and assumptions we made to create a model are as follows:

- Precipitation comes with rainfall and snowfall. If the temperature is below 0°C (32°F), the precipitation is channeled into the snow/ice variable. Otherwise part of it infiltrates into the unsaturated water and the rest goes into the surface water.
- We assume that rainfall infiltrates immediately into the unsaturated layer and only accumulates as surface water if the unsaturated layer becomes saturated or if the daily infiltration rate is exceeded.

> *Always be clear and honest about all the assumptions and simplifications that were made when model was built.*

- Surface water may be present in rivers, creeks, streams or ponds. Surface water is removed by overland flows and by evaporation.
- Surface water flow rates are a function of dynamically varying plant biomass, density, and morphology in addition to surface and water elevation. However, at this point we ignore details of surface water flow.
- Water from the unsaturated layer is forced by gravity to percolate down towards the saturated layer. As it accumulates, the level of the saturated water goes up while the amount of water in the unsaturated layer decreases.
- Transpiration is the process of water removal from soil by the sucking action of roots. Transpiration fluxes depend on plant growth, vegetation type and relative humidity.
- Saturated groundwater can reach the surface and feed into the flow of surface water. This process is what feeds the streams and rivers between the rainfall events – the so-called baseflow.

After looking at individual processes and variables, we can put together the whole model for the hydrologic cycle, assuming that we can single out an area that is more or less independent of the adjacent regions. We assume that we are looking at an area of less than 1 km², located in relatively flat terrain that is not too much affected by horizontal fluxes of groundwater. There is a certain gradient of elevation that is sufficient to remove all the excess surface water that did not get a chance to infiltrate into the ground over one time-step. The groundwater table is rather stable and tends to be at equilibrium at the initial conditions. The climatic data that we have are at a daily time-step, and therefore there is no reason to assume a finer time-step in the model. Thus, we can agree that our time-step is 1 day and our spatial resolution is 1 km².

The model diagram in Figure 6.25 is quite complex, but you will certainly recognize some of the modules and submodels previously considered.

The Globals sector (Figure 6.26) contains climatic data that are input as graphs and the empirical model for

> *Keep your model diagram and code tidy and logical. Explain things wherever possible. Avoid long connections. Try to put the model into submodels or modules.*

solar radiation. Here, we also define the elevation of the area considered. This might not be very important for the unit model, but it will become crucial if we decide to combine the unit models into a spatial simulation.

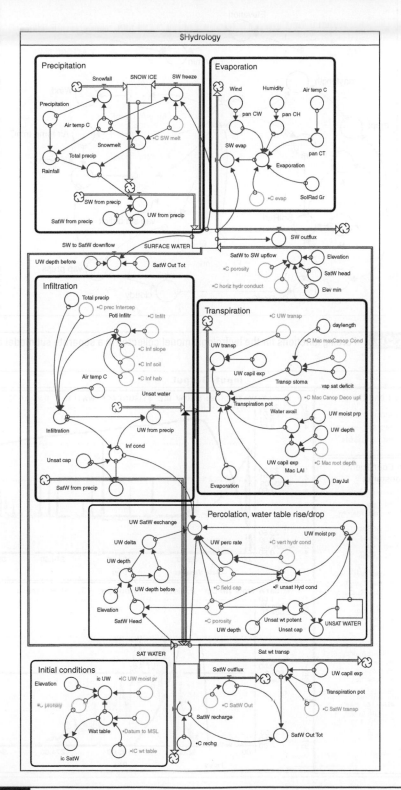

Figure 6.25   Full Stella model for unit hydrology.

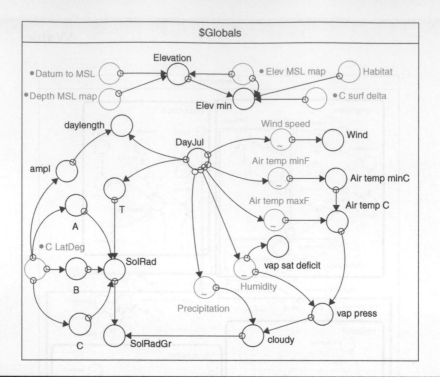

Forcing functions for the hydrologic model collected in a separate submodel called Globals.

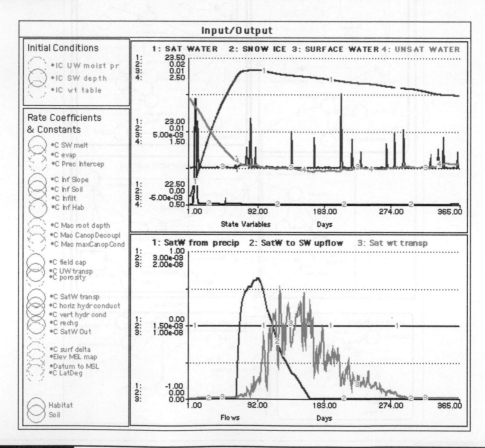

Input/output section for the hydrologic model.
Note that it is easier to manipulate parameters if they are collected in one place using the "ghost" feature in Stella.

The Input/output sector presents all the model parameters and the graphic output generated by the model (Figure 6.27). Notice that it always makes sense to put at least some effort in organizing the interface in the most practical way. In this case we have copied all the parameters spread out across the model diagram into one place, so that when calibrating we can easily modify the parameters without searching for them in different places. The model has already become fairly complex, and it is hard to understand its structure and equations without looking at the real Stella model. At this stage, it would be very useful to download the model from the book website and start exploring it.

Exercise 6.6

1. What modifications should be made to the conceptual model if a timescale of 1 year and a spatial scale of 1 km^2 are chosen for the model? What processes can be excluded, simplified and/or described in more detail?
2. Compare dry-year (halved precipitation) and wet-year (doubled precipitation) dynamics within the model. Does the model produce reasonable estimates for the state variables? Does it tend to equilibrium if such conditions prevail, or it shows a trend over several years?
3. Can you find a parameter in this model, modifying which you can produce a trend over several years that destabilizes the system?

The hydrology model described above is very much process-based. As you can see, much effort has been made to present the processes involved in a lot of detail. We track the pathways of water, describe the processes, and consider the various factors that impact the rate of these processes. In contrast, we may choose an empirical approach where, instead of looking at different details, we compound the statistics of observed processes under various conditions, and then use empirical input to drive the model. This is the approach used in a well-known model of Hydrology of Small Watersheds by the Natural Resources Conservation Service (NRCS), also known as TR-55.

The TR-55 model has been developed for small watersheds, especially urbanizing watersheds, in the United States. It can handle a 24-hour duration rainfall, 10 subwatersheds, and minimum 0.1-hour and maximum 10-hour times of concentration (that is the time that is required for water to travel from the most distant point on the watershed). It has been developed for NRCS type distributions of soils.

The main empirical runoff equation that drives the model is very simple:

$$Q = \frac{(P - 0.2\,S)^2}{(P + 0.8\,S)}$$

where Q is the runoff, P is rainfall and S is the potential maximum retention after runoff begins. S is related to the soil and land cover conditions of the watershed through a parameter CN, which stands for "curve number." CN has a range of 0 to 100, and is related to S by

$$S = \frac{1000}{CN} - 10$$

Here we need to use some caution and remember one of the commandments above: mind the units. As for any product coming from a US Federal Agency, TR-55 is designed in imperial units – in this case, inches. US Federal Agencies do not acknowledge that the rest of the world has adopted metric standards, which causes a lot of confusion and errors. So take care whenever dealing with a product that comes from there! In the case of the equations above, the units did not matter until we arrived at the relationship between S (measured in units of length) and CN (a dimensionless empirical curve number). The curve numbers CN are designed to produce S in inches. So in order to stay within the universally accepted metric conventions, a conversion is needed:

$$S(cm) = \frac{2540}{CN} - 25.4$$

All the complexities of the hydrologic cycle that we have explored become embedded in this magical empirical parameter. If there is no retention capacity of the watershed, CN = 100, S = 0 and Q = P, all rainfall becomes runoff. The larger the retention capacity, the smaller the curve number, the less runoff is seen. Curve numbers are produced from empirical studies for various land covers and soil types. A sample of curve numbers is presented in Table 6.1.

Table 6.1 A sample of runoff curve numbers for urban areas. Similar tables exist for agricultural and other types of land uses. See the full TR-55 publication

Cover description		Hydrologic soil group			
Cover type and hydrologic condition	Average percent impervious area	A	B	C	D
Open space (lawns, parks, golf courses, cemeteries, etc.):					
Poor condition (grass cover <50%)		68	79	86	89
Fair condition (grass cover 50–75%)		49	69	79	84
Good condition (grass cover >75%)		39	61	74	80
Impervious areas:					
Paved parking lots, roofs, driveways, etc. (excluding right-of-way)		98	98	98	98
Streets and roads:					
Paved; curbs and storm sewers (excluding right-of-way)		98	98	98	98
Paved; open ditches (including right-of-way)		83	89	92	93

(Continued)

Table 6.1	(Continued)					
Cover description			Hydrologic soil group			
Cover type and hydrologic condition	Average percent impervious area		A	B	C	D
Gravel (including right-of-way)			76	85	89	91
Dirt (including right-of-way)			72	82	87	89
Urban districts:						
Commercial and business	85		89	92	94	95
Industrial	72		81	88	91	93
Residential districts by average lot size:						
1/8 acre or less (town houses)	65		77	85	90	92
1/4 acre	38		61	75	83	87
1/3 acre	30		57	72	81	86

Infiltration rates of soils vary widely, and are affected by subsurface permeability as well as surface intake rates. Soils are classified into four Hydrologic Soil Groups (HSG) – A, B, C and D – according to their minimum infiltration rate, which is obtained for bare soil after prolonged wetting. Roughly, the HSG soil textures are as follows: A – sand, loamy sand, or sandy loam; B – silt loam or loam; C – sandy clay loam; and D – clay loam, silty clay loam, sandy clay, silty clay, or clay.

Comparing the two models, the process-based Stella model and the empirical TR-55, the simplicity of the latter can be appreciated. Note, however, how little the empirical model tells us about the actual processes – about how various forcing functions (temperature, wind, etc.) affect the system. While it is certainly a useful tool for some particular applications, especially where quick estimates are required, it is unlikely to advance our understanding of how the system works. On the other hand, it is quite easy to become buried in all the complexities of the process-based approach, especially if we consider all the parameters we will need to figure out to make it run, and all the data for forcing functions that we will need to find. In some cases, a bicycle is all you need to get there; in other cases, a Boeing-777 would be a better choice. Note, however, that in most situations when a bicycle is a good solution, a Boeing would be a ridiculous or even impossible option. The same applies with different kinds of models.

Also note that both models have quite limited application, since they assume a very small watershed and no horizontal movement of water. If we want to cover larger watersheds, we need to explore how water gets routed and what spatial algorithms are needed to make the models work.

6.3   Spatial model

In reality, hydrologic processes are very much spatial and their description within the framework of a spatially uniform unit model is quite limited. Water, both on the surface and in the ground, does not stay attached to one place. The water that is accumulated on the surface due to rainfall or fluxes from the ground is removed by gravitation, as it tends to find an adjacent location that has a lower elevation. The flow of surface water is therefore controlled by the head difference between the donor and acceptor locations (Figure 6.28). The rate is also a function of the surface roughness – which is why water will run faster over a concrete bed, a parking lot or a street than over a wetland, a crop field or a forest.

If H_i is the total water head in segment i, $H_i = E_i + D_i$, where E_i is elevation and D_i is water depth, then surface water flow (F_i) is a function of $H_i - H_{i+1}$.

Surface water flow is not a process to consider within the framework of a unit model. It is driven by an elevation gradient, whereas elevation in a unit model, like other spatial characteristics, is assumed to be uniform over the whole modeled area. For surface water to start running we need to compare the elevation of two adjacent cells, and we need a spatial framework to do that.

Groundwater transport, the flow of water in the saturated underground layer, is somewhat similar. Just as on the surface, the water can be accumulated on top of an impervious layer underground. In this case, the water that comes from the top with the infiltration and percolation processes accumulates and reaches saturation concentrations. Once all the pores between the soil particles have been filled, the water can either continue accumulating and increasing the height of the water table under the ground, or start moving horizontally to areas with a lower saturated water head. All this is very similar to the processes on top of the surface except that, instead of a free flow of water, in the case of groundwater we are dealing with the motion of moisture in the porous media. The rate of this flow is considerably slower and is

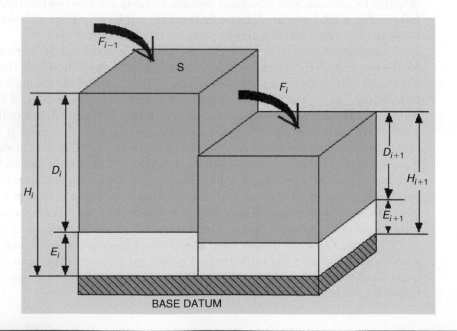

Figure 6.28　Surface water flow between two cells.
Surface water flow (F_i) is a function of head difference $H_i - H_{i+1}$. H_i is total water head: $H_i = E_i + D_i$, where E_i is elevation and D_i is water depth.

defined by horizontal hydraulic conductivity. This rate is very much dependent on the soil type, and can vary by several orders of magnitude.

As with surface water transport, groundwater flow is certainly spatial. It is driven by spatial gradients and the spatial characteristics of soil. In fact, of the four major variables in the unit model considered above, only two (unsaturated water and snow/ice) are attached to a certain area and can be modeled locally. For the other two major actors (surface water and saturated water), we need some representation of spatial dynamics.

As we saw in Chapter 5, Stella is certainly not a proper tool to build spatial models that may become very complex and are likely to require direct links to maps and Geographic Information Systems (GIS). There are two basic approaches used for modeling spatial hydrology (Figure 6.29):

1. Lumped or network-based hydrologic units. Here, the space is represented as a number of hydrologically homogeneous areas that are linked together by a linear network, representing the flow of water in streams.
2. Grid-based units. Here, the space is represented as a uniform or non-structured grid of square, triangular or other cells.

Each of the two approaches has its advantages and disadvantages.

Lumped models

When using network-based segments, the number of individual hydrologic units that are considered spatially may be quite small. The whole area is subdivided into regions, based on certain hydrologic criteria. These may be subwatersheds of certain size, hill slopes, areas with similar soil and habitat properties, etc. In most cases it is up to the researcher to identify the ranges within which factors are aggregated,

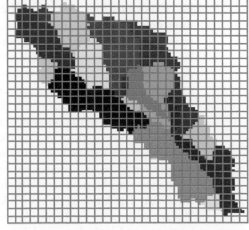

Figure 6.29   Lumped network approach and the grid approach. Each subwatershed or hydrologic unit is presented as a combination of cells.

and therefore decide on the number of spatial units that are to be considered in the model.

This decision is made based on:

- The goals of the model – how much spatial detail do we need about the system, and what are the major processes we want to analyze and understand within the framework of the model?
- The available computer resources – how much memory there is to handle the spatial arrays, and how fast is the CPU to run the full model?
- The available data – how much do we know about the study area, and what is the spatial resolution of the data?

Once the spatial units have been chosen, they are assumed to be homogeneous, and the geometry of the area is fixed. This is also the major disadvantage of the lumped or the unstructured grid approach. If for some reason we need to reconsider the geometry of the watershed and switch to other hydrologic units, it may require a considerable effort to develop a new grid or routing scheme.

Once the routing network is defined, the procedure is more or less the same. Certain empirical or process-based equations are derived to define the amounts of water and constituents that each hydrologic unit may generate. These quantities are then fed into a network model that represents the transport along the river and its tributaries. The network model links together the individual models for the spatial units.

One of the classic examples of this approach is the HSPF (Hydrological Simulation Program Fortran), which is available for download from a variety of sites (http://water.usgs.gov/software/hspf.html). The model was developed in the early 1960s as the Stanford Watershed Model. In the 1970s, water-quality processes were added. HSPF can cover extended periods of time with time-steps ranging from 1 minute to 1 day. It has been used to model various spatial areas, from small sub-catchments of several hundred square meters to the 166,534-km^2 Chesapeake Bay watershed. The model simulates the hydrologic and associated water-quality processes on pervious and impervious land surfaces, and in streams and well-mixed impoundments. It uses standard meteorological records to calculate stream flow hydrographs and pollutographs. The list of processes that are covered by various versions of HSPF is long and impressive: interception, soil moisture, surface runoff, interflow, base flow, snowpack depth and water content, snowmelt, evapotranspiration, groundwater recharge, dissolved oxygen, biochemical oxygen demand (BOD), temperature, pesticides, conservatives, fecal coliforms, sediment detachment and transport, sediment routing by particle size, channel routing, reservoir routing, constituent routing, pH, ammonia, nitrite–nitrate, organic nitrogen, orthophosphate, organic phosphorus, phytoplankton and zooplankton.

Probably one of the best-elaborated versions of HSPF became part of the BASINS suite developed at the US Environmental Protection Agency (EPA) (http://www.epa.gov/OST/BASINS/). A major improvement is the user-friendly interface, which allows users to build a project for a watershed that they are interested in. At this site there may even be data sets that are needed for a model for almost any watershed in the USA. The latest version of BASINS also includes the SWAT model – Soil and Water Assessment Tool (http://www.brc.tamus.edu/swat/index.html) – another well-known spatial hydrology model that is also based on the same lumped subwatershed paradigm. Both models are generally able to simulate stream flow, sediment, and nutrients loading. According to some reports, HSPF simulates hydrology and water-quality components more accurately than SWAT; however, HSPF is less user-friendly than

SWAT, owing to there being even more parameters to control. HSPF is an extremely data-intensive and over-parameterized model, and requires a large amount of site information. SWAT is somewhat simpler; it estimates the surface runoff from daily rainfall using the curve number method we discussed above, and sediment yield is calculated with the Modified Universal Soil Loss Equation (MUSLE).

Yet another two models that are worth mentioning are:

1. TOPMODEL – a classical model that has been used for a variety of rivers and watersheds (see http://www.es.lancs.ac.uk/hfdg/freeware/hfdg_freeware_top.htm).
2. RHESSys – the Regional Hydro-Ecological Simulation System, which is a GIS-based, hydro-ecological modeling framework designed to simulate carbon, water and nutrient fluxes. RHESSys combines a set of physically-based process models and a methodology for partitioning and parameterizing the landscape (see http://geography.sdsu.edu/Research/Projects/RHESSYS/).

Describing any of these models in any decent amount of detail can take as much space as this whole book. However, the basic concept is quite simple and can be illustrated by the same TR-55 model considered above. As we have seen, we can calculate the amount of runoff from a certain drainage area for each rainfall event. By definition, this runoff does not stay in place – it runs. Now we need to look at the horizontal dimension and figure out the factors that can impact this run, since once it starts running it starts to accumulate water from various areas, and that is what configures the flow hydrograph, or the pattern of flow in a stream or river. TR-55 has been developed to estimate the peak flow that an area can generate in response to various rainfall events. It takes the runoff, calculated above, as the potential amount of water that the area can produce, and then takes into account various spatial characteristics of the watershed (such as slope, channelization, surface characteristics, etc.) and the temporal characteristics of rainfall (duration) to estimate the maximal flow that should be expected from this area.

One crucial indicator is the time of concentration (T_c), which is the time for runoff to travel from the hydraulically most distant point of the watershed to a point of interest within the watershed. T_c is computed by summing all the travel times for consecutive components of the drainage conveyance system. Travel time (T_t) is the time it takes water to travel from one location to another in a watershed.

Travel time is affected by several factors, such as surface roughness, channel shape, and slope of surface. For example, undeveloped vegetated areas will have a high degree of roughness and very slow and shallow overland flow. As flow is delivered to streets, gutters and storm sewers, runoff downstream becomes far more rapid. Urbanization will generally significantly decrease the travel time through a watershed. The slope will tend to increase when channels are straightened, and decrease when overland flow is directed through storm sewers, street gutters and diversions. The time of concentration (T_c) is the sum of T_t values for the m various consecutive flow segments:

$$T_c = T_{t1} + T_{t2} + \cdots + T_{tm}.$$

Travel time (in hours) is the ratio of flow length to flow velocity. Water moves through a watershed as sheet flow, shallow concentrated flow, open channel flow, or some combination of these. Sheet flow is the flow over plane surfaces, and usually occurs in the headwater of streams. With sheet flow, the friction value (Manning's n) is an effective roughness coefficient that includes the effect of raindrop impact; drag over the plane surface; obstacles such as litter, crop ridges and rocks; and erosion and transportation of sediment.

Manning's kinematic solution, which works for travel time over 100 m or less, is

$$T_t = \frac{0.007(nL)^{0.8}}{P^{0.5}} \Big/ S^{0.4}$$

where n = Manning's roughness coefficient (Table 6.2), L = flow length (ft), P = 2-year, 24-hour rainfall (in), s = slope of hydraulic grade line (land slope, ft/ft). Note again the confusion with units here.

After a maximum of 100 m, sheet flow usually becomes shallow concentrated flow. It is driven by slope, so for concentration time we have

$$T_t = \frac{L}{3600V}$$

where: L = flow length (m), V = average velocity (m/s) and 3600 = conversion factor from seconds to hours. For slopes less than 0.005 and unpaved conditions,

Table 6.2	Roughness coefficients (Manning's *n*) for sheet flow
Surface description	*n*
Smooth surfaces (concrete, asphalt, gravel, or bare soil)	0.011
Fallow (no residue)	0.05
Cultivated soils:	
Residue cover ≤ 20%	0.06
Residue cover >20%	0.17
Grass:	
Short grass prairie	0.15
Dense grasses	0.24
Bermudagrass	0.41
Range (natural)	0.13
Woods:	
Light underbrush	0.40
Dense underbrush	0.80

$V = 16.1345 \, s^{0.5}$; for paved conditions, $V = 20.3282 \, s^{0.5}$ where s = slope of hydraulic grade line (watercourse slope, m/m). For steeper slopes the equations are similar, but the coefficients will be different.

When flow becomes channelized the equation is different:

$$V = \frac{1.49 r^{2/3} s^{1/2}}{n}$$

So why does 1.49 appear in front of the Manning's equation? What a strange way to write an equation. Why not include the 1.49 in the empirical coefficient n, which is also there? What's so special about 1.49?

Well, your guess is probably correct. Of course, it is the unit conversion. The real Manning's equation is

$$V = \frac{r^{2/3} s^{1/2}}{n}$$

where r is measured in meters and s, the slope, is measured in m/m. While n is an empirical coefficient and is usually presented as dimensionless, in fact it has units. If we want to have V in m/s, we need to have n in $s/m^{1/3}$ – very weird units indeed. But now it is clear that if we wish to use the same empirical values for n, but get the result in ft/s, we'll need some tweaking. Indeed, $s/m^{1/3} = s/(3.281/3 \, ft^{1/3}) = s/(1.49 \, ft^{1/3})$. And there is our 1.49!

The bottom line is, if you really need to use Imperial units, brace yourself for a lot of fun.

Here, r is the hydraulic radius (ft) and is equal to a/p_w, a is the cross-sectional flow area (ft^2), p_w is the wetted perimeter (ft), s is the slope of the hydraulic grade line (channel slope, ft/ft), and n is the Manning's roughness coefficient for open channel flow. This is also known as the Manning's equation.

Finally, the peak discharge (ft^3/s) equation is:

$$q_p = q_u A_m Q F_p$$

where: q_u = unit peak discharge (csm/in), A_m = drainage area (mi^2), Q = runoff (in), and F_p = pond and swamp adjustment factor. Here we know A_m and how to calculate Q from the unit model. F_p is just an adjustment factor if the pond and swamp areas are spread throughout the watershed and are not considered in the T_c computation. The unit peak discharge q_u is what requires most effort to work out. It takes into account T_c, the 24-hour rainfall (in), and once again the curve number, CN. Stepping through a series of tables and graphics, TR-55 finally gets the answer.

There is a Stella implementation of TR-55 developed by Evan Fitzgerald that can be downloaded from the book website or from the "Redesigning the American Neighborhood" project website (http://www.uvm.edu/~ran/ran/researchers/ran55. php). In this simplified version, the standard rainfall–runoff relationships and equations used in TR-55 models have been written into the Stella model to produce near-identical results to the NRCS models. These relationships include the curve number approach as well as the rainfall curve used for the northeast. The time concentration variable was excluded in this version, since the model did not appear to be sensitive to it. A comparative analysis between TR-55 and Stella model results was performed for the time of concentration variable at the fixed scale of 10 acres, and it was determined that the effect of not including this variable in the Stella model was negligible for peak flow rate calibration.

The model also provides a good example of the use of the modeling interface that comes with Stella. In this case, the goal was to explore various alternative management practices for stormwater in a small Vermont neighborhood. There are all sorts of switches and sliders, knobs and graphics that allow the user to define easily the various scenarios and management solutions to compare results in search of a better understanding of the system and an optimal design of management practices.

It is also interesting to note that we have solved a spatial problem by a fairly local Stella model, although we have actually simplified it to the greatest extent possible. In reality, what makes a system really spatially distributed are the variations in data and processes. So far, we are still assuming that all the landscape characteristics (soil and landuse, expressed in the curve number, slope, rainfall pattern, etc.) are spatially uniform. We have provided for some spatial proxies by describing how water gets routed and removed from the unit area, but that is not really spatial.

What the models like those listed above (HSPF, SWAT, RHESSys) and others do is replicate a version of local TR-55 or our Stella Unit Hydrology model for a series of nodes. They then use similar delivery algorithms like the Manning's equation over the network of channels that connects those nodes. This takes care of the delivery mechanism over a large and spatially heterogeneous watershed.

Grid-based models

In grid-based models, the homogeneous spatial units are defined mechanistically, by representing the study area as a grid of cells. The major decision in this case is the size and form of the cell. The size defines the spatial grain – the resolution of the model. Ideally, the smaller the cells, the finer the resolution and the more detail regarding the landscape can be accounted for. However, the reverse side is again the model complexity and the time needed to run the model. The decision about the size and configuration of cells is usually based on pretty similar principles to those above:

- The goals of the study – what is the spatial resolution needed to meet those goals?
- The available computer resources – how much memory is there to handle the spatial arrays, and how fast is the CPU to run the full model?
- The available data – how much do we know about the study area, and what is the spatial resolution of the data?

There is yet one more consideration that may be important. Grid-based models generate huge arrays of output information. They may be quite useless unless there are good data processing and visualization tools that can help to interpret this output. Imagine a model of, say, 10 variables running over a grid of, say, 5,000 cells. And suppose we are running this model for 1 year at a daily time-step. This is probably an average complexity for spatial hydrology models. As an output we will be generating time series of maps, one for each state variable, every day. So potentially we will be obtaining some 3,650 maps for state variables in each of the 5,000 cells, plus as many more as we may want for intermediate variables. What do we do with all this information? Keep in mind that methods of spatial statistics and analysis are quite rudimentary. We also need to remember that it is hardly possible to expect to have anything close to that in terms of experimental data to compare our results and calibrate our model. So chances are that much of the spatial grain that we will be producing will be left unused, and most likely we will be generating some indices and spatially averaged indicators to actually use in our study.

Nevertheless, it is good to have the potential to perform this kind of analysis, and perhaps with the advance of remote sensing techniques and more abundant spatial data we will have more opportunities to test spatial models and improve our understanding of spatial processes. Moreover, spatial output looks

> *Colorful spatial output can be a powerful tool to drive management and planning decisions. Make sure you are not misusing or misinterpreting the results that you get from your model. Be clear about your assumptions and the uncertainties involved.*

so nice in presentations and reports – people like to see colorful maps or animation. Just make sure such output is not being misused or misinterpreted!

In Chapter 5 we visited with the Spatial Modeling Environment – SME – and showed how it can be used to extend local Stella models over a spatial domain. Here, we will take a quick look at a real-life application of this approach to watershed modeling. The Patuxent Landscape Model (PLM – http://giee.uvm.edu/PLM) is a grid-based spatial landscape model that was built upon the SME paradigm. The model uses an ecosystem-level "unit" model built in Stella that is replicated in each of the unit cells representing the landscape (Figure 6.30). For each different habitat type the model is driven by a different set of parameter values (e.g. percolation rate, infiltration rate, etc. are different for a forest vs an agricultural field vs a residential lot) (Figure 6.31). Actually, it is not only one model in Stella but a whole series of them. SME supports modularity in such a way that you can take several Stella models, each representing a certain subsystem, and run them in concert, exchanging information between the different modules.

As a companion tool to SME, the Library of Hydro-Ecological Modules (LHEM – http://giee.uvm.edu/LHEM) has been developed to represent most of the processes important for watershed dynamics and management (Figure 6.32). What is most remarkable with this approach is that it lends ultimate transparency to the model. Unlike the watershed models described above, where the code may not be easily available or indeed available at all (as in some proprietary models), and all the information about the model intestines has to be either figured out from the documentation provided or guessed using common sense, here we have the actual model at our fingertips. We can explore each module, run it as a separate Stella application, understand the dependencies and assumptions, or even make changes if we have better ideas regarding how to present certain processes.

The local hydrology model in LHEM is similar to the unit model in Figure 6.25. In addition to that, there are modules for nutrient cycling, dead organic material, plant growth, etc. Further, there are also spatial algorithms that can be used to move water and constituents between cells. There is a choice of algorithms of spatial fluxing that link the cells together (Figure 6.33). In effect, they are somewhat similar to the procedures discussed above when we were moving water over the network between nodes. Here too we need to decide how far and how fast the water will travel, except, as in the case of PLM, the network is degenerated to a simple case of cell-to-cell piping.

The methods used in LHEM are greatly simplified in order to handle large areas and complex ecological models. They may be considered as an empirical approach to surface-water routing. They are very much based on empirical assumptions and common sense. In a landscape-modeling framework, hydrology is only a part of a much more complex and sophisticated model structure. Therefore we have to try to keep

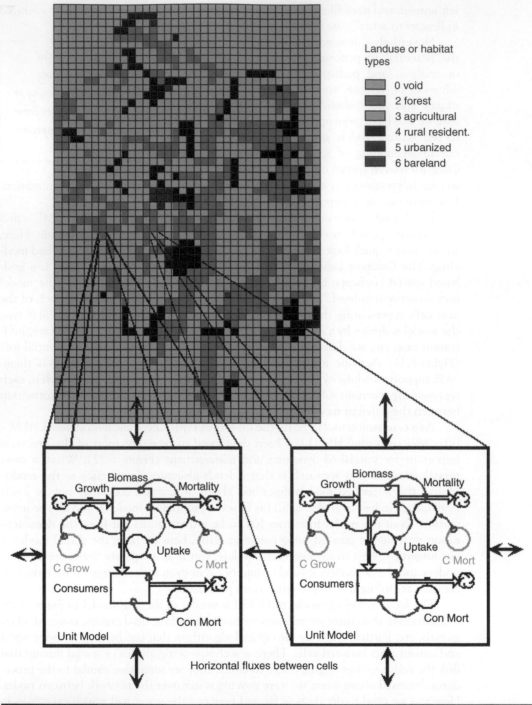

Landuse or habitat types

- 0 void
- 2 forest
- 3 agricultural
- 4 rural resident.
- 5 urbanized
- 6 bareland

Horizontal fluxes between cells

| **Figure 6.30** | Spatial organization of a grid-based model. |

the time-step as large as possible in order to be able to run the models for sufficiently long simulation periods. The methods suggested certainly sacrifice some of the precision, especially in the transfer processes, but they represent the quasi-equilibrium state well and substantially gain in model efficiency in terms of the CPU time required. In

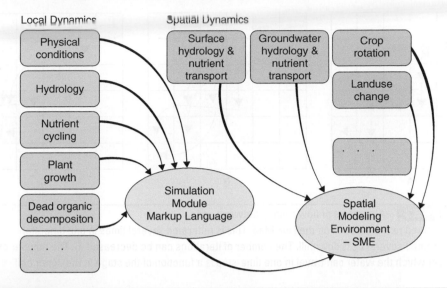

Figure 6.31   Modules and models in the Library of Hydro-Ecological Modules (LHEM). You can choose the ones you need for the goals of your modeling effort. Local dynamics are mostly Stella models. Spatial dynamics are routines in C++ that can be used within the SME framework.

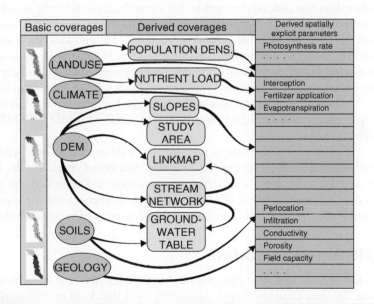

Figure 6.32   Spatial coverages used in the Patuxent Landscape Model (PLM). Basic coverages are used to generate the derived ones, which are stored as input maps. The other spatial data are generated each time the model is run.

this case we have to rely more on the comparison of the model output with the data available, and be ready to switch from the more process-based description to a more empirical one.

Certainly, by choosing this approach, by diverging from the process-based approach and by allowing more parameter and formulae calibration, we decrease the

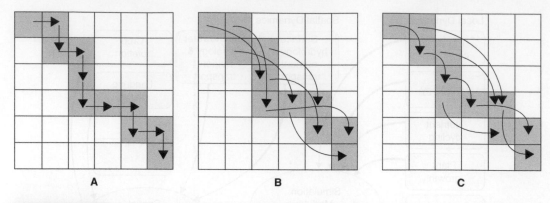

Figure 6.33   Algorithms of flow routing between cells.
A. Water is routed to the next cell on the Link Map. This is reiterated several times in one time-step.
B. Water is routed several cells downhill. The number of iterations can be decreased. C. The number of cells downhill over which the water can travel in one time-step is a function of the stage in the donor cell.

generality of the model, thus requiring additional testing and calibration when switching to other scales or areas. The question is, what is a truly process-based model as compared with an empirical, regression one? In any process-based model there is a certain level of abstraction at which we actually utilize certain empirical generalizations rather than true process description. For example, there is hardly an adequate detailed description of the photosynthesis process to be found among the models of vegetation growth; instead, some version of Michaelis–Menten kinetics is applied, which is already an empirical generalization of the process. Nevertheless, these models claim to be process-based. As we move towards larger systems, such as landscapes, we need to employ even more generalized formalizations, as we did above for modeling hydrology.

Another excellent example of the grid-based approach is the Everglades model at https://my.sfwmd.gov/pls/portal/url/page/PG_GRP_SFWMD_HESM/PG_SFWMD_HESM_ELM?n.

In fact, note that there is not that big a difference in how spatial models are implemented under the lumped and the distributed methods. In both cases we present local dynamics in a series of procedures, and then add the spatial dynamics over a certain network of nodes. The only differences are the number of cells or compartments that we distinguish and the geometry of the network that we allow.

GIS-based models

So far we have looked at dynamic models that describe how water flow changes over time. In some cases we do not necessarily need that kind of information; instead we need a high-resolution spatial model that can tell us what is happening in different parts of our study area. For example, consider a neighborhood that was developed some 15–20 years ago, when there was not much concern about stormwater.

We already know that when houses and lawns replace forests there are quite obvious changes in the hydrology of the area. As seen above, the infiltration rate can change substantially when all the undercover and debris that blanket the ground are removed. Clearly, the infiltration rate of a house roof and a paved driveway (which is close to zero) is very different from the infiltration rate in a forest. The evaporation from the soil also increases dramatically when trees are removed. Suddenly the landscape is producing

Figure 6.34   A standard retention pond such as is built in most new developments to comply with the stormwater regulation. It requires huge investment, and needs to be maintained properly.

far more surface runoff during storm events. Rivers and streams become raging torrents, causing erosion and flooding over vast areas. At the same time, there is less flow in between the storms. The so-called baseflow dries out, since all the water has been already drained and there is not much stored in the ground and wetlands to feed the streams.

The high flows result in highly incised landscapes, with streams digging deeper into the ground, taking out lots of sediment and dumping it into the rivers and lakes. The water quality also dramatically deteriorates. The sediments themselves are a nuisance for adult fish, and can destroy spawning grounds. They also carry large amounts of nutrients. Nutrients also come from fertilizers used to improve residential lawns. The lawns are also treated with chemicals – herbicides and pesticides – which all end up in estuaries and lakes.

The bottom line is that residential neighborhoods have a strong impact on stormwater quantity and quality, and need to start taking care of their runoff. So far, most of the solutions have been quite centralized. In one, the water is captured in large retention ponds, where it is held for a while, losing sediments and partially infiltrating into the soils (Figure 6.34). This solution is quite expensive both

The river network is developed by the landscape. However, it is not just the geology and height that matter; land cover is also a factor. If we have forests, they can absorb most of the rainfall, so there is not much left for runoff. If forests are replaced by impervious or less pervious surfaces, then there is more surface runoff and obviously more streams and rivers are required to conduct all that water. Besides, the more water is channeled through these streams, the wider and deeper they become. It is interesting to realize that perhaps most of the existing river network, especially the smaller streams and rivers, have been developed as a result of our land-cover changing activities.

Figure 6.35   A rain barrel.
This is a simple device to capture water collected from rooftops. It intercepts only the first few centimeters of a rainfall event (depending upon the area of the roof and the size of the barrel). However, it may be quite useful in improving water quality, since it is usually the first flash of runoff that contains most of the constituents, and the more of it we can retain, the better. Check out http://www.likbez.com/AV/barrel/ for how to make your own rain barrel.

to install and to maintain. These super-ponds can be built during the construction phase, when there are clear regulations and controls with which the developers need to comply. However, they are prohibitively expensive to install later on, when the neighborhood is already in place and the homeowners are expected to absorb all the additional costs of redesign.

An alternative solution that is more distributed and does not require huge upfront investment starts right at the door. For example, homeowners can install so-called rain barrels (Figure 6.35), which are simple containers that capture the drainage off the house roofs. However, these can intercept only low- and mid-size storm events, and they can be damaged in winter, when temperatures are below freezing.

Another solution for larger volumes of rainfall are the so-called rain gardens. These are artificial and natural depressions which are planted with vegetation that removes water through transpiration. The concept is not familiar for most homeowners, and it is sometimes hard to persuade them to consider this as an option. A simple spatial model can help in doing that. For example, it might be nice to show what the flows of surface water look like, where they go, how water is accumulated, and where the rain gardens are most likely to work best. This can be accomplished with GIS modeling, provided that we have significantly high-resolution elevation data.

The latest LIDAR (Light Detection and Ranging) point data offer exactly that opportunity. For example, for the whole Chittenden County in Vermont, there are high-resolution data sets. They are collected with aircraft-mounted lasers capable of recording elevation measurements at a rate of 2,000 to 5,000 pulses per second, and have a vertical precision of 15 centimeters. This information can be pulled into a

Figure 6.36   The drainage network as generated by ArcGIS.
The red line is the main stem of the stream, the yellow lines are engineered drainage pipes. The blue lines are the surface flows. We can see individual houses, and how surface flow is channeled from each property.

Geographic Information System (GIS), such as ArcGIS, which has some quite elaborate hydrologic modeling tools embedded in it.

First, we can build a Digital Elevation Model (DEM) using the Inverse Distance Weighted (IDW) interpolation tool from the ArcGis 9.2 ToolBox. Another ArcGIS tool can be used to calculate the stream network and subwatershed delineation on the basis of these DEMs (Figure 6.36). Results of analysis show that the modeled water drainage network follows the stormwater pipelines and street curves – even depressions along the property lines. If we further decrease the threshold, we will generate a micro-drainage network that gives us even more detail about the routing of surface water (Figure 6.37). This kind of information helps us to visualize the fate of stormwater on individual properties and in the neighborhood, and can also serve as a communication tool to help several neighbors to agree on where it will be most efficient and cost-effective to locate the treatment area. In most cases, a bigger shared rain garden will be much cheaper than several smaller gardens on different properties.

It's said that "a picture's worth a hundred words." Indeed, when they look at these images the homeowners can actually recognize their houses and properties and see how the water flows over their land. It is also clear where the rain gardens can be located in order to be most efficient. These models are powerful tools for deliberation and decision-making. In fact, in the "Redesigning the American Neighborhood" project – an attempt to find stormwater management solutions at the scale of small towns, cities and developments in Vermont – such visuals developed by Helena Vladich worked very well in directing the attention of homeowners in two small neighborhoods toward the distributed alternative engineering solutions. After seeing how the rain-garden method could be implemented, and comparing costs with the super-pond option, the citizens agreed that the small-scale distributed approach would be more promising and decided to pursue that technique.

Models do not need to be dynamic if we are mostly interested in the spatial context. The GIS framework offers numerous tools for spatial modeling that are quite simple to implement, and should certainly be considered when the temporal domain is not important or is not supported by any data or information.

Figure 6.37   The micro-drainage network gives more details about surface water flow and can be used to locate rain gardens on the property. In areas where we have a greater density of drainage, it makes sense to have a rain garden.

6.4   Conclusions

It is hard to overestimate the importance of water for human livelihood. Models of various levels of complexity can help us to understand how our lifestyles and economies impact water supply and the quality of this supply. There is always a question about choosing the appropriate modeling effort for the problem being studied. While some of the simple models can be easily assembled at little or no cost, the more complex, especially spatial applications, require many man-hours of work and a lot of expenditure on research and monitoring. Therefore, it always makes sense to start simply and try to obtain as much information as possible from the most straightforward models. This is another reason why close collaboration with the ultimate user of the model can save a lot of money and effort. We will see how this works in Chapter 9.

There is a great deal of concern about water quality in Chesapeake Bay. Excessive nutrient loads to the Chesapeake Bay from surrounding cities and rural counties have led to eutrophication, especially in small harbors and inlets. Several Agreements and Comprehensive Plans call for reductions in nutrients entering the Bay. Most sewage in rural residential areas is treated by onsite sewage disposal systems (septic systems). Almost all of the nitrogen pollution that enters local waters from Calvert County comes from non-point sources, of which the Maryland Department of Planning estimates 25 percent is from septic systems. The Calvert County therefore commissioned a study to estimate the contribution of septics to the amount of nitrogen that enters Solomons Harbor, an estuary on Chesapeake Bay. Despite high population densities, only a small portion of the watershed is serviced by sewers there. The question is whether there are any spatial differences in how septics impact water quality, and whether these spatial variations should be considered when regulating septic improvement or removal.

It seemed a natural task for some spatial modeling. However, even a quick preliminary analysis of data showed that atmospheric pollution, primarily from transboundary sources, contributes the greatest nitrogen load to the watershed and harbor. Fertilizers were proven to be the second most important factor (and, depending on assumed fertilizer usage, could surpass atmospheric pollution). Nevertheless, a spatial model was built which, not surprisingly, showed that the overall input from septic tanks was the lowest among all the anthropogenic nitrogen sources. In addition, the discharge is leached into groundwater, which affects surface water quality in the long term. The large buffering capacity of groundwater means that changes made to septic systems will take longer to materialize as an improvement in surface water quality. The housing density (the total number of homes in the watershed) is the most important criterion to consider when developing septic-related policies. The distribution of these homes surprisingly had relatively no effect on total nitrogen loads to the harbor. Indeed, local governments have little control over atmospheric deposition except through their influence on regional bodies. Managing septic loads is most feasibly implemented at the local level. However, decreasing fertilizer application was certainly the most cost-effective solution that emerged immediately, even without any sophisticated modeling efforts.

The results were certainly not what Calvert County Government was expecting, and would probably have been a hard sell if the project had not been conducted as a participatory modeling effort with much stakeholder involvement and support. This helped significantly to communicate the alternative strategies to the decision-makers.

Was the complex spatial modeling effort needed at all? Couldn't we have singled out the excessive fertilizer application as the most likely source of nitrogen in the runoff? Probably we could have done that. Unfortunately, the reality is that in some cases the more complex model simply looks more convincing than back-of-the-envelope estimates.

Information about water goes way beyond just the quantity and supply of it. The flow of water serves as an indicator of the relief and landscape characteristics on the one hand, and as an integrator of many of the processes occurring within the watershed on the other. The water quantity and quality are indicators of what is happening on the whole watershed. If there is a lot of environmental disturbance on the watershed, that will show in the water at the outlet. If the landuse has changed from forests to agriculture, there will get one set of changes in the water – there will be more flash flooding and lower baseflow, because of lower infiltration and higher runoff. There will be more nutrients, herbicides and pesticides in the water. If the landuse changes from agricultural to residential, there will be even less infiltration, higher peak flow, lower baseflow and lots of different chemicals in the runoff, ranging from caffeine to petrochemicals, antibiotics and detergents. Just by looking at water quality it is possible to figure out what the businesses on the watershed are, and what their level of environmental awareness is.

In one respect, the watershed approach seems to be more versatile than the general ecosystem management view. The system boundaries associated with a watershed approach are objective. Instead of being the result of historical, subjective, oftentimes unfair, voluntary or contradictory processes, they are based on certain geographical characteristics – such as elevation and flow gradient – which it is difficult to change and makes little sense to dispute. The watershed approach is not intended to substitute the existing borders and regions, but rather it offers a supra-administrative viewpoint to exercise consensus across economic, social and administrative bodies.

A hierarchical context is another crucial component of successful management schemes. The implied hierarchical structure of super-watersheds and subwatersheds is instrumental for upgrading and downgrading, zooming in and out, and changing resolution, depending upon the type and scale of the managerial problems to be resolved. This hierarchical approach adds flexibility to management, breaking the usual rigid connection between policy and scale. In most cases the scale is driven by the policy problem, and it is usually unclear who should formulate the policy question and at what scale. With the hierarchy provided by the watershed approach, the scale of the targeted management object becomes less crucial, as long as it is presented as an element of the whole hierarchical structure. The smaller watersheds are embedded in the larger ones, and the various policies formulated can be treated at the appropriate level. The hierarchy in this case is not imposed on the system from the outside, as is the case in administrative divisions, but is embedded in the physical characteristics of the system and offers a much larger variety of scales.

The potential of the watershed management approach may be illustrated by the fact that the US Environmental Protection Agency (EPA) has currently adopted it as its primary approach to addressing water-quality problems. The US Army Corps of Engineers is also advocating the watershed approach. The US Geological Survey (USGS) has defined a multi-digit classification system for watersheds, based on the size of the stem stream and the Hydrological Unit Classification (HUC) system. There are 2,149 watersheds within the continental United States identified as HUC-8 systems, and they are often used as standards for the watershed approach. Groups of stakeholders may apply their efforts to the HUC-8 scale, or may move up or down the scale as appropriate to their local problems and their concerns. More than 20 states are known to be developing or implementing management frameworks that use watersheds as the organizational basis for integrating water resource protection and restoration activities. These frameworks address the process and procedures for coordinating activities – from public outreach to strategic monitoring and assessment, to integrated management.

Further reading

As for other chapters in this book, the perfect companion in fundamentals of ecological economics is the textbook by Daly, H., Farley, J. (2004). Ecological Economics. Island Press. There you can find more about the dichotomy of resources (excludable – non-excludable and rival – non-rival).

For more on basics on hydrology see, for example, Chow, V. T., Maidment, D.R. and Mays, L.W. (1988). Applied hydrology. New York, McGraw-Hill; or Novotny, V. and Olem, H. (1994). Water Quality. Prevention, Identification, and Management of Diffuse Pollution. Van Nostrand Reinhold.

The National Climatic Data Center (NCDC) that is run by NOAA is a great source of climatic data, especially if you are doing a project in the USA. Their web site has plenty of data from metheostations across the country at different temporal resolutions. Most of the data can be downloaded for free: http://www.ncdc.noaa.gov/.

Nikolov, N.T. and Zeller, K.F. (1992). A solar radiation algorithm for ecosystem dynamic models. *Ecological Modelling*, 61: 149–168 – *This describes a model of solar radiation based on the lat-lon coordinate of the location that you are studying.*

The Christiansen model of evaporation is described by Saxton, K.E. and McGuinness, J.L. (1982). Evapotranspiration. *In*: Haan, C. T., H. P. Johnson and D. L. Brakensiek (Eds.) (1982). Hydrologic Modeling of Small Watersheds. St.Joseph, ASAE Monograph. 5: 229–273.

The so-called TR-55 model is documented in "Urban Hydrology for Small Watersheds" Natural Resources Conservation Service, Conservation Engineering Division, Technical Release 55, June 1986.

HSPF is now maintained by Aquaterra, but it is also in the public domain and can be downloaded from several sites. One of the latest HSPF applications developed by the Chesapeake Bay program is at http://www.chesapeakebay.net/temporary/mdsc/community_model/index.htm. *To read about HSPF see* Donigian, A.S., Imhoff, J.C., Bicknell, B.R. and Kittle, J.L. (1984). *Application Guide for Hydrological Simulation Program – FORTRAN (HSPF).* Athens, GA, U.S. EPA, Environmental Research Laboratory; Donigian, A.S., Jr., Bicknell, B.R., and Imhoff, J.C. (1995). Hydrologic Simulation Program – FORTRAN (HSPF). *Chapter 12 in Computer Models of Watershed Hydrology,* V.P. Singh, Ed., Water Resources Publications; *or* Donigian, Jr., A.S., Bicknell, B.R., Patwardhan, A.S., Linker, L.C., Chang, C.H., and Reynolds, R. (1994). *Chesapeake Bay Program Watershed Model Application to calculate bay nutrient loadings.* U.S. EPA Chesapeake Bay Program Office, Annapolis, MD.

See Voinov, A., Fitz, C., Boumans, R., Costanza, R. (2004). Modular ecosystem modeling. *Environmental Modelling and Software.*19, 3: 285–304 – *to learn more about the Library of Hydro-Ecological Modules (LHEM). It also describes in more detail most of the modules from the unit model that we have seen presented in this Chapter.*

Vermont Center for Geographic Information (VCGI http://www.vcgi.org/) *is the source for digital data on hydrologic stream network, roads, houses, landuse, engineered catchments, pipeline network and inlet points that was used in the RAN project. See here for more info about LIDAR remote sensing techniques:* http://www.csc.noaa.gov/products/sccoasts/html/tutlid.htm. *The spatial analysis for this project was performed by* Helena Vladich. *See* http://www.uvm.edu/~ran/ran *for more details.*

For more information about the Calvert project see Brown Gaddis, E. J., Vladich, H., and Voinov, A. (2007). Participatory modeling and the dilemma of diffuse nitrogen management in a residential watershed. *Environ. Model. Software.* 22 (5): 619–629.

7. Adding Socio-Economics

SUMMARY

What models can we build to describe social and economic systems? Economics has developed its own models, and has become one of the most mathematized branches of science. However, most of those models do not take into account the natural side, the ecology. Can we apply some of the models and methods that work good in the natural world to describe economic processes? Would these models then work for ecological economics? In many cases, the answer is yes. We can use population models to describe the dynamics of human populations. We can try to mimic some of the well-known properties of the market economy, such as the dynamics of supply, demand and price. However, we immediately realize that the transition regimes are quite difficult to reproduce. Whereas classic economics operates in the margin, we start considering some substantial changes in the system. This turns out to be somewhat hard to model. Some simple qualitative models can help us to understand processes embedded in our socio-economic and political systems. For example, we can explore how lobbying works to promote big corporations, and how this can allow such corporations to "rule the world." We can even combine some of the processes from the socio-economic field with natural capital and try to consider scenarios of sustainable development. Analyzing these integrated ecological and economic systems, we find a new meaning in the model time-step. It can be related to the efficiency of the decision making process, since this is the time over which the system reacts to change, the time over which processes are updated in the model. If the rates of processes in the system grow, it is essential that the time-step decreases – otherwise the growing system is likely to crash. Similarly, simple analysis of the peak oil phenomenon gives us some insight into the possible future of the end of cheap oil. It seems likely that in the global scale, where we do not have easily available substitutes, the trajectory of oil extraction may extend somewhat further than the peak at one-half of extracted resource. However, the following crash will be steeper and harsher. This could be avoided if sufficient investment were piped into alternative energy resources early enough, while fossil resources are still abundant. Finally, we will look at a different class of models; those that are used to study the dynamics in the global scale. These models contain much information about different processes, and should be treated as knowledge bases of a kind. Some scenarios of futures and applications to ecosystem services are also described here.

Keywords

Population dynamics, natality, mortality, migration, Canada, Malthus, age cohorts, population pyramid, population senescence, Social Security crisis, supply and demand, price, corporations, competition, subsidies, carrying capacity, TerraCycle, Miracle-Gro, lobbying, liquid coal, sustainable development, investment, production, chaos, fossil fuel, non-renewable, biofuel, cheap oil, Hubbert curve, Critical Natural Capital, Energy Return on Energy Invested, alternative energy, conservation, global dynamics, ecosystem services, scenario, futures.

7.1   Demographics

We have already considered several population models earlier in this book. Modeling a human population may be quite similar to modeling a population of woozles, as long as we have the same information about the factors that affect the population dynamics. In most cases, what we need to consider are primarily the growth due to births (natality), the decline due to deaths (mortality), and change due to in- and out-migration.

Consider, for example the data that are available at the Statistics Canada web page (see Table 7.1). This table presents the dynamics of the population of Canada over the past century (in thousands). Based on those data, we have estimated and added to the table the *per capita* natality and mortality rates.

A simple Stella model can be put together based on this data. Let us assume first that there is no migration, and formulate the model of exponential growth with varying birth and death coefficients:

$$\frac{dx}{dt} = (b(t) - m(t))x$$

where x is the population size, $b(t)$ is the birth rate and $m(t)$ is the death rate. Using the "To Graph" option in Stella, it should be easy to insert the data regarding the time-dependent birth and death factors into the Stella model and run it. (Actually, it is not as easy as it should be. Because of a bug in some versions of Stella, it is impossible to copy and paste the numbers from the Excel file column into the Graph description in Stella. For some reason this operation supports only three digits, and all the numbers that are larger than that will be split into two lines. It is important to be aware of this, since it may occur on a line that is not visible in the opened window and therefore all the graph data may be shifted and treated incorrectly. It seems to be much easier to do it in Madonna – so maybe that is how we will do it next time.)

We can either put together the model ourselves, or download it from the book website.

Figure 7.1 gives a comparison of a model run with the data for the total population numbers. The model seems to perform quite nicely for the first 11 decades, but then it consistently underestimates the population growth. If we look at the difference between in- and out-migration in the table, we can see that it has a pronounced

Table 7.1	Dynamics of the population of Canada over the last century							
Period	Census at end	Total growth	Births	Deaths	Immigration	Emigration	Births/ind./year	Deaths/ind./year
1851–1861	3,230	793	1,281	670	352	170	0.04	0.021
1861–1871	3,689	459	1,370	760	260	410	0.037	0.021
1871–1881	4,325	636	1,480	790	350	404	0.034	0.018
1881–1891	4,833	508	1,524	870	680	826	0.032	0.018
1891–1901	5,371	538	1,548	880	250	380	0.029	0.016
1901–1911	7,207	1,836	1,925	900	1,550	740	0.027	0.012
1911–1921	8,788	1,581	2,340	1,070	1,400	1,089	0.027	0.012
1921–1931	10,377	1,589	2,415	1,055	1,200	970	0.023	0.01
1931–1941	11,507	1,130	2,294	1,072	149	241	0.02	0.009
1941–1951	13,648	2,141	3,186	1,214	548	379	0.023	0.009
1951–1961	18,238	4,590	4,468	1,320	1,543	463	0.024	0.007
1961–1971	21,568	3,330	4,105	1,497	1,429	707	0.019	0.007
1971–1981	24,820	3,253	3,575	1,667	1,824	636	0.014	0.007
1981–1991	28,031	3,210	3,805	1,831	1,876	491	0.014	0.007

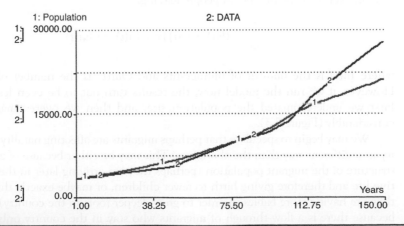

Figure 7.1 Modeling population dynamics with no migration.

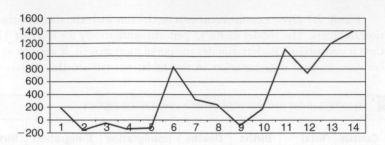

Figure 7.2   Net migration or difference between immigration and emigration rates in Canada.
There is a substantial increase of immigration in the second half of the twentieth century, which explains why it is hard to match the data without taking migration into account.

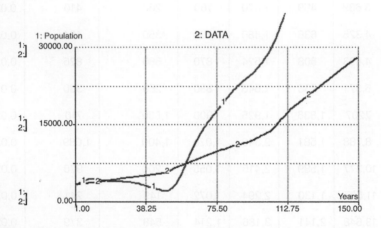

Figure 7.3   Modeling population with migration included.
Actually there is an error in this model. Can you figure out what it is?

growth trend over the years (Figure 7.2). It becomes especially large over the past five decades, which quite clearly matches the period when our model started to fail.

It seems to make perfect sense to bring the migratory processes into the picture and include them in the model. The simplest way is just to add the incoming population and subtract the number of people leaving:

$$\frac{dx}{dt} = (b(t) - m(t))x + In(t) - Out(t)$$

where $In(t)$ is the number of immigrants and $Out(t)$ is the number of emigrants. However, if we run the model now, the results turn out to be even less satisfying. First we underestimated the population size, and then we overestimated it quite considerably (Figure 7.3).

We may begin to speculate that perhaps migrants are affecting natality and mortality in a different way than the aborigines. This may be either because of a specific age structure of the migrant population (perhaps they are arriving later in their reproductive life and therefore giving birth to fewer children, or maybe exactly the opposite – they are having more babies in order to grow deeper roots in the country), or perhaps because there is a flow-through of migrants who stay in the country only for a short

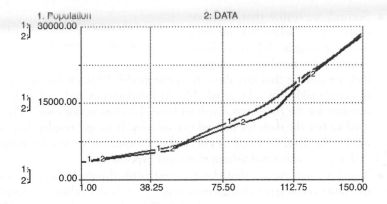

Figure 7.4 Modeling population with migration after calibration with data (or after fixing the error).

period of time without dying or giving birth. We may assume that migrants participate in the population dynamics with certain coefficients:

$$\frac{dx}{dt} = (b(t) - m(t))x + a_1 In(t) - a_2 Out(t)$$

and then use these coefficients, a_1 and a_2, to calibrate the model. By the "educated guess" procedure, we will easily derive that with $a_1 = a_2 = 0.1$ we get a pretty close fit to the data (Figure 7.4). However, we may also go back to Table 7.1 and take a closer look at the numbers, and this shows that the immigration and emigration flows are presented there over 10-year time-steps. This means that in the Stella model we should be using approximately one-tenth of those flows when running the model on a yearly time-step. What a nice coincidence, with the $a_1 = a_2 = 0.1$ calibration result that we have obtained above. Sometimes the model works really well in testing data for internal contradictions and helps us to find errors in our line of reasoning. It is important, though, to make sure we check all our numbers before starting to make new assumptions to explain the inconsistencies in our results!

Exercise 7.1

Try to improve the calibration of the population model by further modifying the parameters a_1 and a_2. You can either continue the trial-and-error exercises or upload the model into Madonna and try the curve-fitting in there. Do not forget to add the Error function to the model, since visual comparison becomes quite hard once you get really close to the optimal solution. Is there a better combination of parameters than $a_1 = a_2 = 0.1$? If there were a better combination of parameters, how would you explain that, bearing in mind that actually they should be equal to 0.1?

By performing this simplified analysis, we have already gained some insight into the population dynamics.

- First, we can see that external factors seem to have affected population dynamics in Canada over the past five decades quite considerably, and that migration patterns need to be taken into account.

- Second, we realize that we have to be careful when adding data into our models. When we have a rate (like natality or mortal-

> Mind the units. If the model time-step is "per day," make sure your data are not "per year."

ity), we can use that rate directly in the model. When we have the actual number of people added or subtracted (like in immigration and emigration), we need to account for the time-step involved (over 10 years in our case). The model can be used to test the data for mutual consistency. If we get results that do not seem to match the data, we need to check the information we use for consistency before making any additional assumptions and hypotheses.

- Third, we may notice that there are established patterns in how the natality and mortality coefficients change over time. They both decline at a similar rate, maintaining a ratio of approximately 1:2. Why does this happen, and what are the other factors that may be related to this? It does not seem to be obvious from looking at the available data. We might need to bring some other information into the model – for example, look at the economic development of the country over this period.

Exercise 7.2

Consider some additional processes, such as economic growth, that may be responsible for the decline of birth rate and mortality. Add them to the Stella model so that the graphs can be replaced by some endogenous processes in the system. Describe your assumptions and results.

Some very basic experiments with the population model immediately show us what exponential growth is, and how populations tend to take that trajectory. As long as the birth rate is higher than the death rate, the population grows – and grows quite quickly. This is what made Malthus worry as early as 1798, when he wrote "The power of population is so superior to the power of the earth to produce subsistence for man, that premature death must in some shape or other visit the human race." The world's current population growth rate is about 1.14 percent, representing a doubling time of 61 years. We can expect the world's population of 6.5 billion to become 13 billion by 2067 if current growth continues. The world's growth rate peaked in the 1960s at 2 percent, with a doubling time of 35 years. Most European countries have low growth rates. In the United Kingdom, the rate is 0.2 percent; in Germany it's 0.0 percent, and in France it's 0.4 percent. Germany's zero rate of growth includes a natural increase of −0.2 percent. Without immigration, Germany's population would be shrinking, as is that of the Czech Republic. Many Asian and African countries have high population growth rates. Afghanistan has a current growth rate of 4.8 percent, representing a doubling time of 14.5 years!

We can also see from the model that it's not the growth or death rate that really matters, but rather the difference between them. In most isolated societies, higher birth rates correspond to higher death rates and *vice versa*. This helps to keep population growth more or less under control. In developed countries, death rates have decreased quite dramatically due to improved life quality, better medical care, lower infant mortality, etc. However, at the same time birth rates have started to decrease as well as a result of better education, especially for women; also, a greater number of children is no longer considered a prerequisite for support of the elderly. As a result the net growth has decreased almost to zero, as in Germany. Problems in developing countries begin when certain natural death-reduction programs, such as vaccination, arrive. The death rate then starts to decrease while the birth rate remains the same and, in Paul Ehrlich's terms, a population bomb explodes.

Does this mean that we should stop humanitarian and medical aid to underdeveloped African countries? Probably not. However, it does mean that interfering with complex systems such as human populations is dangerous without considering all the aspects and relationships. If we decrease mortality, we need to make sure that we also administer programs that will decrease natality. Otherwise populations will grow exponentially, wiping out natural resources, creating civil unrest and refugee problems. Eventually population growth will adjust, but this might be a long and quite an ugly process, involving crises, bloodshed and relocation.

Another possible explanation for the discrepancies that we observe is the age structure of the population. In every population, different age groups function differently. For example, there is a certain age range during which women are most likely to give birth. Migration may also occur more or less often in different age groups — for example, older people are less likely to migrate. These alterations may affect the overall population dynamics. Let us build a model and see what changes we might observe.

The information regarding population age structure is usually presented in the form of population pyramids. For Canada, you can go to http://www.statcan.ca/english/kits/animat/pyone.htm to see how the population structure changed in that country over the twentieth century. It starts with a distribution that really looks like a pyramid, in 1901.

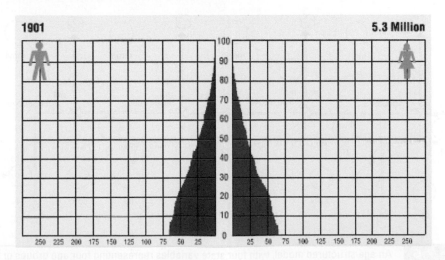

However, as time goes by the distribution graph becomes quite distorted, representing the arrival of the baby-boomers in the 1950s and clearly showing the trend towards a predominantly older population.

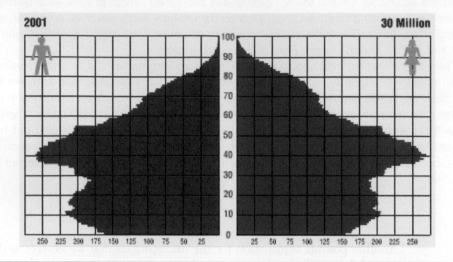

Consider four age groups: children (aged 0–15 years), adults1 (16–40 years), adults2 (41–65 years) and retired adults (over 66 years of age). The goal of looking at these age groups is two-fold. First, we want to separate the childbearing group (adults1); secondly, we wish to distinguish between the working adults (adults1 + adults2) and the rest (the non-working population). In some cases we may need to consider more age groups (also called cohorts), but usually it makes sense to differentiate only between the ones that have different functions. After all, why make the model more complex?

The Stella diagram for this model can be seen in Figure 7.5. We have four state variables with transfer functions, t1, t2 and t3. Each transfer function should

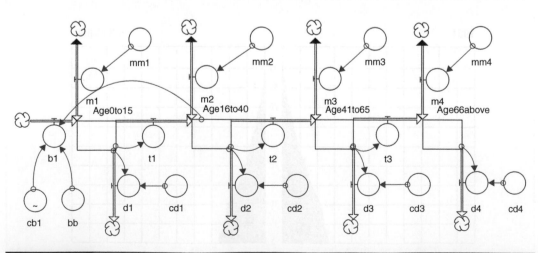

Figure 7.5   An age-structured model, with four state variables representing four age groups or cohorts.

be designed in such a way that it moves all the individuals in one age group to the next over the time period that an individual stays in the age group. For example, if a baby is born and put in the Age0–15 group it will stay in this group for the next 15 years and then be transferred to the next group, Age16–40. This means that every year one-fifteenth of the number of individuals in this age group will be moved to the next group. Therefore, t1 = (Age0–15)/15. Similarly, t2 = (Age16–40)/25, and t3 = (Age41–65)/25. Individuals stay longer in these groups, and therefore only one-twenty-fifth of the group is transferred to the next group annually.

The other processes are similar to those in the standard population model considered above, except that the birth flow is proportional only to the number of individuals in the second childbearing age group, Age16–40, and not the total population, as we saw before. If we want to use the birth rate from the data set that we have, we need to add a scaling factor that will rescale the birth rate for the whole population to the birth rate controlled only by the Age16–40 age class. Similarly, we need to figure out how the total deaths will be distributed among the different age groups.

Common sense dictates that death rates in the younger age groups should be smaller than the population wide average. The death rate in the Age41–65 group should be close to the average, while the death rate in the Age66 above group should be considerably higher than the average. Assumptions should also be made to distribute the overall migration data (= immigration − emigration) among the various age classes.

Overall, we end up with the following set of equations. Definitions of the four age groups are almost identical:

$$[\text{transfer from the younger age group}] - [\text{death}] + [\text{net migration}]$$

In addition, there is birth in the first age group.

[Age0to15(t) = Age0to15(t − dt) + (b1 + m1 − d1 − t1) * dt
INIT Age0to15 = 600
INFLOWS:
b1 = cb1*bb*Age16to40
m1 = mm1
OUTFLOWS:
d1 = cd1*Age0to15
t1 = Age0to15/15

Age16to40(t) = Age16to40(t − dt) + (t1 + m2 − d2 − t2) * dt
INIT Age16to40 = 2100
INFLOWS:
t1 = Age0to15/15
m2 = mm2
OUTFLOWS:
d2 = cd2*Age16to40
t2 = Age16to40/25

Age41to65(t) = Age41to65(t − dt) + (t2 + m3 − d3 − t3) * dt
INIT Age41to65 = 300
INFLOWS:
t2 = Age16to40/25
m3 = mm3

OUTFLOWS:
d3 = cd3*Age41to65
t3 = Age41to65/25

Age66above(t) = Age66above(t − dt) + (t3 + m4 − d4) * dt
INIT Age66above = 130
INFLOWS:
t3 = Age41to65/25
m4 = mm4
OUTFLOWS:
d4 = cd4*Age66above
bb = Total/Age16to40

The following is the distribution of deaths among different age groups:
cd1 = cdd1*C_mort
cd2 = cdd2*C_mort
cd3 = cdd3*C_mort
cd4 = cdd4*C_mort
cdd1 = 0.4
cdd2 = 0.75
cdd3 = 1.2
cdd4 = 1.9

Similarly net migration is distributed among age groups:
M = (In-Out)/10
mm1 = cm1*M
mm2 = cm2*M
mm3 = cm3*M
mm4 = (M-mm1-mm2-mm3)
cm1 = 0.1
cm2 = 0.5
cm3 = 0.2

Some totals and more data for calibration purposes:
Total = Age0to15 + Age16to40 + Age41to65 + Age66above
Total_died = d1 + d2 + d3 + d4
births_p_y = Births/10
deaths_p_y = Death/10

Birth, death rate as well as immigration and emigration numbers are defined as graphic
functions based on the data in Table 7.1.
cb1 = GRAPH (TIME)
C_mort = GRAPH (TIME)
In = GRAPH (TIME)
Out = GRAPH (TIME)

By tweaking the parameters of death rates in different age groups, we can get the
model to closely represent our data. Figure 7.6 represents the dynamics of the total
population. Also we can display the dynamics of individual age groups (Figure 7.7).
Contrary to our expectations, the distribution of migration among age groups does not
play a major role in the system. As long as the total numbers of migrants is correct, we
get almost similar results.

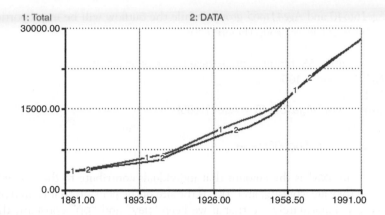

Figure 7.6   Model results for total population in the age-structured model after calibration of birth- and death-rate parameters.

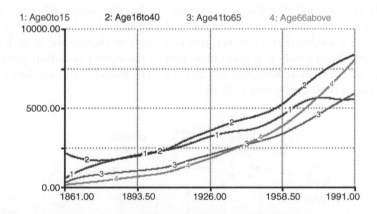

Figure 7.7   Dynamics of different age cohorts in the age-structured model.
Towards the end of the simulation, the numbers in the elderly classes start to grow more rapidly than in other cohorts. A time-bomb for a generational storm is set, unless some drastic measures are taken.

What we can do next with this model is explore the daunting problem of population senescence that is currently looming over most of the societies in developed countries. With the growth of affluence and education, people are less inclined to have children. As a result birth rates are decreasing, while the advances in medicine and health care are decreasing the mortality rate. These changes may not affect the total population numbers (after all, we are decreasing both the inflow and the outflow for the stock of the population number), but they will have a substantial effect on the shape of the age distribution. The number of old and retired people keeps growing, causing an increasing burden on the welfare system. At the same time, the number of people of working age becomes relatively smaller, so there are fewer people contributing to the support of the retirees in the system.

In our Canadian model we already have the decline of birth and death rates, and we can already see that numbers in the eldest age group, *Age66above*, are steadily increasing, making this age group dominant in the population. Let us add the social security system into our model. This can easily be performed by introducing another stock in the model that will have an inflow generated by payments from the

Age16to40 and *Age41to65* groups, while the outflow will be in proportion to the size of the *Age66above* group (Figure 7.8):

```
SS_fund(t) = SS_fund(t − dt) + (taxes − pensions) * dt
INIT SS_fund = 100
INFLOWS:
taxes = pay*(Age16to40 + Age41to65)
OUTFLOWS:
pensions = pp*Age66above
```

The "pay" is the amount that individuals contribute to the Social Security fund while they are working, and "pp" is the size of the pension that retired people receive. We can immediately see that if we keep "pay" and "pp" constant, the "SS_fund" will go bankrupt some time in the near future (Figure 7.9). In this model we have assumed that the social security system has been in place since the beginning of our data set in 1861, that the payments to and from the fund have been constant over these years, and that the age of retirement has also remained constant. This is certainly not realistic, and for a better model we should include all these historical data in our consideration. However, this is unlikely to change the overall trend because, again, qualitatively it is quite clear that as the elder population group grows in size we will need more resources to support it. The model is an excellent tool to quantify some of these qualitative notions.

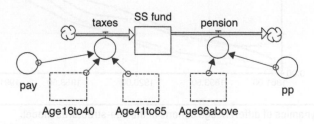

Figure 7.8   A simple submodel of a social security fund.

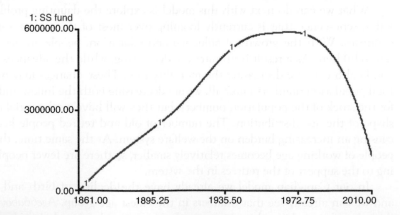

Figure 7.9   Dynamics of the social security fund when population is senescing and there are less people who work and more people who receive pensions.

The problem seems to be quite serious for the USA. According to Kotlikoff and Burns (2005), by mid-century the US centenarian population will exceed 600,000. That's ten times the number of centenarians around today. In 1900, life expectancy at birth was 47 years and the median age of all Americans (half younger, half older) was only 22.9 years. Only 4.1 percent of the population was aged 65 or more. Today, life expectancy at birth is about 76 years – a gain of 29 years – and life expectancy at 65 is now 17 years, up from 12 years in 1900. Gains in life expectancy at 65 seem to be accelerating. As a consequence, those aged 65 and over made up 12.4 percent of the population by 2000 – nearly double the 6.8 percent of population aged under 5 years. At the same time birthrates plummeted from well over 2.1 children per couple (the long-term replacement rate for the population), and are now hovering near the replacement rate. As a result there are more people retiring now, while the number of children coming of age and joining the workforce isn't nearly as large. The forces that would expand the younger (and working) population paying Social Security and Medicare taxes are in reverse, and the result is a kind of perfect demographic storm. Back in 1950, the number of workers per Social Security beneficiary was 16.5; by 2000, this had dropped to 3.4. In the process, most workers started paying more in employment taxes than they pay in income taxes, as employment tax rose fivefold. The wages subject to Social Security tax rose as well, rising from $3,000 in 1950 to $87,000 in 2003.

Between now and 2030 we'll have the last big surge: the retirement of the boomers. By then, we'll be close to having only two covered workers per beneficiary. Instead of having 16 workers chipping in to support each senior citizen, there will only be 2. Where we had 35.5 million people age 65 and older in 2000, we'll have 69.4 million in 2030. During these 30 years, the dependency ratio – the ratio of those aged over 65 to those aged 20–64 – will rise from 21.1 percent to 35.5 percent.

According to Kotlikoff, the situation is so scary that he keeps referring to lower life expectancy estimates as "optimistic." What used to be our goal of increased longevity has suddenly become a huge danger for the society. Interestingly, we have been getting quite similar projections with our very simplified model. Then numbers will need to be changed to describe the US situation, but there is good reason to expect that, qualitatively, we will be getting exactly the same scenario that Kotlikoff portrays.

Exercise 7.3

How is it possible to prevent the Social Security fund from crashing? Consider changing the rate of contributions, lowering the pension rate and delaying the retirement age. What should be the retirement age to make the fund sustainable under existing birth and death trends?

As before, let us take a closer look at some of the equations we are dealing with. Without migration, we can write the system equations as follows:

$$\frac{dx_1}{dt} = bx_2 - t_1x_1 - d_1x_1$$

$$\frac{dx_2}{dt} = t_1x_1 - t_2x_2 - d_2x_2$$

$$\frac{dx_3}{dt} = t_2x_2 - t_3x_3 - d_3x_3$$

$$\frac{dx_4}{dt} = t_3x_3 - d_4x_4$$

Here, $x_1 = Age0to15$, $x_2 = Age16to40$, $x_3 = Age41to65$ and $x_4 = Age66above$; b is the birth rate, t_i are the transfer coefficients, and d_i are the mortality rates – just as in the Stella model above. When running Stella, we did have some problems identifying the correct values for the coefficients. Sometimes the variables were growing too fast, or alternatively they were diminishing to zero. Are there any relationships that we should keep in mind when looking for suitable combinations of parameters?

First, let us check for an equilibrium. Making the left-hand side of the equations equal to zero, we get a system of algebraic equations:

$$0 = bx_2 - t_1x_1 - d_1x_1$$

$$0 = t_1x_1 - t_2x_2 - d_2x_3$$

$$0 = t_2x_2 - t_3x_3 - d_3x_3$$

$$0 = t_3x_3 - d_4x_4$$

The first equation yields: $x_1 = bx_2/(t_1 + d_1)$. Substituting this into the second equation, we get $(bt_1/(t_1 + d_1) - (t_2 + d_2)) \cdot x_2 = 0$. This means that we get an equilibrium only if $x_2 = 0$, which then automatically makes all the other variables equal to zero. Or if $(bt_1/(t_1 + d_1) - (t_2 + d_2)) = 0$, in which case x_2 can be any, and

$$x_1 = \frac{bx_2}{t_1 + d_1}, \quad x_3 = \frac{t_2x_2}{t_3 + d_3}, \quad \text{and} \quad x_4 = \frac{t_3x_3}{d_4} \qquad (7.1)$$

Neither of these states is interesting, since the first is trivial, when there is no population, and the second is extremely unlikely because it requires that there is an exact relationship between model parameters. Equality-type relationships are unrealistic for any real-world situations, where there will always be some uncertainty about model parameters and it is impossible to guarantee they will be exactly equal to some combination between other parameters, as we require in this case by asking that $bt_1/(t_1 + d_1) - (t_2 + d_2) = 0$.

However, this analysis is not without merit. What we can see is that when this condition does not hold and, say, $bt_1/(t_1 + d_1) > t_2 + d_2$, then $dx_2/dt > 0$. This means that in this case x_2 will be growing. Keeping in mind (7.1), we can see that all the other variables will also be growing. If, otherwise, $bt_1/(t_1 + d_1) < t_2 + d_2$, all the model variables will be declining. So we have found a simple condition that quickly tells us when the population becomes extinct and when it survives. Interestingly, none of the parameters from the third or fourth equations in the model are involved. Not surprisingly, this means that, for survival of the population, only the first two age groups matter. The remaining two are a tail that can be cut to any length. The population still persists, as long as the childbearing group is in place. Once it gives birth to progeny, it can disappear.

This simple analysis is quite helpful when looking for the right combination of parameters to make the model run. Instead of the trial-and-error method, we can identify certain parameter domains where the model behaves as we would want it to.

If we bring in migration, we get a slightly modified system of equations:

$$\frac{dx_1}{dt} = bx_2 - t_1x_1 - d_1x_1 + m_1$$

$$\frac{dx_2}{dt} = t_1x_1 - t_2x_2 - d_2x_2 + m_2$$

$$\frac{dx_3}{dt} = t_2 x_2 - t_3 x_3 - d_3 x_3 + m_3$$

$$\frac{dx_4}{dt} = t_3 x_3 - d_4 x_4 + m_4$$

Here, m_i are the net migration rates into the four age groups. They can be positive or negative. The other parameters are always positive. This time, we can see that there exists an equilibrium in the model: substituting $x_1 = (bx_2 + m_1)/(t_1 + d_1)$, which comes from the equilibrium in the first equation ($dx_1/dt = 0$), into the second equation at equilibrium, we immediately get a solution for x_2:

$$x_2 = \frac{m_1 t_1 + m_2 (t_1 + d_1)}{(t_2 + d_2)(t_1 + d_1) - bt_1} \tag{7.2}$$

This can be then substituted back into the equation for x_1, to produce

$$x_1 = \frac{bm_2 + m_1(t_2 + d_2)}{(t_2 + d_2)(t_1 + d_1) - bt_1} \tag{7.3}$$

Substituting x_2 into the third equation, we can calculate

$$x_3 = \frac{t_2 x_2 + m_3}{t_3 + d_3}$$

Then, similarly, this value for x_3 can be used to calculate

$$x_4 = \frac{t_3 x_3 + m_4}{d_4}$$

which follows from the fourth equation.

Obviously, these equilibria have to be positive. If the migration coefficients m_1 and m_2 are positive, it follows from (7.2) and (7.3) that when

$$b < \frac{(t_2 + d_2)(t_1 + d_1)}{t_1} \tag{7.4}$$

we have all the equilibria in the positive domain; otherwise we move into the negative domain. If this condition holds, the other two equilibria for x_3 and x_4 will also be positive.

If we now run the Stella model under these conditions, it appears that the equilibrium is stable: we can start modifying the initial conditions, and still will converge to the values that we have identified above. However, if the equilibrium moves into the negative domain, either when (7.4) no longer holds or when migration becomes negative, we get exponential growth or exponential decline patterns. Though analytical analysis can become quite cumbersome, without it it may be hard to figure out that the model can produce all three types of dynamics: exponential growth, exponential decline, or stable steady state. It all depends upon the parameters we choose.

We may once again conclude that looking at the equations can be quite helpful. Unfortunately, the algebra becomes rather tiresome even when we have only four equations and some fairly simple interactions. However, when we have many more equations and parameters, it is still important to run and rerun the model for as many combinations of parameter values and initial conditions as we can afford. This is the only way to attain confidence and understanding of the results we are producing.

Before we continue with some linked models of demography and economics, let us consider a few examples of simple economic and socio-economic models.

7.2   Dynamics on the market

Let us see if dynamic modeling is an appropriate tool to model some economic systems. Consider the basic demand–supply–price theory that is discussed in most classical books on microeconomics, and at a variety of web pages (e.g. http://hadm.sph.sc.edu/COURSES/ECON/SD/SD.html or http://vcollege.lansing.cc.mi.us/econ201/unit_03/lss031.htm).

In essence, we are looking at a system of two state variables, one representing the quantity of a given commodity (G) and the other one representing its price (P). In a market economy, the two are supposed to be determined by the relationship between supply and demand. Let us look at Figure 7.10 to see how we derive the relationship between price and the amount of commodity on the market. Suppose that the price of the commodity is set at p_1. In Figure 7.10A, we will graph the relationship between price and the amount of the commodity on the market; in Figure 7.10B, we will show the change in price over time. Let us first draw the graph of Supply. The Law of Supply states that the higher the price for a commodity, the more products will be offered by the producer on the market. So S should be an increasing function of P. (Note that, mathematically, this is somewhat dubious, since we have just replaced the independent variable in the graph. Nevertheless, this is the way economists do it.)

On the graph, we see that the quantity g_1 corresponds to the price p_1. This projects the first price point in Figure 7.10B. However, there is also the Law of Demand that states that the price of a commodity is inversely related to the amount demanded per time period. In our case, the Demand curve stipulates that at a quantity g_1 the commodity can be sold only at a price as low as p_2 (the second point in Figure 7.10B). With such high supply there is simply not enough demand to keep the price up, so competition among producers increases and they have to push the price down to sell all the g_1 stock that was produced. However, at price p_2 the Supply curve tells

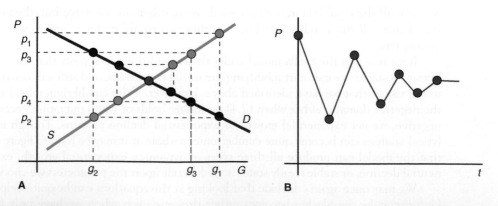

Figure 7.10   Converging of supply towards demand and equilibrating of price after several cycles. A. The demand (D) and supply (S) curves. B. Dynamics of price as defined by the demand and supply.

us that producers will only be willing to produce and ship to the market g_2 commodities. Producing these goods at such a price is not that profitable and lucrative for the producer, so only a few will remain, and they will produce much less.

Once again, projecting the amount g_2 to the Demand curve we realize that now the demand for the commodity is so high (there is an excess demand on the market) that it can sell at a price as high as p_3. For such a price the producer is once again eager to produce more and, according to the Supply curve, will deliver more commodities to the market. We continue this process, observing that the price and the amount of commodity gradually converge to a certain equilibrium state, when the price will be just right for the available quantity supplied, and the quantity supplied will match the amount demanded. This is what is called "market equilibrium." Economic theory considers that markets come to equilibrium in one shot – i.e. when both producers and consumers know exactly the equilibrium price and amount of a market good which will be sold on this market. However, in the real world it always takes some time for supply to adapt to demand and *vice versa*.

How can we describe this process in a dynamic model? Consider a system with two variables: P and G. According to the Supply Law, the production of the commodity G is in proportion to its price. According to the Demand Law, the consumption of the commodity is in reverse proportion to the price. Therefore, we can assume the equation for goods in the following form:

$$\frac{dG}{dt} = c_{g1}P - \frac{1}{c_{g2}P} \tag{7.5}$$

Based on similar considerations, we assume that the price increases in reverse proportion to the amount of the commodity available for consumption and decreases in direct proportion to this amount:

$$\frac{dP}{dt} = \frac{c_{p1}}{G} - c_{p2}G \tag{7.6}$$

We can find the equilibrium for this model by assuming that there are no changes in the system, so

$$0 = c_{g1}P - \frac{1}{c_{g2}P}$$

$$0 = \frac{1}{c_{p1}G} - c_{p2}G$$

The feasible (positive) equilibrium point is ($P = 1/\sqrt{(c_{g1}c_{g2})}$, $G = 1/\sqrt{(c_{p1}c_{p2})}$). If we now put these equations into a Stella model and run it, we find that the equilibrium point is not stable. Instead, if we start anywhere away from the equilibrium point we generate an economic cycle that is quite similar to that seen in the price–commodity oscillations above, except that these oscillations do not dampen out (Figure 7.11).

For any initial conditions and any combination of parameters (except the ones that crash the model, taking the trajectories to the negative quadrangles), the trajectory continues to cycle around an ellipsoid, with no indication of convergence to the stable state. It is not quite clear how to modify the model in such a way that the trajectories lead to equilibrium. Apparently the supply/demand curves that we are choosing (see 7.5 and 7.6) are symmetrical, so we keep cycling around the equilibrium

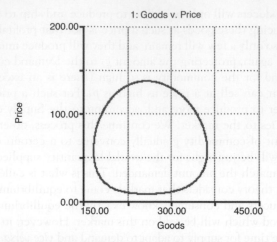

Figure 7.11   Cycles in the Commodity–Price model. Dynamics in the phase plane (P,G).

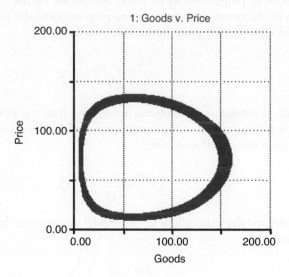

Figure 7.12   Converging cycles in the modified Commodity–Price model. Dynamics in the phase plane (P,G).

without approaching it. One possible modification of the model that seems to make it converge is if we describe the commodity dynamics as:

$$\frac{dG}{dt} = c_{g1}P^{1.5} - \frac{1}{c_{g2}P^{0.5}}$$

In this model, with $c_{g1} = 0.01$, $c_{g2} = 0.02$, $c_{p1} = 0.0055$, $c_{p2} = 0.05$, we can generate a slowly converging trajectory. Note that it took 6,000 iterations to generate the curve shown in Figure 7.12. Besides, this converging model appears to be structurally unstable, since even slight modifications in the formulas used or in the parameter set result in non-convergence or a crash.

Exercise 7.4

Put together the Price–Goods model in Stella, or download it from the book website. Try to find another function or set of parameters that would make it converge faster.

Let us consider another formulation for the same system. Instead of looking at just the price and commodity, let us consider three variables: price (P), supply (S) and demand (D). The

> *If changing parameters doesn't help, change the equations. Perhaps there was something wrong in the assumptions.*

supply is assumed to be somewhat identical to the amount of commodity considered earlier. The demand will be treated as the reverse of supply. The Stella equations can then be as follows:

Demand(t) = Demand(t − dt) + (D_up − D_down) * dt
INIT Demand = 90
INFLOWS:
D_up = 1/C_d1/Price
Just as in the previous model, the higher the price of the commodity gets, the slower the demand grows.
OUTFLOWS:
D_down = C_d2*Price
The higher the price the faster the demand will actually decrease.
Price(t) = Price(t − dt) + (P_change) * dt
INIT Price = 100
INFLOWS:
P_change = C_p * (Demand-Supply)

If the demand exceeds supply, then the commodity becomes scarce and the price goes up. It goes down if more of the commodity is supplied than is demanded.

Supply(t) = Supply(t − dt) + (S_up − S_down) * dt
INIT Supply = 110
INFLOWS:
S_up = C_s1 * Price
There is more incentive to produce a commodity if its price is high.
OUTFLOWS:
S_down = 1/C_s2/Price
If the price is high the commodity is less likely to be consumed.
C_d1 = 0.008
C_d2 = 0.01
C_p = 0.01
C_s1 = 0.01
C_s2 = 0.008

When C_s1 = C_d2; C_s2 = C_d1 we get dynamics, which are identical to those previous: stable oscillations for all initial conditions. However, if these conditions do not hold then the dynamics are different. While price is still displaying stable oscillations,

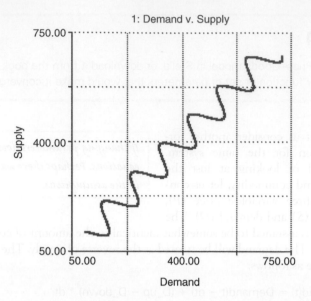

Figure 7.13   Oscillating and growing (or declining) dynamics in the (D,S) phase plane.

Supply and Demand start to oscillate along an increasing (if C_d1 < C_s2 or C_s1 > C_d2) or decreasing (if C_d1 > C_s2 or C_s1 < C_d2) trajectory (Figure 7.13). This is a very crude analysis of the system; however, it already shows that by adding another variable to the system we have modified the behavior quite significantly and generated some new previously unavailable trajectories. We can now represent a situation when both the demand and supply change in a similar way, either growing or decreasing. The price dynamics, however, remain unchanged. Whether this corresponds to reality or not is yet to be figured out. We still cannot make the system converge to an equilibrium state.

Let us further modify the system assuming that Supply and Demand can also interact directly, not necessarily only by means of Price. For the outflow part in the dynamics of S and D we will use the same assumption as above – that is, that the price P will define their value. However, we will now assume that the growth of supply S is decided directly from the knowledge regarding the demand D for the commodity, without the price dynamics being involved. Similarly, the growth of demand D will be directly determined by the supply of the commodity, and will be in reverse proportion to this supply. As a result, we will get the following system of equations:

$$\frac{dD}{dt} = \frac{1}{c_{d1}S} - c_{d2}P$$

$$\frac{dP}{dt} = c_p(D - S)$$

$$\frac{dS}{dt} = c_{s1}D - \frac{1}{c_{s2}P}$$

for the model with direct effects between Supply and Demand.

We can either put together this model ourselves, or download it from the book website.

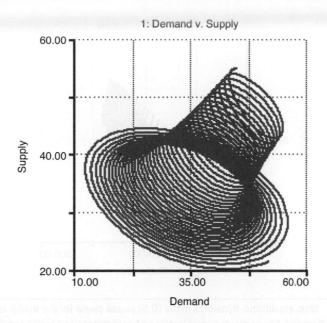

1: Demand v. Supply

Figure 7.14 Dynamics in the (D,S) phase plane for the model with direct effects between Supply and Demand.

By just playing with the Stella model it would be hard to find the equilibrium in this model; however, some simple calculations with the equations will show that if $c_{s1}c_{s2} = c_{d1}c_{d2}$, then there is an equilibrium for any $S = 1/(c_{d1}c_{d2}P)$ and $D = S$.

However, the equilibrium is unstable; if the initial conditions are displaced even slightly, we embark on a spiraling trajectory like the one in Figure 7.14. This eventually brings one of the variables to zero and crashes the model. Some other interesting regimes can be obtained by playing with the parameters and initial conditions. For instance, there is a trajectory (Figure 7.15) that starts on a growing trend but then for some reason reverses and brings the system back downwards towards an inevitable crash. It is yet to be figured out whether this kind of behavior may be found in any real-life economic systems. Most likely, this is quite irrelevant to a real economy.

We still cannot get any closer to the type of dynamics that the economic theory assumes for our system. We have already generated several models that seem to comply quite well with our assumptions about the system; they have produced a wide variety of dynamics, but we still cannot get on the converging path that we are trying to model. Let us give it another try and build yet another model.

Let us further shorten the information links and connect Supply and Demand directly, with Price generated only as a product of the relationship between the two. Suppose there is some direct interaction between Supply and Demand that is not mediated by price. Indeed, we know that if we are offered one glass of water it may have a very high (perhaps even infinite) value for us and will be in very high demand. When we get the second glass, we will probably also take it with thanks. After the fourth, fifth and sixth glasses, our interest will quickly decrease and even become negative. We will no longer want any more water; our demand will become negative (we may even want to throw up that water). This is what economists call diminishing marginal utility. Perhaps we can assume something similar

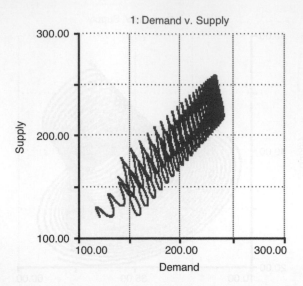

Figure 7.15 Non-equilibrium dynamics in the (D,S) phase plane for the model with direct effects between Supply and Demand. For particular combinations of parameters and initial conditions we may get some weird trajectories. Here $S = D = P = 120$; C_d1 = 0.008, C_d2 = 0.01, C_p = 0.1, C_s1 = 0.01, C_s2 = 0.01.

for the whole market scale, and formulate a Stella model with the following set of equations:

Demand(t) = Demand(t − dt) + (Dgrowth) * dt
INIT Demand = 120
INFLOWS:
Dgrowth = 1/C_d1/Supply-C_d2*Supply
Price(t) = Price(t − dt) + (Pgrowth) * dt
INIT Price = 100
INFLOWS:
Pgrowth = C_p*(Demand-Supply)
Supply(t) = Supply(t − dt) + (Sgrowth) * dt
INIT Supply = 90
INFLOWS:
Sgrowth = C_s1*Demand*(1-Supply/Demand)
C_d1 = 0.009
C_d2 = 0.02
C_p = 0.01
C_s1 = 0.01

As you may see, we have Demand growing in reverse proportion to Supply, and decreasing in proportion to Supply. We also assume that Supply grows in proportion to Demand as long as Supply is less than Demand. When Supply overshoots and becomes larger than Demand, it starts to decrease. For Price, we assumed that it grows if Demand is larger than Supply and vice versa.

A quick analysis of the model equations shows that there is an equilibrium $S = 1/\sqrt{c_{d1}c_{d2}}$: $D = S$. P will also stabilize, but it is hard to say where. Running the Stella

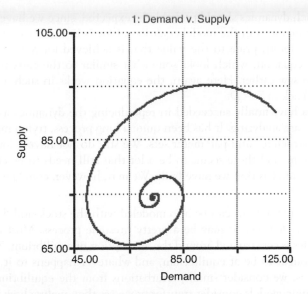

Figure 7.16   Stable focus in the Demand–Supply model.
Dynamics in the (D–S) phase plane.

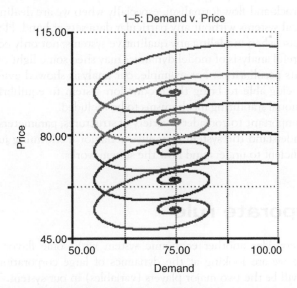

Figure 7.17   Dynamics in the Demand–Price phase plane.
The equilibrium for demand and supply is independent of initial conditions; however, the price equilibrium is decided by the initial conditions for price.

implementation, we see that we get a stable focus (Figure 7.16); after a number of oscillations the trajectories equilibrate at one point in the (S,D) plane. The equilibrium is stable; no matter how we modify the initial conditions, we still arrive at the same point in the (S,D) plane or return to the same line in the (D,P) plane (Figure 7.17). This is still not quite a perfect solution, since the equilibrium price depends upon the initial conditions that we chose for the price.

Such dynamics should probably be expected, since we have built in two stabiliz-ing formulations in the model equations. One is in the Price equation, which always tends to return price to the value that is achieved for $S = D$. The other is in the Supply equation, which looks somewhat similar to the carrying capacity formaliza-tion we saw earlier. Here again, the equation works in such a way that S is always driven back to $S = D$.

We have finally succeeded in reproducing the dynamics assumed in the system that we are analyzing. It has been quite a long process, trying numerous descriptions, model structures and parameter sets. We still do not have a lot of understanding of the system, and there seems to be a lot that still needs to be checked and explored with the models that we have built. We may, however, conclude that:

- Economic systems can be also modeled with the stock-and-flow formalism used in Stella. However, it may be a pretty tiresome process. Most of conventional eco-nomics is constructed around the assumption of equilibrium. The economic system is thought to be at equilibrium, and whatever happens to it is "at the margin," – that is, we consider small perturbations from the equilibrium. In contrast, most dynamic models consider transfer processes that analyze how to reach equilibrium, or how to jump from one equilibrium state to another.
- The systems dynamics language is not very well suited for conventional economic analysis. The language of economics may be somewhat difficult to translate into the stock-and-flow formalism, especially when we are dealing with qualitative the-oretical systems without any particular data sets at hand. However, this is probably the case when modeling any qualitative systems, not only economic ones.
- A careful analysis of model dynamics may shed some light on the system operation and its peculiarities. For example, our analysis showed evidence of price by itself not being able to bring the production system to equilibrium. We needed some additional stabilizing mechanisms to be included.
- It is important to consider a variety of structures, parameters and initial conditions to understand the system dynamics behavior. Performing just a few model runs is insufficient to understand how the system works.

7.3   Corporate rule

Let us consider another economic system with some flavor of social policy in it. Suppose we are looking at the dynamics of large corporations vs small businesses. These will be the two major players (variables) in our system. The main difference in how they operate is that there is hardly any competition between the corporations, which manage to divide their spheres of interests without employing market forces. The small businesses compete with each other and with the corporations. They also try to limit the growth of corporations by legislative means, which is also a non-market mechanism. However, corporations also compete with the small businesses for influ-ence upon the legislators. Let us see how such a system can develop in dynamic terms.

The variables of our system are the corporations (we refer to them as Bigs, B) and the small businesses (Smalls, S). We suppose that B and S are measured in their total value (say, in billions of dollars). Both B and S are assumed to grow exponentially, so that the larger their size the more their absolute growth will be. The Smalls are controlled by self-competition. We think that their total growth in value is mostly because of the growth in their numbers. Therefore, the larger the number, the higher the competition will be. Besides, the Smalls are suppressed by the Bigs: the larger the size of the Bigs, the more they limit the Smalls

$$\frac{dS}{dt} = bS - dB - cS^2$$

where b is the growth rate, c is the self-limitation coefficient and d is the rate of competition with the Bigs. The equation for the Bigs will be:

$$\frac{dB}{dt} = aB\left(\frac{1-B}{MM}\right).$$

Here, a is the Bigs growth rate and MM is a certain carrying capacity, some maximal limit set for the total size of the Bigs. In such a system with unfair competition the only result is gradual elimination of the Smalls while the Bigs reach their carrying capacity. If the carrying capacity MM is set at a high level, the Smalls are entirely wiped out. If MM is small, then coexistence is possible. This leads to a possible way to control the Bigs in a democratic society. The MM should be set at such a level that allows the Smalls to exist and develop. This should be done outside of the economic system, by a specific political process. The allowed size of the Bigs' development then determines the size of both the Bigs' and the Smalls' development.

Note that the so-called self-competition, in mathematical terms, is actually identical to carrying capacity. We can rewrite the equation for Smalls as:

$$\frac{dS}{dt} = bS - dB - cS^2 = bS\left(1 - \frac{S}{b/c}\right) - dB$$

Here, we have carrying capacity equal to b/c. Similarly, rearranging the equation for the Bigs, we can have:

$$\frac{dB}{dt} = aB - \left(\frac{a}{MM}\right)B^2$$

So the whole asymmetry of the system is in the fact that the Smalls are impacted by the Bigs (the $-dB$ term), while the Bigs feel no pressure from the Smalls.

This may be just about the right time to put these equations into Stella and start experimenting with the model. Just to make sure that we are on the same page, let us compare our Stella equations:

```
Bigs(t) = Bigs(t − dt) + (B_in − B_out) * dt
INIT Bigs = 100
INFLOWS:
B_in = a*Bigs
OUTFLOWS:
B_out = a*Bigs*Bigs*m
Smalls(t) = Smalls(t − dt) + (S_in − S_out) * dt
INIT Smalls = 300
```

INFLOWS:
S_in = b*Smalls
OUTFLOWS:
S_out = d*Bigs+c*Smalls*Smalls
a = 0.2
b = 0.2
c = 0.0003
d = 0.001
m = 1/MM
MM = 30000

The dynamics become much easier to understand if we check out the equilibria. Resolving the equilibrium equations, we get

$$B = MM$$

$$S_{1,2} = \frac{b \pm \sqrt{b^2 - 4dcMM}}{2c}$$

There are two points, and one of them seems to be stable. There could be a coexistence of the two, but note that the equilibrium for the Smalls can exist only if the expression under the square root is non-negative:

$$MM < \frac{b^2}{4dc}$$

We can see that for the Smalls to exist, they have to make sure that MM is sufficiently small (Figure 7.18). The decisions about such external controls are made in a political process, which may be assumed to be democratic. In this case, since the number of Smalls is always larger than the number of Bigs, we may hope that the control over MM will be successful. However, in reality the "democratic" process is largely influenced by lobbying, which in turn is defined by the amount of moneys spent to influence the politicians. Let us add the lobbying process into the model.

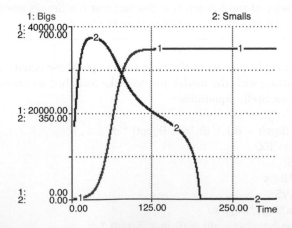

Figure 7.18   Crash of Smalls if carrying capacity established for Bigs is not small enough.

Suppose that both the Bigs and the Smalls spend a certain portion of their wealth on lobbying (*e* and *f*, respectively). The size of MM will be then determined by who spends more.

The equation for Smalls will now be:

$$\frac{dS}{dt} = bS - fS - dB - cS^2$$

For Bigs, we have:

$$\frac{dB}{dt} = aB\left(1 - \frac{B}{MM}\frac{fS}{eB}\right) - eB$$

Here, we have added the loss of wealth for lobbying (*fS* and *eB*) and modified the carrying capacity, assuming that it is now a function of *B/S* – it grows when *eB* > *fS* and declines otherwise. The model dynamics seem to be more complex now. By simply running the Stella model, it may be hard to figure out what is going on. If we have not put together the model ourselves, we can download it from the book website. Playing with the model, we find that there does not seem to be a state of coexistence any longer. If the Smalls prevail, in most cases by increasing their spending on lobbying, the Bigs can turn around the dynamics and wipe out the Smalls entirely, as in Figure 7.19.

A small company called TerraCycle has started to produce fertilizers from worm droppings. Organic waste is fed to worms and the worm poop compost tea is bottled as ready-to-use plant fertilizer, using soda bottles collected by schools and other charities. Started by college students, after five years in business TerraCycle was expecting to reach $6 million in sales in 2007, finally making some profit. This did not look good to the $2.2 billion giant Scotts Miracle-Gro Company, which has 59 percent of the plant food market.

Scotts claims that the two companies' products look similar and will confuse customers, because some TerraCycle plant foods have a green-and-yellow label with a circle and a picture of flowers and vegetables on it. Scotts also objects that TerraCycle says its plant food is as good or better than "a leading synthetic plant food."

Clearly, the expectation is that a small company will not be able to survive a major lawsuit and will go out of business. The Bigs compete with the Smalls.

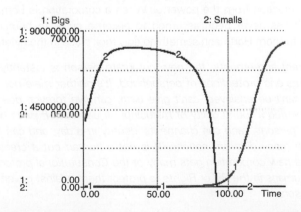

Figure 7.19   Crash of Smalls caused by Bigs increasing their lobbying efforts.

By looking at the model equations, we get some understanding of what is actually happening here. First, we can easily find one trivial equilibrium:

$$B = 0; \quad S = \frac{(b - f)}{c}$$

This one is feasible only if $b > f$, which makes perfect sense, since we would not normally spend more on lobbying (f) than we grow (b). If we use this equilibrium point with the Stella model, we can see that it is stable. The other, non-trivial equilibrium is just a little harder to find:

$$S = \frac{MM \, e(a - e)}{(af)}; \quad B = \frac{S(b - f - cS)}{d}$$

For this point to be in the positive quadrant, we need that $a \geq e$, and

$$MM \leq \frac{af(b - f)}{ce(a - e)} \tag{7.7}$$

Testing this point in Stella, we find that it is unstable. If we displace the Bigs from the equilibrium, making them slightly larger, they shoot to infinity, immediately eliminating the Smalls. If they are moved to below the equilibrium value, the Bigs fall to zero. This point is probably of little interest, except its feasibility condition turns out to be quite useful, since it tells us when to expect the model to switch from the die-off of the Bigs to the elimination of the Smalls. When (7.7) holds, the Smalls win; if (7.7) fails, then the Bigs are the champions. Note that for any reasonable MM, by increasing the lobbying efforts (e) the Bigs can easily make (7.7) fail. While MM is something discussed and probably voted for during the "democratic" process, e is set by the Bigs quietly and secretly.

Corporations were originally designed as a way to stimulate economic growth by protecting the entrepreneurs from some of the risks. Corporations were given certain rights; however, it has always been assumed that they are different from people. Apparently this is not quite so. When a person is born, he or she is protected by the power of the Constitution and has human rights. This is the way to protect the weak from the powerful. The same idea was applied to businesses that also required protection from the powerful. When a corporation is born, it too is assigned certain rights, but it has never been assumed to become equal to humans in these rights. However, according to Thom Hartmann something ugly took place in the nineteenth century.

"Under our current agreements, the new corporate person is instantly endowed with many of the rights and protections of personhood. It's neither male nor female, doesn't breathe or eat, can't be enslaved, can't give birth, can live forever, doesn't fear prison, and can't be executed if found guilty of misdoings. It can cut off parts of itself and turn them into new "persons," and can change its identity in a day, and can have simultaneous residence in many different nations. It is not a human but a creation of humans. Nonetheless, the new corporation gets many of the Constitutional protections America's founders gave humans in the Bill of Rights to protect them against governments or other potential oppressors:

* *Free speech, including freedom to influence legislation;*
* *Protection from searches, as if their belongings were intensely personal;*

- *Fifth Amendment protections against double jeopardy and self-incrimination, even when a clear crime has been committed;*
- *The shield of the nation's due process and anti-discrimination laws;*
- *The benefit of the Constitutional Amendments that freed the slaves and gave them equal protection under the law.*

Even more, although they now have many of the same "rights" as you and I – and a few more – they don't have the same fragilities or responsibilities, either under the law or under the realities of biology."

(Thom Hartmann, 2004)

As we can see from the model we have built, these are all developments towards loss of control over corporations and their total dominance over the governments and people.

Our conclusions from this simple study are:

- By allowing lobbying, we destabilize the system. The Bigs are pretty much forced into active lobbying, since otherwise they may perish entirely. However, their lobbying can result only in eventual total elimination of the Smalls. If the diversity of the Smalls and fair competition is really of value, some campaign reform is clearly in order.
- By creating simple models of processes and systems, we do not intend to analyze the in-depth dynamics that may occur. We are interested in investigating the possible behaviors, especially the ones that are caused by the dynamic structure of the system. These are most likely to show in other scales, as long as the main drivers and actors are similar.
- It certainly does not necessarily mean that a real-life political–economic system will behave in this way. There are plenty of processes and factors that we did not include in our model. However, the fact that the simple dynamic system with skewed competition and embedded self-controls behaves in a certain way gives us reason to expect similar behavior in real systems with similar properties.

As peak oil is approaching and energy costs increase, America is looking at alternative sources of energy. Renewable energy from sun, wind and biogas are small players. The coal industry is BIG. But producing energy from coal can only increase the amount of CO_2 in the atmosphere and further exacerbate global warming. Coal-to-liquid fuels produce almost twice the volume of greenhouse gases as ordinary diesel. In addition to the carbon dioxide emitted while using the fuel, the production process creates almost a ton of carbon dioxide for every barrel of liquid fuel. Besides, mining will devastate landscapes and water, and will hardly last for too long if instead of petroleum we start to burn as much coal. According to some reports, the 200-year supply of coal is vastly exaggerated; it is more likely to last only for 20–30 years if we start mining at the rate needed to replace the oil supply. Research at MIT points out some economic risks. It estimates that it would cost $70 billion to build enough plants to replace 10 percent of American gasoline consumption.

Nevertheless, prodded by intense lobbying from the coal industry, lawmakers from coal states are proposing that taxpayers guarantee billions of dollars in construction loans for coal-to-liquid production plants, guarantee minimum prices for the new fuel, and guarantee big government purchases for the next 25 years. Among the proposed inducements winding through House and Senate committees are: loan guarantees for 6–10 major coal-to-liquid plants, each likely to cost at least $3 billion; a tax credit of 51 cents for every gallon of coal-based fuel sold through to 2020; automatic subsidies if oil prices drop below $40 a barrel; and permission for the Air Force to sign 25-year contracts for almost a billion gallons a year of coal-based jet fuel. Coal executives say that they need government help primarily because oil prices are so volatile and the upfront construction costs are so high. Executives anticipate potentially huge profits. Gregory H. Boyce, Chief Executive of Peabody Energy, based in St Louis, which has $5.3 billion in sales, told an industry conference nearly 2 years ago that the value of Peabody's coal reserves would skyrocket almost tenfold, to $3.6 trillion, if it sold all its coal in the form of liquid fuels.

So the lobbying machine has kicked in. Coal companies have spent millions of dollars on the issue, and have marshaled allies in organized labor, the Air Force and fuel-burning industries like the airlines. Peabody Energy, the world's biggest coal company, urged in a recent advertising campaign that people "imagine a world where our country runs on energy from Middle America instead of the Middle East." Coal-industry lobbying has reached fever pitch. The industry spent $6 million on federal lobbying in 2005 and 2006 – three times what it spent each year from 2000 through 2004, according to calculations by Politicalmoneyline. com. Peabody, which has quadrupled its annual lobbying budget to about $2 million since 2004, recently hired Richard A. Gephardt, the Missouri Democrat who was House majority leader from 1989 to 1995 and a candidate for the Democratic presidential nomination in 1988 and 2004, to help make its case in Congress.

Do you think it will work? You bet.

Based on Edmund L. Andrews, "Lawmakers Push for Big Subsidies for Coal Process," *The New York Times*, May 29, 2007

7.4   Sustainability

There is a considerable number of publications about "sustainability." Most of the research on sustainability considers an integrated system of natural and man-made processes in an attempt to figure out the conditions for this system to persist in perpetuity.

Unfortunately, people tend to define sustainability in ways that suit their particular applications, goals, priorities and vested interests, and often use the term with no explicit evidence and recognition of the exact meaning being implied. Just like biodiversity, sustainability has become more of a political issue than a scientifically supported concept.

Most sustainability definitions originate from the relationship between humans and the resources they use. Originally the Bruntland Commission defined sustainable development as that which meets the needs of the present without compromising the ability of future generations to meet their own needs (WCED, 1987). Most of the later definitions are somewhat similar except that they started to apply to scales other than the global.

Wimberly (1993: 1) states that "to be sustainable is to provide for food, fiber, and other natural and social resources needed for the survival of a group – such as a national or international society, an economic sector or residential category – and to provide in a manner that maintains the essential resources for present and future generations." This already implies scales other than global.

Costanza (1992: 240) emphasizes the system's properties, stressing that "sustainability... implies the system's ability to maintain its structure (organization) and function (vigor) over time in the face of external stress (resilience)." Solow (1991) says that the system is sustainable as long as the total capital of the system is equal or greater in every next generation. Costanza and Daly (1992) argue that sustainability only occurs when there is no decline in natural capital. Whatever the flavor of the different definitions, there is one common component; all of them talk about maintenance, sustenance, continuity of a certain resource, system, condition and relationship, and in all cases there is the goal of keeping something at a certain level, of avoiding decline. This is also how Google's definition tool defines sustainability: as a state or process that can be maintained indefinitely, to keep in existence, to maintain or prolong, to use resources in a manner that satisfies current needs while allowing them to persist in the long term.

No wonder sustainability has become a welcome concept in the Western developed world. People are quite happy with what they've got, and it is a sweet idea to figure out how to preserve the *status quo* indefinitely. It is also no surprise that sustainability is not that easy a sell in developing countries, where people are much more interested in growth of their economies. Clearly such growth is an unsustainable goal, bearing in mind the limited resources and declining natural capital. Also note that sustainability in the West is far from giving up the growth paradigm.

Let us now consider an integrated ecological economic system and see if we can derive some simple patterns from its behavior.

For the integrated model we will consider a system with four variables:

1. Population – the size of the human population in a certain area
2. Development – the size of the human created economy
3. Capital – the amount of capital that is available for investment
4. Resources – the amount and state of natural resources available.

We have seen the related conceptual model on p. 43.

For the human population, let us use the model that we developed above with a few additional simplifications. We are assuming that there is no migration in or out of the system, and using the average values for the birth and death rate parameters.

The economic growth will be controlled by a Cobb-Douglas function:

$$D\_growth = Population^{0.3} * Resources^{0.4} * C\_dev^{0.3}$$

The idea here is that economic growth is stipulated by the Population (i.e. labor force), Resources and Capital. In our case C_dev is the amount of Capital spent for economic development. Without any growth the economic structure depreciates at a constant rate, so that

$$D\_decay = C\_decay*Development$$

where C_decay is the rate coefficient. Economic development generates Capital:

$$C\_in = C\_cap\_gr*Development$$

C_cap_gr is the rate coefficient, which tells how much investment capital the economy can generate. The capital will be spent for two purposes. One is to fund further economic growth, C_dev; the other is to restore the resources, C_env. The equation to spend capital will be:

$$C\_out = \frac{C\_cap\_sp*Capital}{DT}$$

C_cap_sp is the proportion of capital spent. If C_cap_sp = 1, all capital is reinvested. Note the division by DT that makes sure that this is independent on the time-step. C_out is divided between C_dev and C_env according to the control parameter F_dev, 0 ≤ F_dev ≤ 1.

F_env = 1 − F_dev indicates that whatever is left from the economic investments is spent on environmental restoration. The amount of resources will then grow in proportion to the restoration efforts:

R_in = C_restor*C_env + C_self*Resources, where C_restor is the efficiency of restoration and C_self*Resources is the process of self-rehabilitation, self-restoration. The resources are used for economic growth at a rate of

$$R\_out = C\_env\_des*D\_growth.$$

Of course this is a very simplistic model, very much along the lines of neo-classical economic theory. We assume that resources can be always regenerated or substituted. However, for the moment let us assume that this is indeed possible and see what behavior such a system can display.

We have already considered some preliminary dynamics in this model, when discussing the different integration methods (Chapter 3). We have observed dynamics that do not seem very sustainable: after an initial rise in development, the resource base is quickly depleted and the economy crashes. The population continues to grow, which is obviously unrealistic and begs for some improvement.

There are many obvious additions that can and should be made to the model, but before we go into any further details let us analyze the model that we have already put together. First, let us play with some of the parameters. We assume that the model has been put together in Stella or another modeling package, or downloaded from the book website.

First, if the resources crash, how can we sustain them? In the model we have the parameter F_dev, which defines what fraction of the capital is spent for development. What is left is spent on restoration. We had F_dev = 0.9. If we decrease the coefficient to F_dev = 0.6 we will get a perfect growth pattern, where development is generating enough revenue to provide for resources recovery – a world vision of a technologic optimist (Figure 7.20).

However, note that the growth trajectories are in place because of a very high efficiency of our restoration procedures (C_restor = 0.5). If we decrease it to, say, 0.1, we will be back to the rise-and-crash scenario. There seem to be only two ways the system can possibly develop: one is runaway growth, where all the elements grow to infinity; the other is rise and crash, where after a period of initial fast development the system variables decline to zero.

Since there is no feedback at this time from the other system variables to Population, let us single it out and see how the system behaves if population is assumed constant and at equilibrium. Now we find that the infinite growth behavior becomes

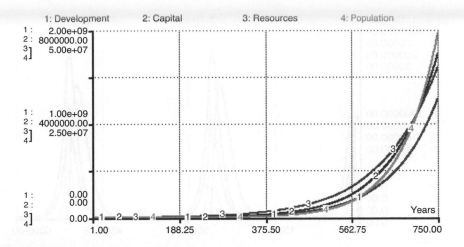

1: Development   2: Capital   3: Resources   4: Population

Figure 7.20   A technologic optimist world view.

very robust. Even if there is no restoration available, C_restor = 0., and almost everything is reinvested in development, F_dev = 0.8, we still have the system evolving along the growth curve. It does crash when F_dev = 0.9. One curious conclusion already emerges from this: Apparently the human component is extremely important when analyzing sustainability. With a small and fixed number of people, the development growth is controlled only by Capital and Resources. This allows development to grow gradually, based on the self-recovery of resources. Sustainability is possible when development is based only on the existing resource base. It is really the growing human influence in production that destabilizes the system.

If the Population is so important, let us bring it back into consideration and also look at some of the obvious feedbacks that the rest of the system should have with respect to the population. One thing that seemed quite strange in the original model was that Population continued to grow *ad infinitum* even when all the resources were gone and the economic system had crashed. Actually it was this infinite growth of Population that broke up the model, both for the Euler and the Runge–Kutta methods.

Let us assume that when resources are depleted, mortality increases (this can be due to, say, a decrease in air and water quality):

$$C\_mortality = C\_mortality + \frac{1}{1 + C\_mor\_env*Resources}.$$

C_mor_env is the rate of environmental effects on mortality. Note we wrote $1 + C\_mor\_env*Resources$ to make sure that we do not get a division by zero in case Resources become very small. If Resources are plentiful, this equation returns a value almost equal to the original mortality coefficient. However, as Resources decrease, mortality rate starts to grow. As a result we get the oscillating "grow-and-crash" type of dynamics, where all the elements of the system initially display rapid growth, followed by an equally rapid decline as resources become scarce (Figure 7.21).

If we further increase the development by allocating more capital to economic growth, the pattern becomes somewhat chaotic, with sudden outbursts of development followed by even steeper declines (Figure 7.22).

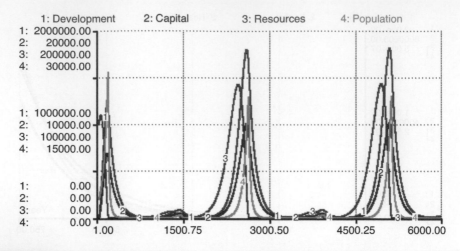

Figure 7.21   The grow and crash pattern of dynamics, F_dev = 0.6.

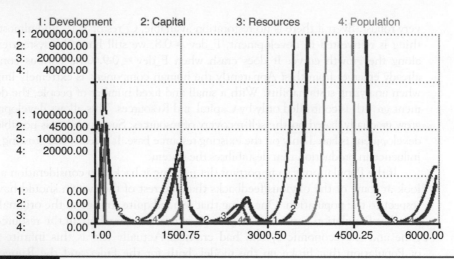

Figure 7.22   The grow and crash pattern of dynamics, F_dev = 0.9.

In any case, this is definitely not the type of dynamics we would call sustainable. Let us try to introduce some self-limitations into the system that could potentially dampen the oscillations. We will make the decision about the investments based on the current availability of Resources. If Resources are plentiful, F_dev is unchanged. When Resources decline, F_dev decreases, so that F_env = 1–F_dev can increase and more will be reinvested in restoration. The s-shaped function, discussed among other functions, seems to be a perfect choice to provide this type of behavior:

$$F\_dev = \frac{0.7 * Resources^2}{C\_half^2 + Resources^2}$$

where C_half – is the half saturation parameter, which in this case is the amount of resources at which F_dev is to be half of the original. This is some sort of an adaptive management that is embedded into the system. We are trying to make the system react to the changing conditions and adapt accordingly. As a result, we get a

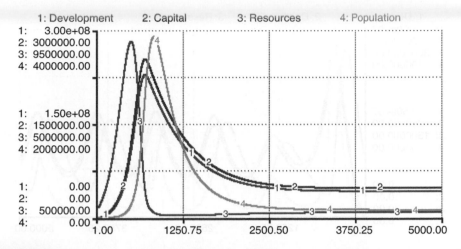

| | Figure 7.23 | "Sustainability" in a model with "adaptive management," F_dev = 0.6. |

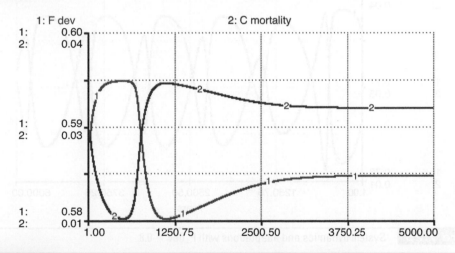

| | Figure 7.24 | Changes in mortality and investments that stabilize the system. |

behavior of the system that may be called sustainable. After an initial peak in economic development, the system returns to a non-zero condition which persists. The resources are not depleted, the population is not too large, and the economic development is such that it sustains the population and the regeneration of resources (Figure 7.23).

The adaptation is provided by changes in the mortality rate and in the investment strategy, as shown in Figure 7.24. Is this the scenario humans might follow, where adaptations and adjustments are made only when it is too late and the Resources have declined to a relatively low level?

We could claim that we have built a model of a sustainable system, if it were not for the fact that the model turns out to be structurally quite unstable. If we start from a different initial investment strategy and make F_dev = 0.8, we put the system into stable oscillations, as displayed in Figure 7.25.

If we further increase the initial investment into development, the system oscillations become chaotic. Figure 7.26 presents the cycles in the phase plane for Resources

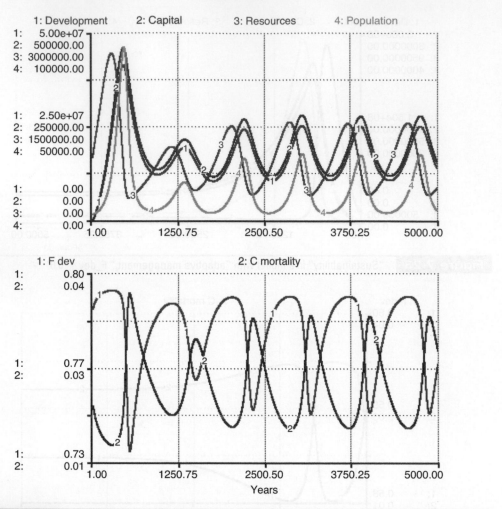

Figure 7.25   System dynamics and adaptations with F_dev = 0.8.

and Population. Development and Capital display similar chaotic oscillations. It is not quite clear what the future of this system will be. If instead we decrease F_dev and make it equal to 0.5, we get yet another entirely new behavior: steady growth of the economic subsystem with a very low resource base (Figure 7.27). Apparently all the resources are very efficiently being used for economic development, with the population entirely careless about the state of the environment as long as development is ensured.

If F_dev is further decreased, the growth becomes so rapid that the system quickly falls into discontinuous jumps and falls, clearly indicating the insufficiency of the numerical accuracy of the computer calculations. In reality, it is simply because numbers become too large for the computer to handle properly.

This numerical insufficiency deserves some further consideration. In Figure 7.28 we present the model trajectories achieved when, instead of the quadratic switching function for F_dev, we use a function of the Michaelis–Menten type:

$$F\_dev = \frac{0.7 * Resources}{C\_half + Resources}$$

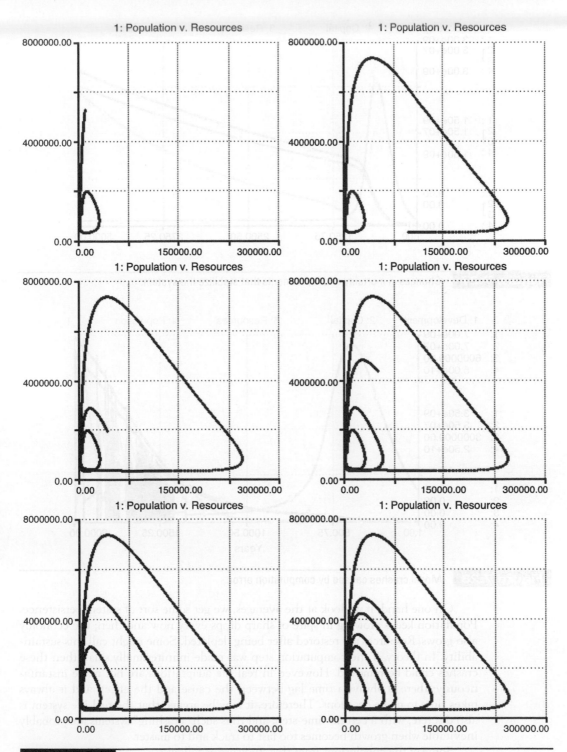

Figure 7.26 Chaotic cycles of Population and Resources when F_dev = 0.9.

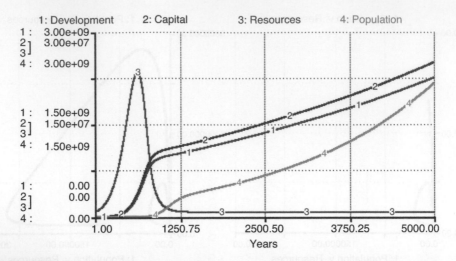

Figure 7.27 Changes in mortality and investments that stabilize the system.

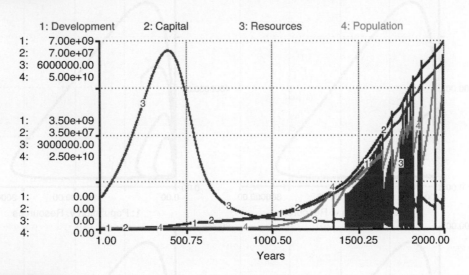

Figure 7.28 Model crashes caused by computation error.

On one hand, if we look at the averages, we get some sort of system persistence. Population keeps growing in spite of sharp drops every now and then; the economy also grows. Resources are restored after being depleted. Some might call this sustainability. In theory, if the computation step was made infinitesimally small then these crashes could be removed. However, in real life adaptations are not made instantaneously; there is always a time lag between the cause and the effect, and it always takes time to make decisions. Therefore, it may be argued that the real-life system is also discrete, with a certain time-step, and thus such "crashing" systems are probably inevitable when growth becomes too fast to track and to master.

Some of the conclusions from this study are as follows:

- The behavior of an ecological–economic system is quite complex and hard to control. We may create some behavior which might resemble sustainable development; however, it seems to be very much dependent upon the particular parameterization of the model.

- It is important to test the model with a variety of parameters and formalizations to make sure that we have really captured the essence of the system dynamics. It is wrong to jump to conclusions about system behavior based upon only one model realization.
- The neo-classical paradigm quite often results in system behavior that is focused on economic development, where ecological resources are used only to provide for further economic growth. This may be well in conflict with other human priorities, such as environmental quality and human health.

7.5   The end of cheap oil

And we ought not at least to delay dispersing a set of plausible fallacies about the economy of fuel, and the discovery of substitutes for coal, which at present obscure the critical nature of the question, and are eagerly passed about among those who like to believe that we have an indefinite period of prosperity before us.

W.S. Jevons, 1865

Water and energy are the two renewable resources that are essential for human livelihood. Whereas we have been mostly concerned with non-renewable resources as the human population grows in size and in terms of the impact that it has on the biosphere, renewable resources become equally important. Renewable resources may become limiting if the rate of their renewal is not fast enough. Renewal of water is dependent on energy. Production of energy, especially of renewable energy (biofuel and hydro), is dependent on water. In both cases, for energy and water, we compensate the lack of flow by digging into the stocks. The fossil fuels are the non-renewable reserves that we are quickly depleting. It is actually the stocks that have allowed humans to develop into a geological force (Vernadskii, 1986) which may very well bring itself to extinction, unless we find alternative development goals and paradigms. As with energy, we are compensating for a lack of water by extracting from fossil groundwater reserves. In both cases this is an unsustainable practice that leaves future generations dry, with no safety net to rely upon.

We looked at water in some detail in Chapter 6. Let us now focus on energy. There has been much discussions lately about the so-called "peak oil." Back in the 1950s, a USGS geologist, King Hubbert, was observing the dynamics of output from individual oil wells and noticed that they seemed to follow a pretty similar pattern. At first their productivity was low, then it gradually grew, until it peaked and then followed a pattern of steady decline. He has generalized these observations over multiple oil wells in various regions, and for the contiguous US he came up with a projection that said that oil production across the whole country will peak. He even estimated when it would happen – in the early 1970s. It turned out that his projection was remarkably close to what happened in reality (Figure 7.29). The next obvious step was to apply this same methodology to world oil production. According to those projections, the peak is supposed to happen some time really soon – by some estimates, it has actually already happened.

Why is peak oil such a big issue? Primarily because the demand for oil continues to grow exponentially, which means that as soon as oil production peaks there will be an increasing gap between demand and supply. For such an essential resource as energy, this gap may result in catastrophic outcomes.

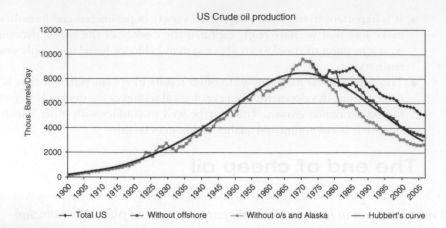

US Crude oil production

Figure 7.29 US oil production 1850–2050, as predicted by the peak oil theory of King Hubbert and in reality. The dashed line is Hubbert's prediction. The solid line is the actual extraction. Note that the timing of the peak was predicted almost exactly.

Both energy and water belong to the so-called Critical Natural Capital category, which means that they are essential for human survival. As they become scarce, they exhibit high price-inelasticity of demand, so that a small reduction of quantity leads to a huge increase in price.

A small decrease in supply will lead to an enormous increase in price, so that total value (price × quantity) paradoxically increases as total quantity declines. This is true for any resource that is essential and non-substitutable. As there is less water or energy available, the price quickly increases towards infinity. This creates havoc with markets and pretty much puts the whole system out of control – as we saw during the energy crisis of the 1970s. While energy and water are abundant, their value is low; it may seem that we have an infinite supply, and there is nothing to worry about. However, as depletion accelerates, even small perturbations due to unforeseen climatic events or technical malfunction may result in disproportionate changes in price.

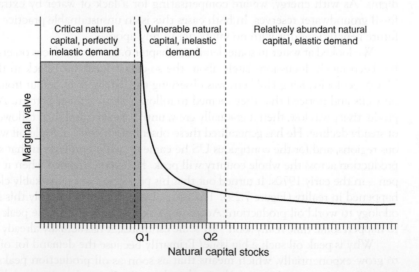

As pointed out in Chapter 6, as long as we rely upon purely renewable energy and water, they are non-rival and non-excludable. However, as we need to dip into reserves of fossil water or energy, or even into the temporary reserves (lakes, reservoirs, or forest and crop biomass), immediately the resources become excludable and rival. As resources become scarcer we easily create conflict situations (water and energy wars, one of which we are waging right now).

(Farley and Gaddis, 2007)

While most official sources have been quite reluctant to discuss this issue, in 2007 several publications appeared indicating that there is a growing concern even in circles closely related to governments. In July 2007 the International Energy Agency (IEA), an arm of the Organization for Economic Cooperation and Development (OECD), published the "Medium-Term Oil Market Report." The report predicts that world economic activity will grow by an average of 4.5 percent per year during the next several years, driven largely by strong growth in China, India, and other Asian countries. Global oil demand will, as a result, rise by about 2.2 percent per year, pushing world oil consumption from an estimated 86.1 million barrels per day in 2007 to 95.8 million barrels by 2012. If there are no catastrophes and there is ample new investment, the global oil industry may be able to increase output sufficiently to satisfy this higher level of demand – but if so, barely. Beyond 2012, the production outlook appears far grimmer. And remember that this is the best-case scenario.

Let us see what we can find out about the future of oil supplies using some simple dynamic modeling. Suppose we have a stock of oil. Since it is a non-renewable resource, it is safe to assume that it is limited. There will always be oil in the ground, but it is quite clear that eventually we will run out of the energetically profitable resource. So with this stock comes just an outflow, which we will call Extraction. Let us assume that Extraction is driven by Demand. Demand is exponentially growing, just as it has been over the past years:

$$\text{Demand}(t) = \text{Demand}(t - dt) + (\text{Growth}) * dt$$

$$\text{Growth} = \text{C\_grow} * \text{Demand}$$

Besides satisfying Demand, Extraction should also produce enough to power Extraction itself. This is what is known as the EROEI (Energy Return on Energy Invested) index. If e_{out} is the amount of energy produced and e_{in} is the amount of energy used in production, then EROEI, $e = e_{out}/e_{in}$. In some cases the net EROEI index is used, which is the amount of energy we need to produce to deliver a unit of net energy to the user: $e' = e_{out}/(e_{out}-e_{in})$. Or $e' = e/(e - 1)$.

To account for EROEI, we put:

$$\text{Reserves}(t) = \text{Reserves}(t - dt) + (-\text{Extraction}) * dt$$

$$\text{Extraction} = \text{Demand} * \left(1 + \frac{1}{\text{eroei}}\right)$$

It also makes sense to assume that EROEI is not constant. In fact, at some point we had oil fountaining out of the ground, so we just needed to collect and deliver

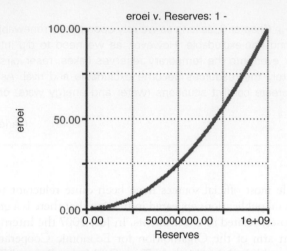

Figure 7.30   The parabolic dependency between the EROEI index and the amount of reserves still available. The fewer reserves are left, the more we need to invest in production.

it; now we need to drill kilometers deep into the ground and pump the oil out, then pump water or CO_2 in to push some more oil out, and so on. The energy return has declined from over 100:1 in the 1930s to 30:1 in the 1970s to around 10:1 in 2000. EROEI is a battle between technology and depletion, and depletion is winning. In the future, more energy investment will be needed, taking energy out of a non-energy society.

Let us assume that EROEI drops with Reserves decreasing, according to the parabolic function shown in Figure 7.30. Then

$$\text{eroei} = \text{e\_ini} * \left(\frac{\text{Reserves}}{\text{r\_ini}} \right)^2$$

assuming that e_ini = 100 is the original EROEI and that r_ini = 1000000000 is the original stock of oil in the Reserves.

If we run this model, we will get an expected result: the growing demand will certainly deplete the resources (Figure 7.31). What is noteworthy about this graphic is the power of exponential growth. While we have very slow, almost negligible change over a long initial period of time, things start to accelerate tremendously by the end of the season. Most of the change is compressed into a rather short period of time, when action is really needed, but there is very limited time to do something. Also, note how much faster we need to pump out our reserves to supply the demand as reserves become depleted.

It is also noteworthy that the values of the parameters that we used in this model do not really matter. The exponential growth or decline has a vivid trace that shows through any modifications in parameters. We can even try another function

You can increase your confidence in model results if the model is structurally stable. It is hard to prove structural stability, but it is always good to search for models that have a good deal of robustness to structural modifications.

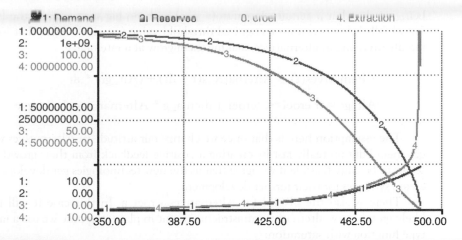

Figure 7.31   System dynamics shows very slow dynamics at first, followed by a period of very high growth rate and eventual crash of the system due to depletion of resources.

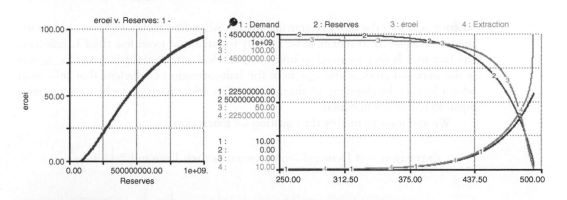

Figure 7.32   The s-shaped EROEI function produces very similar results.

We may argue that the model is structurally quite stable. With qualitatively similar assumptions about the driving forces and processes, the exact formulations and parameter value do not matter that much.

for the EROEI. Suppose we choose an s-shaped one (remember that which we discussed in Figure 2.20?):

$$eroei = \frac{e\_ini * Reserves^2}{\left(\dfrac{r\_ini}{2}\right)^2 + Reserves^2}$$

For this function we also get a similar pattern, with very slight changes in the trajectories (Figure 7.32). Once again, the last drop of oil is extracted at an exceedingly high rate.

However, it could certainly be argued that there is other energy out there, and there is really no reason to expect that we are so ignorant not to realize the imminent crash and not to start exploring alternatives. Let us add alternative sources of energy to our model. Let us assume that the infrastructure for alternative energy is being produced at a certain slow rate (a_g_c) with no big success until the EROEI for oil falls below a certain recognized threshold value (eroei_t). After that we start rapidly investing in alternatives, making them grow at a rate of a_g:

$$\text{Alternatives}(t) = \text{Alternatives}(t - dt) + (\text{Alt\_gr}) * dt$$

$$\text{Alt\_gr} = \text{if eroei} < \text{eroei\_t then a\_g} * \text{Alternatives else a\_g\_c}$$

The assumption here is that once we change our attitude to Alternatives we can get them built up really fast by creating a positive feedback from their growth. This seems to be quite feasible if we agree that as the new technologies get developed they create synergies for their further development.

There is also the EROEI for Alternatives, eroei_a. In this case it will mostly likely grow as new alternative infrastructure is put in place. Suppose we use a monod-type function with saturation:

$$\text{eroei\_a} = \text{e\_a\_min} + \frac{\text{e\_a\_max} * \text{Alternatives}}{\text{Alternatives} + \text{e\_a\_hs}}$$

where e_a_min is the minimal starting eroei_a, when new technologies are only starting to be deployed. It makes sense initially to have it at even less than 1, reflecting the fact that at first we need to invest a great deal with very little return. e_a_max is the maximal eroei_a and e_a_hs is the half-saturation coefficient that tells us at which level of development of alternative energy (Alternatives stock) we get eroei_a equal to half of the maximal.

We also want to modify the equation for Extraction.

$$\text{Extraction} = \text{if Demand} > \text{Alternatives then Demand} * \left(1 + \frac{1}{\text{eroei}}\right)$$

$$- \text{a\_eff}*\text{Alternatives}*\left(1 - \frac{1}{\text{eroei\_a}}\right), \text{ else } 0$$

The logic here is that if all the demand can be covered by alternative energy (Demand < Alternatives), then there is no need to continue extraction of fossil energy, and Extraction = 0. Otherwise, we need to extract enough to cover the demand. The alternative infrastructure chips in with the efficiency a_eff (a negative term in the Extraction equation), but to produce this alternative energy we need to invest 1/eroei_a (a positive term in the Extraction equation). The higher the eroei_a, the less we need to run the alternative infrastructure.

Let us run the model with the following parameter values:

a_eff = 10, a_g = 0.2, a_g_c = 100, c_grow = 0.03

eroei_t = 20, e_a_hs = 500000, e_a_max = 10, e_a_min = 0.5

e_ini = 120, r_ini = 1000000000

We will mostly be concerned with general qualitative behavior, and will not try to figure out what the real values for these parameters are (which is also a very worthwhile effort). For now, let us explore what the overall system dynamics are. With the

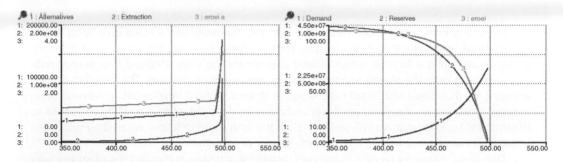

Figure 7.33   If we start investing in Alternatives too late, we only accelerate the crash of the system.

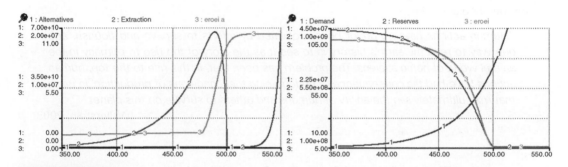

Figure 7.34   Early investment in Alternatives, while there is still ample supply of conventional energy, allows for a smooth transition to renewable energy.

values above, Alternatives make almost no change to the system (Figure 7.33). On the contrary, investing in the alternative sector when we are already pumping out the second half of our reserves only accelerates the crash. Changing different parameters related to Alternatives efficiency does not seem to help. The system still crashes.

What does help is changing parameters related to the timing of the switch to alternatives. If we start developing alternatives when the EROEI of traditional energy is still as high as about 60 or more (eroei_t > 58), we get a completely different picture (Figure 7.34). The same opportunity exists if we have been slowly developing alternatives since the very beginning (a_g_c = 1000). In these cases we have a pretty smooth transition from fossil-based energy to alternative energy, with extraction going down to zero while there is still plenty of oil left in the ground.

Obviously other factors will kick in, such as limited land resources, so it is a major simplification to think that indeed we will be always able to provide for the exponentially growing demand. However, in terms of energy we can do it (or, more likely, could have done it). That is if we had started the transition early enough to provide for the new alternative infrastructure. To find out more exactly how late we are arriving at the show, we will need to find some more realistic values for the parameters. In this case, there are several parameters that do matter. However, qualitatively it seems that the half depletion threshold is an important factor in these dynamics. If we start the transition to the alternatives well before we have half depleted the resource, there is enough to fund the development of the new alternative infrastructure. If we procrastinate any longer, the crash is imminent.

It is becoming increasingly clear that, in the long run, humanity can survive only living within the limits of resources that it has. In this context, fossil fuels appear as winning a lottery ticket, or inheriting a fortune, which indeed is precisely the case. Without this energy subsidy, we have only the steady supply of energy coming from the sun. The windfall of high energy gain that came with fossil fuels was our chance to learn how better to harness solar energy. Whatever we learn and whatever technologies we build while we still have access to cheap fossil fuels is what we will be left with in the long years to follow when they will be no more. And we certainly will run out of them, sooner or later, but quite inevitably.

One clear reason for caution is the very disturbing statistics in biographies of lottery winners. Too often they end up in bankruptcy, poverty or suicide. So far humanity has been living off lottery winnings – that is, digging out the treasure buried in previous eras. This windfall of cheap and abundant energy provided for the fast population and economic growth of the previous two centuries.

"Human actions have no doubt been motivated by efforts to survive and flourish, and one way to read the earth's history is to see it as the story of the rise to primacy in the animal world of homo sapiens. The problem has been that, in this rise to the top, human actions have had the consequence of undermining the 'conditions of production' in ways that may ultimately sap the ability of humans and others to survive on this planet."

(Wallerstein, 2003)

There are few more observations that can be made.

- The depletion pattern is likely to be different on the global scale than it is locally. The smooth sigmoid curve predicted by Hubbert and observed for most of the local and regional operations may not hold for global extraction. When individual wells started to show high decline rates, we found a replacement for them. When regional, say USA, production went beyond the peak, we moved to other locations to extract oil relatively cheaply, with a higher EROEI. We could use the same technology, the same equipment; we just moved to a different place. It will be different in the global scale, where we have really nowhere else to go. We have to move to a different technology and create new infrastructures for production and distribution. That will require considerable investment in energy that we will try to get from the traditional sources, even if we will have to continue pumping at lower EROEIs. This can somewhat delay the peak, but significantly increase the rate of fall. So the more-or-less symmetrical sigmoid curve of peak oil is likely to be shifted to the right, with the second half extracted much faster than the first half. We will climb higher to fall down harder.
- Another no less significant factor is global climate change. There is no way we can stop it at this time, so we need to adapt to it. However, adaptation also requires new infrastructure, which will be yet another burden on the dwindling resources. We are in a race against time that may become quite ugly. And the price for further procrastination may be exceedingly high.
- Curbing the demand is probably the cheapest and fastest solution. The problem is that it moves the focus from the technical, engineering arena primarily to the socio-psychological domain. There is an urgent need for a paradigm shift from promoting growth to achieving sustainability. Supply should not be dictated by demand. On the contrary, we should develop demand in those places where we

have supply, and to a level that can be sustained. Ironically, in many places we have exactly the reverse: population is growing most rapidly where water and energy are least available. Conveying energy creates more losses: currently up to two-thirds of electric energy is lost in transmission. Conveying water requires much energy, and also results in significant losses due to evaporation and seepage.

There is a clear correlation between energy consumption and economic development (Figure 7.35). At the same time, there is no obvious correlation between GDP and such indicators as life satisfaction or life expectancy (Figure 7.36). We can see that with no sacrifice to life quality indices we can at least halve the *per capita* GDP, and therefore energy consumption. It is really a matter of choice, social attractiveness, and cultural priorities. These can be changed only with a strong leadership that should be advanced and promoted by the federal government.

Decreasing consumption may be an unpopular measure that makes federal involvement especially important. So far, most of the advertising industry is working towards increasing consumption, buying things that we do not need, wasting more energy and water. Only federal action can stop that and help to shift awareness of the population towards conservation and efficiency. Increasing efficiency in all areas – industrial, residential and agricultural – is another clear focus begging for action.

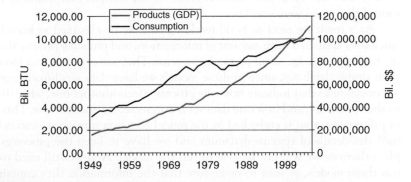

Figure 7.35   Energy consumption from all sources and GDP in USA (EIA, 2006).

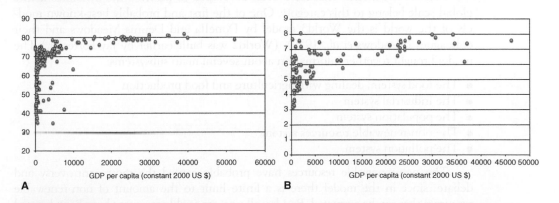

Figure 7.36   A. Relationship between GDP *per capita* and life expectancy based on 2005 World Bank data for 184 countries (http://devdata.worldbank.org/query/default.htm). B. GNP and Life Satisfaction Index for 2000 (Veenhoven, 2004; another great source of this kind of information is http://www.gapminder.org/world/).

Compare productivity *per capita* in the USA and Japan. They are at a comparable level, and are actually the best in the world. Yet Japan needs only half of the energy that the USA needs! Japan emits 9.5 tons CO_2 *per capita*, whereas the USA emits 19.7 tons CO_2 *per capita* – roughly proportional to the energy consumption ratio of the two countries. That shows an obvious way to cut GHG emissions.

7.6   The World

So far we have been trying to focus on some very simple models, the dynamics of which we can carefully explore to reveal some of the emergent properties and surprises in systems' behavior. These models easily tend to become more and more complex. As we find more connections, processes, factors and parameters that seem to be important for the overall system's dynamics, the enticement is very strong to add them to the model, because, indeed, they seem important and the model would not look relevant without them. Sometimes we promise ourselves that we will try to simplify the model later on, after running sensitivity analysis, and finding parameters and processes that are not really making much of a difference. Quite often we forget about that, especially if we are happy with the results that we are getting, and we tend to care less about the more elaborate model analysis that would be nice to perform if the model were simpler.

As a result we tend to build models that may be classified as knowledge bases, since they contain a huge amount of information, and probably present the best state-of-the-art knowledge about particular systems. They are certainly way more advanced than simply databases, since in these models we have data sets linked together: there are casual links that indicate how one process affects another one, what the feedbacks in the system are, and how one data set is connected to another one. This abundance of information that is embodied in the model comes at a price: we can no longer dig into the details of systems dynamics and we have to keep the processes quite simple, otherwise we will not be able to run the model. And we still need to be able to run these models, at least to make sure that the information they contain is consistent, that the logic of the links and relationship works, and that we get a meaningful, coherent picture of the modeled system.

Most of the models that have been developed to describe the dynamics in the global scale belong to this category. One of the first and probably best-known models of the world is the World3 model by Donella and Dennis Meadows and their colleagues (Meadows *et al.*, 1979). (World2 was built earlier by Jay Forrester.) The model brought together information about several main subsystems:

- The food system, dealing with agriculture and food production
- The industrial system
- The population system
- The non-renewable resources system
- The pollution system.

The non-renewable resources have probably caused the most controversy and debate, since in the model there is a finite limit to the amount of non-renewable resources that can be extracted. Besides, all non-renewable resources have been lumped into one. This allowed immediate and costless substitution of one non-renewable resource (coal) for another (say, gas), but excludes the substitution by other resources through new technology that science and engineering are yet to discover.

This single resource assumption was supposed to be a simplification but, as with all simplifications in models of this sort, this opens the door to much debate. The overall complexity of the model does not help to communicate the assumptions and approximations to users, causing more controversy. The model comes as part of the standard distribution of Vensim – one of the systems dynamics modeling packages that we mentioned in Chapter 2. You can download a free educational Vensim PLE version and explore the World3 model for yourself to find out that, even with less than 20 state variables, it is not very simple to figure out all the ideas and functions that went into it. However, it certainly gives an appreciation of the system complexity and of all the different factors that we would like to take into account. There is also a Stella version (although poorly documented), which can be downloaded from the book website.

The issue of resource substitutability is an important one in ecological economics. It has led to two definitions of sustainability: weak and strong. The technological optimists argue that there are no real limits to resources, because as resources dwindle people will find other, better ones to use instead. Neumeyer (1999) defines sustainable development as one that "does not decrease the capacity to provide non-declining per capita utility for infinity." Those items that form the capacity to provide utility are called capital, which is then defined as a stock that provides the flow of services. For weak sustainability it is then necessary to preserve the value of the total aggregate stock of capital. This obviously implies that components of capital and utility function are substitutable. "Strong sustainability instead calls for preserving the natural capital stock itself as well." As models show us, this is probably a crucial assumption in deciding what kind of future we are facing.

The finite resources are probably not so serious a concern in this model, especially if we remember the Jevon's paradox, well known in ecological economics theory. After all, so far new technology has not really decreased overall resource consumption. A different type of criticism came from David Berlinski, as mentioned earlier in Chapter 3. Indeed, the model is pretty much entirely constructed based on exponential growth equations. These could not possibly give a different type of output than what the model produces – that is, a system crash (Figure 7.37). Once again, we wonder to what extent the model runs and their results are actually important, or whether they are pretty much artifacts of the model structure.

William Stanley Jevons, a British economist, noticed a strange phenomenon while studying the coal industry in England: as more efficient technology is developed, overall consumption does not decrease. Actually, 20 years earlier Williams had written: "The economy of fuel is the secret of the economy of the steam-engine; it is the fountain of its power, and the adopted measure of its effects. Whatever, therefore, conduces to increase the efficiency of coal, and to diminish the cost of its use, directly tends to augment the value of the steam-engine, and to enlarge the field of its operations." (C.W. Williams (1841), *The Combustion of Coal*, p. 9.) Indeed, the more efficient Watt steam-engine brought about only greater coal consumption, since it was more efficient than the Newcomen engine, and therefore it was put into more widespread use and coal combustion increased. We see this happening all the time. As cars became cheaper and more efficient, we got more cars. As electricity and construction became cheaper and more efficient, we got bigger houses. Even now, when a new super energy-efficient refrigerator is installed in our kitchen, the old one is moved to the garage, so we have two refrigerators, with the consequent higher energy consumption.

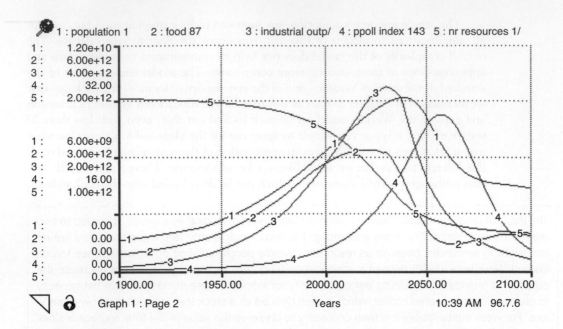

1 : population 1   2 : food 87   3 : industrial outp/ 4 : ppoll index 143 5 : nr resources 1/

Figure 7.37   A typical output from the World3 model. The system crashes when non-renewable resources are consumed.

The main result of this model was that it stimulated much discussion on several global problems, such as population growth, depletion of natural capital, pollution, etc. According to some estimates, the number of lines of text contributed to these debates has exceeded the size of "The Limits to Growth" by two or more orders of magnitude. The marketing of the World3 model has drawn much attention to applications of models in politics and policy-making. The unfortunate outcome is that nothing or very little has been actually accomplished to solve or mitigate the problems that were brought to light by the model.

A more recent reincarnation of a world systems dynamics model is the Global Unified Metamodel of the Biosphere (GUMBO), developed by Roelof Boumans and other scientists at the Gund Institute for Ecological Economics to simulate the integrated Earth system with the implicit goal of assessing the dynamics and values of ecosystem services. The model is presented as a synthesis and simplification of several existing dynamic global models in both the natural and social sciences, and claims to aim for the intermediate level of complexity. With 234 state variables, 930 variables in total, and 1715 parameters, this may be a bit of a stretch. We are certainly dealing with a beast of a different kind than that we have seen in other chapters of this book. If somebody thought that some of those models were complex – think again. However, indeed, there are certainly more complex models available.

GUMBO is the first global model to include the dynamic feedbacks among human technology, economic production and welfare, and ecosystem goods and services within the dynamic earth system. GUMBO includes modules to simulate carbon, water, and nutrient fluxes through the Atmosphere, Lithosphere, Hydrosphere, and Biosphere of the global system. Social and economic dynamics are simulated within the Anthroposphere (Figure 7.38). GUMBO links these five spheres across

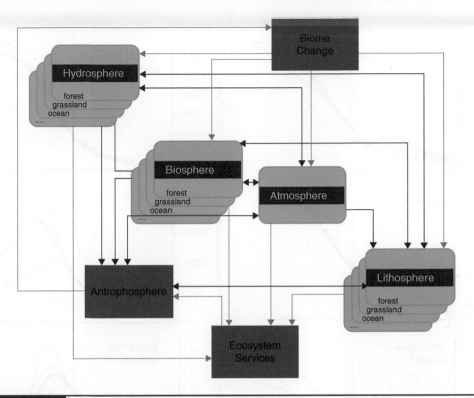

Figure 7.38   Overall structure of the GUMBO model. Using the Stella array functionality, all the main "spheres" are replicated over the 11 biomes assumed in the model.

eleven biomes (Open ocean, Coastal ocean, Forests, Grasslands, Wetlands, Lakes/Rivers, Deserts, Tundra, Ice/rock, Croplands, and Urban), which together cover the entire surface of the planet.

The Stella version of the model can be downloaded from http://ecoinformatics. uvm.edu/GUMBO/GUMBO.zip. Perhaps it would be most useful to download the model and do some clicking on the diagram to understand what it looks like and what it is doing.

The dynamics of 11 major ecosystem goods and services for each of the biomes are simulated and evaluated. Historical calibrations from 1900 to 2000 for 14 key variables for which quantitative time-series data were available produced an average R^2 of 0.922. For a model of this level of complexity, this level of correlation with data is very unusual and quite astounding. The only possible explanation is that we are working at a very aggregated level and there is not much variability in the data (Figure 7.39). As we can see, most of the dynamics are really still in the future, so it will take at least a couple of decades to find out whether the model projections are right or wrong.

However, it needs to be stressed that, for these knowledge-base type models, forecast precision and accuracy of the results are really not the point. They are not built to reproduce the exact day of collapse. What we are looking for are the trends, the understanding of how the overall system performs. This seems to be very well captured by the model, which was used to analyze the four scenarios of future human development proposed by Robert Costanza in 2000.

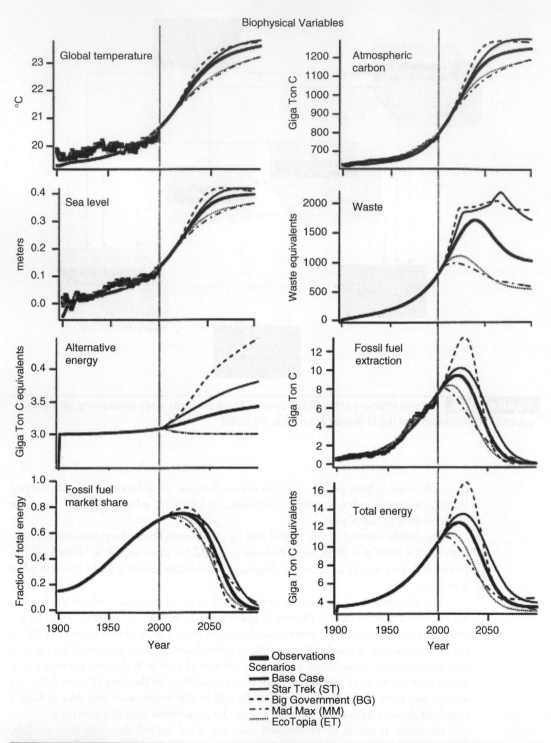

Costanza described these four possible futures in Costanza (2000). Briefly here is what they are about.

1. Star Trek: The Default Technological Optimist Vision

"Warm fusion" was discovered and powered humanity to the stars. By 2012, natural resources were very strained. The warm fusion allowed a rapid reduction of global fossil fuel burning, with eventual reversal of the greenhouse effect. The air pollution problem was essentially eliminated over the period from about 2015 to 2050. Electricity came increasingly from warm fusion, nuclear fission reactors were decommissioned and some hydropower stations were eliminated. The world was still getting pretty crowded. The solution was space colonies, built with materials taken from the moon and asteroids and energy from the new warm fusion reactors. Since food production and manufacturing are mainly automated and powered by cheap warm fusion energy, only about one-tenth of the population actually needs to work for a living. Most are free to pursue whatever interests them. Often the biggest technological and social breakthroughs have come from this huge population of "leisure thinkers." People also have plenty of time to spend with family and friends, and the four-child family is the norm.

2. Mad Max: The Skeptic's Nightmare

The turning point came in 2012, when the world's oil production finally peaked, and the long slide down started. There were no cheaper alternatives for oil, only more expensive ones. Oil was so important in the economy that the price of everything else was tied to it and the alternatives just kept getting more expensive at the same rate. The greenhouse effect was really kicking in and the earth's climate and ecological systems were in a complete shambles. The pollution crisis came next. Rising sea level inundated all low-lying coastal areas by about 2050.

The financial bubble really burst. Both the physical infrastructure and the social infrastructure have been gradually deteriorating, along with the natural environment. The human population was declining since the global epidemic killed almost 25 percent in 2025–2026. The population was already weakened by regional famines and wars over water and other natural resources. Since then death rates have exceeded birth rates almost everywhere, and the current population of 4 billion is still decreasing by about 2 percent per year. National governments have become weak, almost symbolic, relics. Transnational corporations run the world, making the distribution of wealth even more skewed. Those who work for global corporations lead comfortable and protected lives in highly fortified enclaves. These people work 90- or 100-hour weeks with no vacation. The rest of the population survives in abandoned buildings or makeshift shelters built from scraps. There is no school, little food, and a constant struggle just to survive. The almost constant social upheavals and revolutions are put down with brutal efficiency by the corporate security forces (governments are too broke to maintain armies anymore).

3. Big Government: Reagan's Worst Nightmare

The turning point came in 2012, when the corporate charter of General Motors was revoked by the US Federal Government for failing to pursue the public interest. Even though "warm fusion" had been discovered in 2015, strict government regulations had kept its development slow while the safety issues were being fully explored.

Warm fusion's slowness in coming on line was balanced with high taxes on fossil energy to counteract the greenhouse effect and stimulate renewable energy technologies. Global CO_2 emissions were brought to 1990 levels by 2005, and kept there through 2030 with concerted government effort and high taxes, after which the new fusion reactors eliminated the need for fossil fuels. The worst predicted climate-change effects were thus averted. Government population policies that emphasized female education, universal access to contraception, and family planning managed to stabilize the global human population at around 8 billion, where it remained.

The income distribution has become much more equitable worldwide. Governments have explicitly advocated slow or no-growth policies, preferring to concentrate instead on assuring ecological sustainability and more equitable distribution of wealth. Stable human population also took much of the pressure off other species.

4. Ecotopia: The Low Consumption Sustainable Vision

The turning point came in 2012, when ecological tax reform finally was enacted almost simultaneously in the US, the EU, Japan and Australia. Coincidentally, it was the same year that Herman Daly won the Nobel Prize for Human Stewardship (formerly the prize for Economics). People realized that governments had to take the initiative back from transnational corporations and redefine the basic rules of the game. The public had formed a powerful judgment against the consumer lifestyle and for a sustainable lifestyle. A coalition of Hollywood celebrities and producers got behind the idea and began making a series of movies and TV sit-coms that embodied the "sustainable vision." It suddenly became "cool" to be sustainable, and un-cool to continue to pursue the materialistic, consumer lifestyle.

All depletion of natural capital was taxed at the best estimate of the full social cost of that depletion, with additional assurance bonds to cover the uncertainty about social costs. Taxes on labor and income were reduced for middle- and lower-income people, with a "negative income tax" or basic life support for those below the poverty level. The QLI (Quality of Life Index) came to replace the GNP as the primary measure of national performance. Fossil fuels became much more expensive, and this both limited travel and transport of goods and encouraged the use of renewable alternative energies. Mass transit, bicycles and car-sharing became the norm. Human habitation came to be structured around small villages of roughly 200 people. The village provided most of the necessities of life, including schools, clinics and shopping, all within easy walking distance. People recognized that GNP was really the "gross national cost," which needed to be minimized while the QLI was being maximized. By 2050 the workweek had shortened in most countries to 20 hours or less, and most "full-time" jobs became shared between two or three people. People could devote much more of their time to leisure, but rather than consumptive vacations taken far from home, they began to pursue more community activities (such as participatory music and sports) and public service (such as day care and elder care). Unemployment became an almost obsolete term, as did the distinction between work and leisure. The distribution of income became an almost unnecessary statistic, since income was not equated with welfare or power, and the quality of almost everyone's life was relatively high. With electronic communications, the truly global community could be maintained without the use of consumptive physical travel.

GUMBO could handle these scenarios to produce the results in Figures 7.39 and 7.40. Again, the exact numbers on those graphs are hardly important, and may be difficult to justify. What really matters is that the model took into account much of the existing knowledge about global processes and translated that knowledge into meaningful trends that can be discussed, compared and evaluated.

The further development of the model was for valuation of ecosystem services. Ecosystem services in GUMBO are aggregated to 10 major types. These are: gas regulation, climate regulation, disturbance regulation, water use, soil formation, nutrient cycling, waste treatment, food production, raw materials, and recreation/cultural. These 10 services together represent the contribution of natural capital to the economic production process. They combine with renewable and non-renewable fuels, built capital, human capital (labor and knowledge), and social capital to produce economic goods

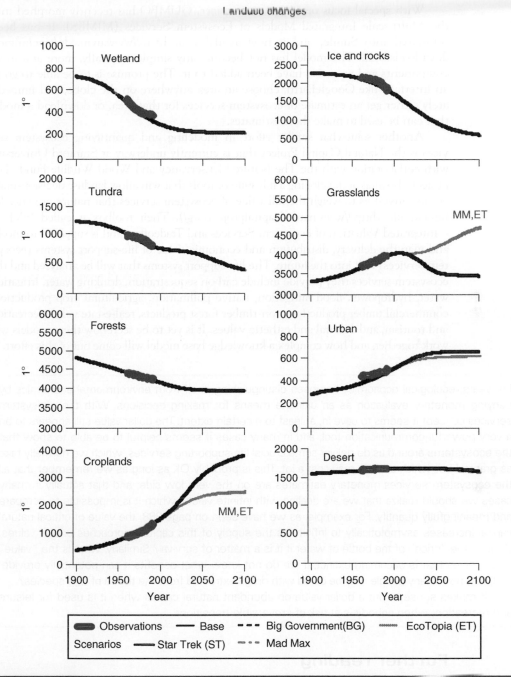

Landuse changes

Figure 7.40 Change in landuse composition under various future development scenarios.

and services. They also contribute directly to human welfare. Several different methods to value ecosystem services are implemented in the model, allowing users to observe all of them and compare the results. Historical data on landuse, CO_2 concentration in the atmosphere, global mean temperature, economic production, population and several other variables are used to calibrate the model.

With special focus on ecosystem services, GUMBO has recently morphed into the Multi-scale Integrated Models of Ecosystem Services (MIMES). It has been translated into Simile, and is now available at http://www.uvm.edu/giee/mimes/downloads.html. The model has not become any simpler; actually, more and more components and processes have been added to it. The promise is to be able to go to an interface like GoogleEarth, choose an area anywhere on the globe, and immediately either get an estimate of ecosystem services for that area, or download a model that can be used to make these estimates.

Another somewhat similar effort in modeling and quantifying ecosystem services is the Natural Capital Project that is currently underway at Stanford University, with collaboration with the The Nature Conservancy and World Wildlife Fund. The project also aims at developing a full suite of tools that will allow landuse decision-makers and investors to weigh the full value of ecosystem services that nature provides for human life (http://www.naturalcapitalproject.org/). Their toolbox is called InVEST – Integrated Valuation of Ecosystem Services and Tradeoffs – and is supposed to model and map the delivery, distribution and economic value of life-support systems (ecosystem services) well into the future. The life-support systems that will be analyzed and the ecosystem services they provide include carbon sequestration, drinking water, irrigation water, hydropower, flood mitigation, native pollination, agricultural crop production, commercial timber production, non-timber forest products, real-estate value, recreation and tourism, and cultural and esthetic values. It is yet to be seen how these models will work together, and how complex a knowledge base model will come out of this effort.

For years, ecological economics has been distinguishing itself from environmental economics by denying monetary evaluation as an ultimate means for making decisions. With the ecosystem services concept it seems to cave in, at least to a certain extent. The dollar value still appears to be a very powerful communication tool, and in many cases it seems helpful to be able to show that the ecosystems around us do deliver some crucial life-supporting services, which we normally take as granted but which actually may cost a lot. This is probably OK as long as we remember that all the ecosystem services monetary estimates are on the very low side, and that actually in many cases we should realize that we are dealing with infinite values which it is impossible to compare and meaningfully quantify. For example, as we have seen on page 288, the value of critical natural capital increases asymptotically to infinity as the supply of this capital approaches critical values. What is the "price" of the bottle of water if it is a matter of survival? Similarly, what is the "value" of a species that is becoming extinct, if we do not know what benefits it can potentially provide, and, say, how many people may be cured with drugs extracted from the tissue of that species?

It makes sense to put a dollar value on abundant natural capital when it is used for leisure and recreation, when nobody is at risk of irreversible transitions.

Further reading

Ehrlich, P.R. (2000). Population Bomb. Random House. 202 pp. – *First published in 1969, the book gives an excellent account of what exponential growth of population means for this planet. A later book by* Ehrlich P.R. and Anne H. Ehrlich, 1991. *The Population Explosion.* Touchstone Books, 320 pp. – *Digs deeper into causes and consequences of population growth.*

Kotlikoff, L.J. and Burns, S. (2005). *The Coming Generational Storm: What You Need to Know about America's Economic Future.* The MIT Press. 302 pp. – *Gives a vivid account of the current trends in US population, explores the future of a country with an increasingly older population and with a welfare and social security system on the verge of collapse.*

For scrupulous mathematical analysis of population models, including models with age structure, see Logofet D.O. (1993). *Matrices and Graphs: Stability Problems in Mathematical Ecology*, CRC Press. *For even more on matrix modeling of population dynamics see* Caswell, H., 2001. *Matrix population models: construction, analysis, and interpretation*, 2nd ed. Sinauer, Sunderland, MA, 722 pp.

Hartmann, T. (2004). *Unequal Protection: The Rise of Corporate Dominance and the Theft of Human Rights*. Rodale Books. 360 pp. – *A history of corporate takeover in the USA and now globally. It shows how very small, insignificant events in the past (like a clerical error) can result in tremendous consequences for all. See* http://www.terracycle.net/ *and* http://suedbyscotts.com/ *for more information about the TerraCycle story. Another very well known book on this subject is* David Korten, 2001. *When Corporations Rule the World*. Berrett-Koehler Publishers, 385 pp.

Much of the sustainability talk started after the famous report of the Bruntland Commission: WCED (World Commission on Environment and Development), 1987. *Our Common Future*. Oxford University Press, Oxford, 400 pp. *For various definitions of sustainability and the relevant scales and hierarchies, see* Voinov, A., 2007. Understanding and communicating sustainability: global versus regional perspectives. *Environ. Dev. and Sustain.* (http://www.springerlink.com/content/e77377661p8j2786/). *The definitions quoted here are taken from:* Wimberly, R. C., 1993. Policy perspectives on social, agricultural and rural sustainability. *Rural Sociology* 58: 1–29; Costanza, R. 1992. Toward an operational definition of ecosystem health. *In*: Costanza, R., Haskell B.D., Norton B.G., (Editors). *Ecosystem Health: new goals for environmental management*. Island Press, Washington, DC, pp. 239–256; Costanza, R., Daly, H. E. 1992. Natural capital and sustainable development. *Conservation Biology*. 6, 37–46; Solow, R. M., 1991 *Sustainability: an economist's perspective*. Marine Policy Center, WHOI, Woods Hole, Massachusetts, USA. *For further analysis of sustainability and its various economic implications see* Neumayer, E., 1999. *Weak versus Strong Sustainability*. Edward Elgar, Cheltenham, UK, 294 pp.

Our observation on management problems with rapidly growing and complex socio-economic systems echoes with a book by Joseph Tainter, 1990. *The Collapse of Complex Societies (New Studies in Archaeology)*. Cambridge University Press, 260 pp.

Vladimir Vernadskii is a Russian scientist of great importance, who unfortunately is still very little known in the West. One of his major findings was the theory of the noosphere, the new state of the biosphere characterized by the dominance of intellectual and moral powers of humans. Vernadskii, V.I., 1986. *Biosphere*. Synergetic Press, 86 pp. – *This is a reprint of the 1929 paper that contains some of these ideas.*

There are several classic books on the Peal Oil issue that could be recommended, such as Kenneth Deffeyes, 2001. *Hubbert's Peak: The Impending World Oil Shortage*, Princeton University Press, 224 pp; Richard Heinberg, 2005. *The Party's Over*, Temple Lodge, 320 pp; David Goodstein, 2005. *Out of Gas: The End of the Age of Oil*, W. W. Norton & Company, 148 pp.

Paul, Robert (2005). *The End of Oil: On the Edge of a Perilous New World*. Mariner Books. 416 pp.

Perhaps the best latest account of the situation with oil and how far we are from its "peak" can be found on The Oil Drum *blog at* http://www.theoildrum.com/.

The "Medium-Term Oil Market Report" from the International Energy Agency (IEA) can be purchased at http://omrpublic.iea.org/mtomr.htm.

For an interesting discussion on how humans make decisions, and why there is a disjoint between ecological and economic reasoning see Wallerstein, I., 2003. The Ecology and the Economy: What is Rational? Paper delivered at Keynote Session of Conference, *World System History and Global Environmental Change*, Lund, Sweden, 19–22 September 2003. http://www.binghamton.edu/fbc/iwecoratl.htm

The Critical Natural Capital concept is discussed by Farley, J. and E. Gaddis, 2007. An ecological economic assessment of restoration. In J. Aronson, S. Milton and J. Blignaut (Eds). *Restoring Natural Capital: Science, Business and Practice*. Island Press: Washington, DC. *The perfect inelasticity* of essential goods and services when they become increasingly scarce is analyzed in Daly, H., Farley, J., 2004. *Ecological Economics*. Island Press (p.197)

The World3 model is described in Donella H. Meadows, Dennis L. Meadows, Jorgen Randers, William W. Behrens III, 1979. *The Limits to Growth*. Macmillan, 208 pp. *For a more recent analysis of the model and the various discussions and controversies that followed, see* Donella H. Meadows, Jorgen Randers, Dennis L. Meadows, 2004. *Limits to Growth: The 30-Year Update*. Chelsea Green, 368 pp. (*For World2 model see:* Jay W. Forrester, 1972. *World Dynamics*. Cambridge, MA: Wright-Allen Press, Inc.)

The Jevon's paradox is well presented in the visionary book by W. S. Jevons, *The Coal Question; An Inquiry Concerning the Progress of the Nation, and the Probable Exhaustion of Our Coalmines*, 1865. URL of an E-Book: http://oll.libertyfund.org/EBooks/Jevons_0546.pdf. *The book would probably not be that remarkable if it was not written almost 150 years ago. What do you think about this quote in the context of the recent Hydrogen Economy hype:* "The fallacious notions afloat on the subject of electricity especially are unconquerable. Electricity, in short, is to the present age what the perpetual motion was to an age not far removed. People are so astonished at the subtle manifestations of electric power that they think the more miraculous effects they anticipate from it the more profound the appreciation of its nature they show. But then they generally take that one step too much which the contrivers of the perpetual motion took – they treat electricity not only as a marvellous mode of distributing power, they treat it as a source of self-creating power."

The GUMBO model is described in R. Boumans, R. Costanza, J. Farley, M. A. Wilson, R. Portela, J. Rotmans, F. Villa and M. Grasso, 2002. Modeling the dynamics of the integrated earth system and the value of global ecosystem services using the GUMBO model. *Ecological Economics*, Volume 41, Issue 3, p. 529–560.

To learn more about the value of ecosystem services, see the special issue of *Ecological Economics*, Volume 25, Issue 1, April 1998. In particular, the famous paper by R. Costanza, R. d'Arge, R. de Groot, S. Farber, M. Grasso, B. Hannon, K. Limburg, S. Naeem, R, V. O'Neill, J. Paruelo, et al. *The value of the world's ecosystem services and natural capital*. p. 3–15.

For more analysis, including philosophical and economic issues of valuation see: G. Daily (Ed.), 1997. *Nature's Services. Societal Dependence on Natural Ecosystems*. Island Press, 412 pp. *For more case studies see:* G. Daily, K. Ellison, 2002. *The New Economy of Nature. The Quest to Make Conservation Profitable*. Island Press, 250 pp.

8. Optimization

SUMMARY

Running a model, we get a glimpse of the system behavior for a given set of parameters and forcing functions. This set is called a scenario. We run a scenario and learn how the system may behave under certain conditions. Suppose we know how we want the system to behave. Can we make the computer sort out through various scenarios to find the one that would bring the system as close as possible to the desired behavior? That is exactly what optimization does for us. If we have some parameters that we can control, the computer will look at various combinations of values that can make the result as close as possible to the desired one.

The software that can help us do it is Madonna. We will look at a couple of simple models to learn how optimization can be performed. For more complex systems, especially if they are spatially explicit, simple methods do not work. We will need to invent some tricks to solve the optimization tasks. Furthermore, in some systems it seems as though the system itself involves an optimization process that is driving the system, as if the system is seeking a certain behavior that is optimal, in a sense. When modeling such systems, it makes sense to embed this optimization process in the model.

Keywords

Objective function, control parameter, constraints, Madonna software, global and local optimum, Monte Carlo method, optimality principle.

8.1 Introduction

In many cases, we want to do more than understand how a system works. We want to figure out how to improve its performance, or, ideally, find the best way it can possibly perform. In these cases we will be talking about *optimization*. We briefly came across this concept in Chapter 4, when we were exploring model calibration. Remember, in that case we also wanted the model to behave in a certain way. There were the data and there was the model output. The goal in that case was to bring the model output as close as possible to the data. The model was the system that we were optimizing.

To optimize, we need two things. First, there should be a goal in mind. We can optimize system performance only with respect to a certain goal. There is no such thing as general, overall or absolute optimality; there is no optimality *per se*. System optimality is always relative. It is relative to the goals that we choose. So before we optimize a system we need to identify the *goal or objective*.

Second, optimization of a system requires that we have some *controls*, some dials that we can turn or levers that we can move. If there is nothing that can be changed in the system, there is no way to improve its performance. Therefore, we cannot make it perform "better" with respect to the goal that has been identified. These controls may be static or fixed, or they may change with time. We may choose a target, take aim, steady the gun and shoot, or we may need constantly to modify the controls – especially if we have a moving target. In addition, there are also certain limits within the system on how much the controls can change. For example, if we optimize for speed, it would be great if an airplane could accelerate at 5 g – that is, five times gravity. However, if this were a passenger plane we would not want to expose humans to such levels of overload. Therefore we would impose a limit, a *restriction*, which would say that the acceleration should not be more than, say, 2 g. We would then be finding ways to move the plane as fast as possible, but ensuring that it still never goes beyond the allowed acceleration.

Optimizing a real system may involve a lot of pain and investment. Imagine, if we do not exactly know what the reaction of a system will be to changed control, we may inadvertently cause serious damage or even complete destruction to the system while searching for its optimal behavior.

On April 25, 1986, prior to a routine shutdown, the reactor crew at Chernobyl-4 began preparing for a test to determine how long turbines would spin and supply power following a loss of main electrical power supply. Similar tests had already been carried out at Chernobyl and other plants, despite the fact that these reactors were known to be very unstable at low power settings. They wanted to find out how to better optimize for reactor operations at low power levels. To carry out the tests under extreme conditions, the operator disabled the automatic shutdown mechanisms. As the flow of coolant water diminished, the power output increased. We do not know exactly how it all happened, but apparently when the operator moved to shut down the reactor from its unstable condition arising from previous errors, a peculiarity of the design caused a dramatic power surge.

The fuel elements ruptured and the resultant explosive force of steam lifted off the cover plate of the reactor, releasing fission products into the atmosphere. A second explosion threw out fragments of burning fuel and graphite from the core and allowed air to rush in, causing the graphite moderator to burst into flames. There is some dispute among experts about the character of this second explosion. The graphite burned for 9 days, causing the main release of radioactivity into the environment. A total of about 14 EBq (10^{18} Bq) of radioactivity was released, half of it being in the form of biologically inert noble gases.

Some 5,000 tonnes of boron, dolomite, sand, clay and lead were dropped onto the burning core by helicopter in an effort to extinguish the blaze and limit the release of radioactive particles. Really nasty things can happen when we try to optimize real-life systems.

Using models of real systems to find optimal regimes makes a lot of sense. If we have a good model of a system, we can define all sorts of sets of controls and subject the system to all sorts of experiments at no risk. We just need to make sure that the model is still adequate within the whole domain of changing control factors.

An optimization task in a general form can be formulated as follows.

Suppose we have a model of a system:

$$X_{t+1} = F(X_t, P, t) \tag{8.1}$$

where $X_t = (x_1(t), x_2(t), \ldots)$ is the vector of state variables of the system at time t, and $P = (p_1, p_2, \ldots)$ is the vector of parameters. We assume a dynamic model that describes the system in time. Therefore, we define each next state of the system, X_{t+1}, as a function, F, of its previous state, X_t, a vector of parameters, P, and time, t.

Suppose we have identified a goal or an *objective* function, which tells us where we want the system to be. The objective function is formulated as a function of model parameters and state variables. It is formulated in such a way that we can then try to minimize or maximize it.

For example, we can be studying an agricultural system that produces grass, $x_1(t)$ and sheep, $x_2(t)$. Our goal could be to maximize the output of goods produced by the system. The objective function would then be based upon the sum of biomasses of sheep and grass, $x_1(t) + x_2(t)$. However, a sum of these two variables does not give us a value to maximize. We should either decide that we want to track the total biomass over the whole time period [0.T]:

$$G = \int\limits_{t=0}^{T} (x_1(t) + x_2(t))dt$$

or agree that it is only the final biomass that we are interested in, because that is when we take the products to the market:

$$G = x_1(T) + x_2(T)$$

If sheep are the only product we are concerned with, we may not care about grass, let sheep eat as much grass as they wish and maximize only the biomass of sheep. Then

$$G = x_2(T)$$

Alternatively, if we are maximizing for the farm profits, we may be getting more revenue from grass than from sheep and we thus want to include both, but with weights that will represent the market values of both goods at the time we sell them:

$$G = p_1 x_1(T) + p_2 x_2(T)$$

where p_1 and p_2 are the prices of grass and sheep, respectively. Clearly, defining the right objective function is a very important part of the optimization task. If we do not do a good job describing what we want to optimize, the results will be useless.

In Chapter 4, as you may remember, the objective function was the difference between model trajectories and the observed dynamics given in the data available. We were then trying to minimize this function by choosing the right set of parameters. Similarly, with sheep and grass, the total biomass produced is a function of climatic conditions, soil properties, fertilizers applied, grazing strategies, etc. Some of these parameters can be changed while others cannot. For instance, we will not be able to change climatic conditions; however, we can change the amount of fertilizers used. The parameters that are at our disposal, that we can change, are called *control factors*. Those are the ones that we can control when trying to maximize (or minimize) the objective function.

Again, in Chapter 4 we had certain parameters that were measured in experiments and we knew their values quite well. We did not want to change those when tweaking the model output. There were other parameters that were only estimated, and those were our control factors – those we could change anywhere within reasonable domains to bring the objective function to a minimum.

Let us put some more formalism into these descriptions. We have a model (8.1), and define an objective function $G(X,P)$, where X is the vector of state variables $X = (x_1, \ldots, x_n)$, and P is the vector of parameters, $P = (p_1, \ldots, p_k)$. For this objective function we then find a

$$\min_{R \in P} G(X,P) \qquad (8.2)$$

subject to $S(X,P) = 0$, and $Q(X,P) \geq 0$. This should be read as follows: we minimize the *objective function* $G(X,P)$ over a subset $R = (p_k, \ldots, p_l, \ldots)$ of parameters P, which are the *control parameters*, provided that the *constraints* (or restrictions) on X hold.

If the controls are scalars and constant, they are also called decision variables. If they are functions and allowed to change in time, they are known as control variables. As we will see below, in many cases we may want to describe our control variable in terms of some analytical function with parameters – say, a polynomial or a trigonometric function. Then your time-dependent control variable becomes formulated in terms of constant parameters, and we can say that a control variable is expressed in terms of several decision variables.

The constraints bound the space where the model variables can change. There may be two types of constraints: equality type ($S(X, P) = 0$) and inequality type ($Q(X, P) \geq 0$).

Note that it really does not matter whether we are minimizing or maximizing the objective function. If we have an objective function G, which we need to maximize, we can always substitute it with a function $G^* = 1/G$, or $G^{**} = -G$, which you can now safely minimize to get the same result.

Since the model trajectories are a result of running the model (8.1), the model becomes part of the optimization task (8.2). Here is how it works.

First, we choose an objective function, which is formulated in terms of a certain model trajectory. We identify the parameters that we can control to optimize our system. For a combination of these control parameters, we run the model and figure out a trajectory. For this trajectory we calculate the objective function, then we choose another combination of control parameters, calculate a new value for the objective function, and compare it with the previous one. If it is smaller, then we are on the right track, and can try to figure out the next combination of parameters such that it will take us further down the minimization path.

For example, in our Grass–Sheep example, suppose we have built a model that describes the growth of both species in terms of a system of differential equations:

$$\frac{dx_1}{dt} = f_1(x_1, x_2, P)$$

$$\frac{dx_2}{dt} = f_2(x_1, x_2, P)$$

where P is the vector of parameters, which include mean air temperature and precipitation, fertilizer application rate, grass harvest schedule, initial number of sheep, market price of sheep and market price of grass.

The objective function G is the profit we make on the farm by harvesting and selling grass several times over the vegetation period, and selling sheep at the end of season.

The control parameters – that is, the ones that we can change – are the fertilizer application rate, the grass harvest schedule, and the initial number of sheep.

The constraint is that the biomass of grass must not drop below a certain value, otherwise we will be putting the long-term vitality of our farm at risk: if we wipe out all the grass it will not be able to regenerate and the sheep will destroy the soil cover, causing erosion and massive loss of topsoil. Something we certainly want to avoid by all means.

Exercise 8.1

1. Imagine that your system is an airplane and you need to land it. Formulate an optimization task that will allow you to safely touch the ground at the airport nearby. What is the objective function? What are the control parameters, and the restrictions?
2. Think of an optimization problem of your own. Define the system, the objective function, the controls and the restrictions.

In order to perform the optimization task, we need to set up a procedure that will browse through various combinations of control parameters; for each set calculate the trajectory of the model, making sure that the trajectory stays within the limits defined by the constraints; then, based on this trajectory, define the value for the objective function and compare the current value with the ones that have been calculated previously for different parameter sets. Figure 8.1 presents this algorithm that we are to follow. The real question is, how do we choose the next set of control parameters based on the results that we have generated for the current one?

This is what the various optimization algorithms are designed to do. There are many different methods, the details of which are clearly out of the scope of this book. Here, we can only briefly discuss some of the basic ideas that may be useful in understanding how these algorithms may work.

It should be easier to understand some of the optimization techniques if we look at the simplest case. Suppose we have a function of one variable, $F(x)$, over an interval $x \in [a,b]$. We need to find the value of x where $F(x)$ is minimal: $\min_{x \in [a,b]} F(x)$. How do we do it? It depends largely upon the function F that we have. The real problem is to find the *global* minimum, instead of being trapped in one of the multiple *local* minima that may exist for this function.

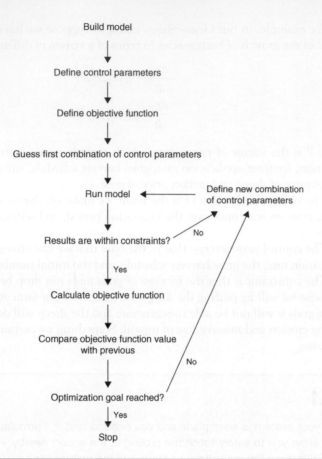

Figure 8.1   The optimization process.
When solving an optimization task, we normally go through all these steps. The real challenge that is solved by optimization methods is where to make the next step, how to find the next combination of control parameters.

If $F(x)$ is a linear function (Figure 8.2A), obviously we get a minimum on one of the ends of the $[a, b]$ interval. Either $x = a$, or $x = b$ delivers a minimum to the function. When generalized to several independent variables, we find ourselves in the realm of a special branch of optimization called linear programming. Since for linear functions the minimum is always on the boundary, that is where the linear programming methods search. They are designed in such a way that they go over all the possibly complex boundaries of the function domain in the multivariate space.

If $F(x)$ is non-linear but nice and smooth (Figure 8.2B), the most common minimization technique is the so-called gradient method, or the "steepest descent" method. In our case of one independent variable, we can choose a point, x_1, calculate the value $F(x_1)$, and then choose the next point, x_2, such that $F(x_2) < F(x_1)$. We may need to try several directions before we find such a point. If we have several variables, we will move in the direction of the variable that delivers the lowest value to the function in the vicinity of x_1. We then move to this next point, x_2, and repeat the same procedure to find x_3. And so on, until we realize that, whichever direction we go, we are only increasing the value of the function.

This algorithm works really well unless the function we are dealing with has several local minima, like the function in Figure 8.2C, which has a minimum on the

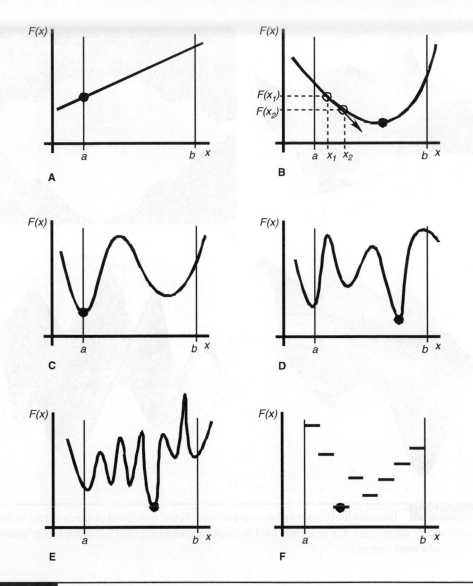

Figure 8.2   Searching the minimum in some simple functions.
We can see that a local minimum can be quite different from the global one.

boundary, or the function in Figure 8.2D, which has two local minima inside the domain for x. If we follow the algorithm above, chances are that we will find the other minimum and will stay there, never realizing that there is yet another minimum which is even smaller. The solution in this case could be to try several starting points for the gradient search algorithm and see where we end up going downhill. Then we can compare the values we get and choose the minimal one.

As the function $F(x)$ becomes more poorly behaved, with strong non-linearities as in Figure 8.2E, the gradient search becomes almost impossible. The chances that we will hit the global minimum are becoming very low. In this case we might as well do a random search across the whole interval $[a, b]$, picking a value for x, x_i, finding the value of $F(x_i)$, then picking the next value x_{i+1} again at random and comparing $F(x_i)$

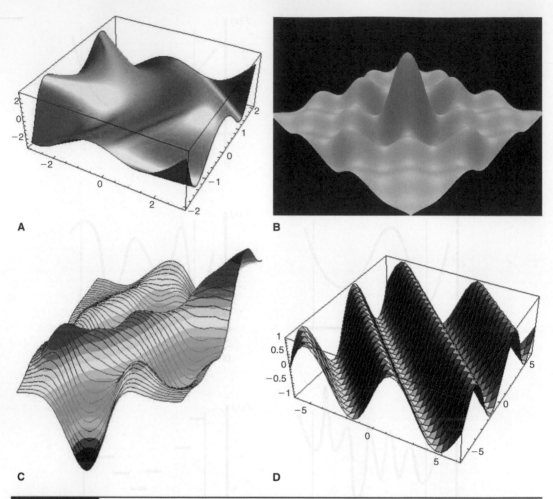

A

B

C

D

Figure 8.3   Objective functions may have very unusual forms, which makes it only harder to find good methods for optimization. It is especially hard to do global optimization, yet that is the kind of optimization which is usually most desired.

and $F(x_{i+1})$, keeping the lowest value for further comparisons. If we are lucky playing this game of roulette, we may eventually get pretty close to the real global minimum for our function. Like other methods based on random search this method is known as the Monte Carlo method of optimization, after the famous casino town in Europe. Just as when playing roulette we do not know what the result will be (except that most likely we will lose!), here too we keep randomly picking a set of control parameters from their domain of change, hoping that eventually we will aim somewhere close enough to the global minimum. It also may be helpful to combine the random walk algorithm with the gradient search, when for each randomly chosen value of x we also make a few steps in the direction of the steepest decent. In this way we avoid the unpleasant possibility of being lucky enough to pick a point somewhere really close to the global minimum and then moving away from it, only because we were not close enough.

The gradient search is entirely inappropriate for piecewise linear or categorical function, like the one in Figure 8.2F. In this case we cannot even define the direction of the steepest decent by exploring the vicinity of a point of our choice – we get the same results for $F(x)$ unless we jump over to the next segment. For such functions, it is only the random walk or some variations of it that are appropriate.

There are numerous other optimization methods available these days. Among them are the Genetic Algorithms and other evolution strategy methods that try to mimic the way genes mutate in search for an optimal configuration. There is the Simulated Annealing algorithm, which, by analogy with the physical process of annealing in metallurgy, on each step replaces the current solution by a random "nearby" solution, chosen with a certain probability. The optimization problem is not an easy one; the objective function can become very complex, especially when it involves multiple variables, and it becomes quite hard to find the minimum in functions like the ones shown in Figure 8.3. It takes a lot of mathematical creativity and computer power to find these optima. Still, in many cases this kind of computer simulation is a much safer and cheaper alternative than many other kinds of optimization.

8.2   Resource management

Let us consider a simple example of a system where optimization can help us find the best way to manage it. Suppose there is a natural resource that we wish to mine to sell the product to generate revenue. The resource is limited; there is only a certain amount of this resource that the mine has been estimated to contain. How do we extract the resource in order to generate the most profit? There are also a few economic considerations that we need to take into account. First, clearly our profits will be in direct proportion to the amount of resource sold. But then also our costs of production will be proportional to the amount produced and inversely proportional to the amount of resource left. That is, the more we mine and the less resource is left, the harder and more expensive it will be to get the resource. In addition, we should realize that the price we charge for the product can go down if we dump too much of it on the market. The law of demand tells us that the more goods are produced and offered, the less will be the price that we can sell these goods for.

Since we know that we will be doing optimization, let us use Madonna to put the model together. Just like Stella, Madonna uses a set of icons in its interface to formulate the model. The model as outlined above is presented in Figure 8.4. For anyone familiar with Stella, it should be quite easy to understand this diagram. The cylinder tanks are the reservoirs, with flows taking material in and out. The balls are the parameters and the intermediate variables. It is almost like Stella, but drawn in 3D. Even the clouds that represent the exterior of the system in Stella are now replaced by the infinity signs ∞, which look almost like clouds. As in Stella, if we double click on any of the icons, we open up a dialogue box that allows us to specify the formulas or parameter values to use. For example, for the price variable you get:

As you draw the diagram you put together the equations of the model. Below are the Madonna equations generated for this model with comments added in brackets.

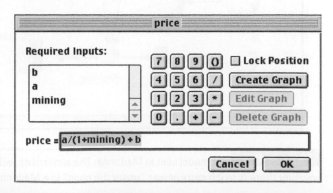

{Top model}

  {Reservoirs}
   d/dt (Resource) = –mining
      INIT Resource = 10000
   d/dt (Profit) = +to_profit
      INIT Profit = 0
{Flows}
mining = if (qq <= 0) then 0 else if (Resource > qq) then qq else Resource
         {We are checking that there is enough resource to extract}
to_profit = price*mining-costs
         {Proceeds from sales of products minus costs of operations}
{Functions}
   qq = d*TIME^2 + e*TIME + f
         {This is the amount extracted. We define it as a function of time to be able
           to find the optimal extraction strategy, as explained below}
e = 0.1
f = 20
d = 10
         {Parameters of the extraction function}
price = a/(1 + mining) + b
         {Price of goods modified by the amounts of goods produced. Price can
           increase substantially if there is very little supply, mining is small}
costs = cc*mining/Resource
         {Costs of mining are in proportion to the volumes extracted. Costs grow as
           the resource becomes scarcer}

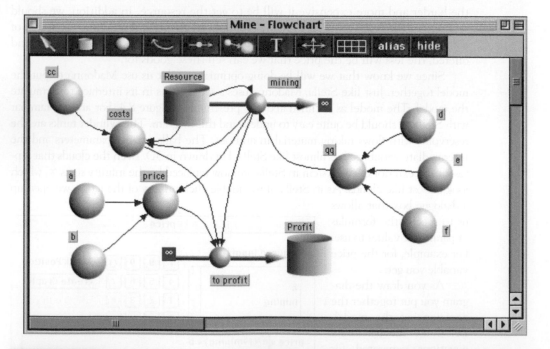

Figure 8.4   A model built in Madonna. The similarities with the Stella interface are quite clear;
however there is much more power "under the hood" in a Madonna implementation.

a = 100
b = 10
cc = 0.3

In addition to the equations, Madonna assembles the parameters window, which is convenient to manage all the parameters in the model:

This window contains all the parameter values from the equations, as well as the initial conditions and the simulation control parameters: STARTTIME – when to start the model run; STOPTIME – when to stop; DT – the time-step for the numerical method; and DTOUT – the time-step for output. In this window we can also choose the numerical method to solve the equations. The (STOPTIME-STARTTIME) in our model actually tells us what the lifetime is for the mine that we have in mind – that is, for how long we plan to operate it.

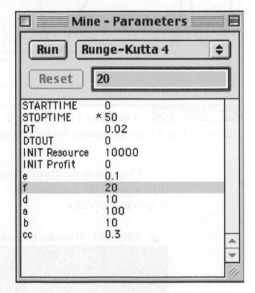

Once the model is defined we can set up a couple of graphics for output and run the model. Note that the pattern of extraction is defined by a parabolic function described in

$$qq = d * TIME^2 + e * TIME + f \qquad (8.3)$$

If d = e = 0, we get constant rate of extraction throughout the lifetime of the mine. By changing f, we can specify how fast we wish to extract the resource. By changing d and e, we can configure the rate of extraction over time, making it different at different times. We can set up some sliders and start our optimization by manually changing the values for the parameters.

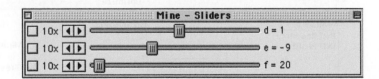

As noted above, every time we move a slider, Madonna will calculate a new set of trajectories, so we can get some idea of how changes in parameter values impact the model dynamics. Figure 8.5 gives a sample of model output for the parameter values defined in this slider window. Note the spike of price around year 5, when mining was very low and supply of the resource plummeted. The Profit at the end of the simulation is around $103,817, which is quite high, as we can easily see by trying to adjust the parameters in the slider window. The question is, can we further increase it by finding the very best combination of control parameters?

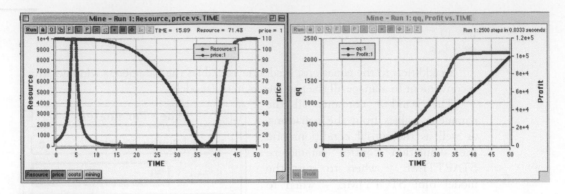

The reason we chose Madonna for this analysis is because it can run optimization algorithms automatically. Indeed, in the "Parameters" menu let us choose the "Optimize" option:

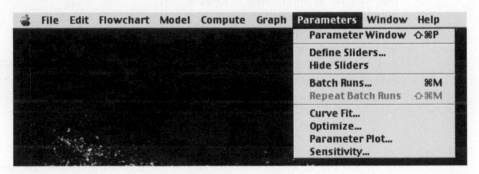

Another dialogue box opens up, where we are advised to choose the parameters that are allowed to change (control parameters), and to specify the function that is to be minimized (the objective function):

In our case, the control parameters will be the d, e, and f in (8.3); these define the pattern of resource extraction over time. Madonna can only minimize, so our previous consideration of maximization as the reverse of minimization comes in quite handy. The objective function to minimize can be the negative of the Profit, since our goal is to maximize Profit. Obviously, minimizing −Profit is the same as maximizing Profit. This is one of the ways we can convert a maximization task into a minimization one. Note that Madonna, when doing the optimization, conveniently uses the ending values of the variables in the objective function. So in our case we will be indeed optimizing for the profit at the end of the simulation, and not at some intermediate steps.

For each of the control parameters we are asked to set the limits of their allowable change. The smaller the intervals we choose, the easier and quicker it will be for the algorithm to find a solution. On the other hand, we need to keep the ranges broad enough to account for a variety of different scenarios of resource extraction.

When choosing these restrictions on parameters, it is essential to take into account the ecological meaning of the parameters we specify. For example, if some of the control parameters were presenting, say, rates of change, or values related to biomass or other stocks, it would be clear that they need to be clamped to be positive. There is no need to search for optimal solutions that would include negative growth rates of populations. These controls are not possible, so there is no need to consider them as options.

In the case of (8.3) there are no obvious ecological conditions for d, e, and f, except that we do want to make sure that the resulting scenario will produce a positive flow of resource, qq ≥ 0, for 0 < TIME < 50. We may take a closer look at (8.3) and come up with some relationships between parameters that would keep qq ≥ 0, or we might play with the sliders in Madonna and see what combinations of parameters make qq become negative and then try to exclude them from the optimization. However, in our case this may not be so important, because we have built the condition of qq being positive into the model formulation.

Indeed, when describing the flow for "mining" we have put:

$$\text{mining} = \text{if (qq} <= 0) \text{ then } 0$$

Effectively, we have implemented a constraint that is usually part of a general optimization task (see the definition in section 8.1), but which has no special place in the optimization procedure in Madonna. In our particular model it means that we do not necessarily have to limit the control parameters to such values that would guarantee that qq ≥ 0. This will be taken care of by the model.

Anyway, we still want to set some limits to these parameters, making sure that the rate of extraction is not overly high. For d we choose –2 < d < 4. This is because larger values of d cause very big differences in the extraction rate over the lifetime of the mine – something we probably want to avoid. For e, we choose the interval −20 < e < 30. The qq is not very sensitive to changes in e, as we can see from playing with the sliders. For f, let us choose a larger interval of 0 < f < 300. This is to allow high enough rates of extraction if the algorithm chooses a constant rate.

Next, to run optimization we are required to specify a couple of "guessed" values for each parameter. Remember, when discussing how most optimization algorithms work, we mentioned that in most cases we solve the equations for a couple of fixed parameter values and then compare which of the solutions brings a lower value to the objective function. That helps us to determine in which way to go in search of the optimum. While we do not know how exactly the optimization algorithm works in Madonna, most likely it also needs some values to initialize the process, and probably those are the guesses that we need to specify. Certainly the guesses need to be within the parameter domains (that is, larger than the Minimum value and less than the Maximum). They should also be different. Other than that, they can really be quite arbitrarily set. However, it may help to rerun the optimization with a set of different guesses. This may help us step away from a local minimum and find a better solution and a different combination of control parameters.

Finally, we can press the "OK" button, sit back and watch the optimization magic happen. The model will be rerun multiple times with different combinations of parameters in search for the one that will make the objective function the smallest. While optimizing, a box will appear reporting how many model runs have been made and what the current value of the objective function is:

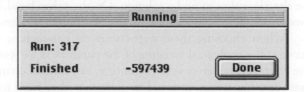

We will also see that apparently there are several algorithms involved in the optimization process and a combination of them is used.

Running optimization with the chosen settings returns a set of control parameters, d = –2.26029e–6, e = 1.14093e–4, f = 214.536, with which the ending profit is $104918. If we try other combinations of parameter values, we do not seem to get anywhere better than that. We can round up these parameter values to d = 0, e = 0 and f = 214.54, and see that we are talking about a constant extraction rate as an optimal strategy of mining (Figure 8.6).

Note that there is yet another parameter in the Optimize dialogue box, which is called "Tolerance." This specifies the accuracy of the optimization, telling us when to stop looking for a better combination of parameter values. By default Tolerance is set to 0.001, and that was the value we used in our computations above. If we try

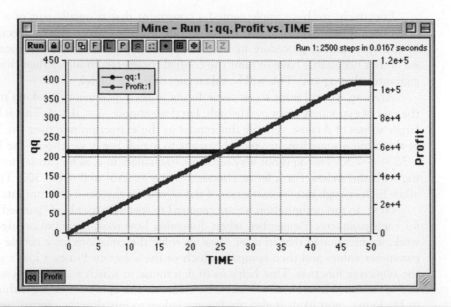

Figure 8.6   Model output for the optimized set of parameters that turns out to stand for a constant extraction scenario.

to make Tolerance large enough, say 0.1, we will notice that it takes less time to run the optimization, fewer model runs will be required, but the results will become way more sensitive to the initial guesses that we made for the parameters. Suppose we take parameter d, $-2 < d < 4$, and use the following two guesses, $d_1 = -1$, and $d_2 = 4$. If we now hit the OK button, we will get back the following values for the control parameters: $d = 0.27$, $e = -6.69$, $f = 180.53$. The resulting pattern or the resource extraction (Figure 8.7) looks quite different than what we had above, while the Profit with these parameters is almost the same as before: Profit = \$104918.

Now, with the same tolerance, let us make another guess for parameter d: $d_1 = -1$, and $d_2 = 3$. Now the solution to the optimization process comes with control parameters: $d = -0.12$, $e = 6.17$, $f = 155.08$. Again there is yet another different pattern for extraction (Figure 8.8), and what is most surprising the Profit is again \$104918, which is the same as with a constant extraction rate. So what is going on?

Apparently there are several local minima that are quite close to the global one. To stop the optimization process Madonna calculates the deviation of the objective function between different model runs, and once the change becomes less than the Tolerance it stops. When the tolerance is large enough it is more likely to stop at a local minimum, instead of continuing to search for a better solution elsewhere. That is how we get into the Figure 8.7 or Figure 8.8 solutions. The Figure 8.8 solution is probably still a little worse than what we generated when running the model with the smaller tolerance (Figure 8.6), but we cannot see it because profit is reported with no decimals. In any case the difference is probably negligible, but it is good to know that the very best solution is very simple: just keep mining at a constant rate, and the model can tell us what that rate should be. However, if the constant extraction is not an acceptable solution for some other reasons, we can still come up with alternative strategies (Figures 8.7 and 8.8) which will produce results quite identical to the optimal strategy. Actually, if we compare the pattern of resource depletion for all these strategies, it is clear that the differences are quite small.

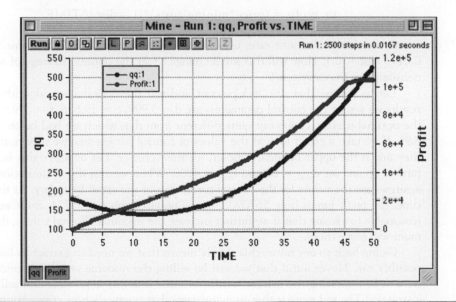

| **Figure 8.7** | Optimization results for a different tolerance parameter. |

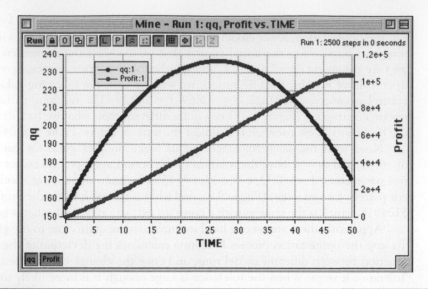

Figure 8.8   Yet another quasi-optimal strategy produced from a different initial guess of parameters.

Let us now slightly modify our system and introduce a discounting rate into our calculations. The way the economic system works today is that $1 today is worth more than $1 tomorrow. When we have a growing economy, the idea is that if we have this $1 today we can always invest it into something that will have a positive return and therefore tomorrow we will have $1+Δ. As a result, in our economic calculations we have to take this discounting in account when we calculate our future profits: the money that will be coming in later on will be worth less. This can be easily taken into account if we add a small modification to our model:

$$to\_profit = (price*mining\text{-}costs)*(1 - disc)\text{\textasciicircum}TIME$$

where disc is the discount rate, usually varying between 1 and 10 percent. Hence $0.01 < disc < 0.1$. How will this small change affect the optimal strategy of resource consumption in our system?

Let us assume that disc = 5 percent and run the optimization algorithm. The results we get for the control parameters are d = 3.999, e = 29.656, f = 299.998, and the optimal strategy comes out quite different from that which we had before (Figure 8.9). If we take a closer look at the values of control parameters, we may notice that they are at the upper boundary chosen for their change. Let us move this boundary further up and set d_{max} = 20 and f_{max} = 4000. When we run the optimization, once again we get the values for the control parameters at the upper boundary. At the same time, the profit jumps from $67090.1 in the previous run to $93774. So it would be reasonable to assume that if we further increase the maximal allowed values, the optimum will move there.

Going back to our mine, this simply means that we need to extract as fast as we possibly can. Never mind that we will be selling the resource at a lower price since we will be saturating the market – still just mine it as fast as possible and sell. Quite a strategy! The sad news is that in many cases that is exactly how conventional economics deals with natural resources. The economic theory tells us to mine them

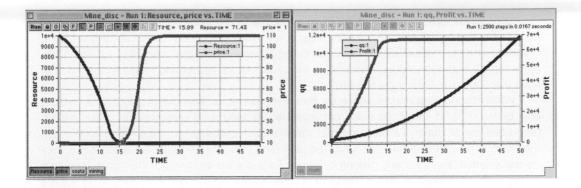

Figure 8.9   Optimization results from a model with discounting.
The optimal strategy is dramatically different, and now it is better to extract as fast as possible.

as fast as we can. If you have a forest, cut it down, then invest the profits on Wall Street. If you have a coal mine, dig it out, then do the same – not a very sustainable way to manage the resources! However, if optimality in terms of profits is all that drives the process, then this is the solution.

While discounting is a major part of the backbone of conventional modern economics, it is clear that there are quite a few problems associated with it. An obvious justification for discounting is the so-called marginal opportunity cost of capital. If a firm can keep reinvesting its assets in activities that generate 7 percent returns (which is the average real rate of return on long-term stock market investments), $1 invested today will become $114 in 70 years time. Conversely, a firm should count $114 in 70 years as worth only $1 today (an investor will presumably be dead in 70 years, and might count $114 dollars at that time as nothing today). These calculations make sense only as long as there are continuous investment opportunities that average 7 percent. However, over the very long term, no investment opportunity can offer returns greater than the growth rate of the economy, or else the part must become greater than the whole. So if the discount rate is assumed at 7 percent while the economy grows at 2 percent in a year, this is already cheating. But then if there is a steady state economy, or a declining economy, then the discount rate should be zero or negative!

Among the more serious problems, conventional exponential discount rates, even very low ones, treat catastrophic events that occur far enough in the future as essentially irrelevant today. Global warming is a case in point. Some economic analyses of global warming discount the future at rates as high as 6 percent (IPCC, 1995). At such rates, we would not spend $2,500 today to prevent a $30 trillion dollar loss (the approximate gross global product today) in 400 years. People discount due presumably to both impatience and uncertainty – we have short and uncertain life spans, and may not even be alive when the future arrives. Empirical evidence shows that individual discount rates can be very high in the short term – how many people in America (a country with low interest rates) are holding credit-card debt at 18 percent or higher? However, while economists use such evidence to justify exponential discounting, empirical studies actually find that while an individual might prefer $100 today over $200 next year, he or she is likely to prefer $200 in 10 years over $100 in 9 years. This suggests that people discount at a fairly high rate over a short time horizon, but at a lower rate over a longer time horizon. In other words, if we plot the discount function (the weight placed on payoffs received at different points in time) on the y axis against time on the x axis, it takes the shape of a hyperbola. Economists refer to such discount rates as "hyperbolic".

Curiously, another justification for discounting the future value of natural resources is that new technologies will provide superior substitutes, making existing resources less valuable. The irony is that, more often than not, technology seems to develop new uses for resources more rapidly than substitutes, increasing their value – which if anything would suggest a negative discount rate. Oil is a case in point.

While discounting makes sense for purely economic or financial projects (one person, one generation), it becomes quite controversial for long-term projects and when natural resources are at stake. We assume that these resources bring us higher value today than they will bring in the future, to future generations. This intergenerational discounting may be really unfair, since we are making decisions for our children in which they have very little say at present. Moreover, if the supply of non-renewable resources decreases while we are assuming a growing economy (a must to justify discounting) that obviously implies a growing demand for these resources, we should be then also assuming a growing price of these scarce resources. Forget the moral issues; pure economic logic should say that the future flows of profits may not be lower (as calculated by discounting) but might well be higher (because of the growing price per unit of resource).

An individual may have a very high discount rate as an individual, but a lower discount rate as part of a family or part of a country. For example, parents might smoke or over-eat, valuing the present well over the future, but will simultaneously invest heavily in their children's education, even when the net present value of such investments is negative (Sumaila and Walters, 2004). People give up their lives for their families or for their countries, sacrificing everything now for future gains they will never see. Government and religious institutions (larger hierarchical levels of society) have repeatedly invested in infrastructure projects with life spans of hundreds or even thousands of years, some of which have taken several generations to build.

Exercise 8.2

Change the model to take into account the growing future demand and therefore a higher future price for the resource. Introduce another parameter which will describe the growing demand for the resource. Make the price a function of both demand and supply. How does this change the results of optimization? Does the discounting still play an important role?

8.3   Fishpond

Let us take a look at another, more complex model where optimization can be useful. The system is a fishpond, an artificial aquatic ecosystem designed to grow fish.

Since ancient times, humans have used fishponds to meet their nutritional – and particularly protein – requirements. Fish breeding was highly developed in ancient China. High efficiency is achieved only with optimum values of the control parameters, such as input of feed and fertilizers, and aeration of the water body, and with optimum choice of the seed piece characteristics. The management affects the entire fishpond ecosystem, often resulting in unpredictable and not always desirable changes in the ecodynamics of the pond. A model can help figure out the relationships between different variables and control factors, and hopefully avoid the unwanted regimes.

Let us start with the ecosystem part of the model. Here we will present a simple version of a predator—prey, or rather resource—consumer, model. Let us assume a pond that is entirely driven by artificial feed. A Madonna model can be put together as in Figure 8.10. "Feed" is a variable that presents the accumulation of artificial feed in the pond. "Fish" is the biomass of fish. Note that there is yet another state variable, "Detritus." An important condition of fish survival is that there is a sufficient level of dissolved oxygen in the pond. Oxygen is consumed for fish respiration, but is also, very importantly, utilized for decomposition of dead organic material – detritus. Detritus is formed from the products of fish metabolism, excreted by fish, as well as from the remains of the feed that are not utilized by fish and stay in the pond.

Detritus is an important factor in the pond ecosystem because as its concentration grows, anoxia or anaerobic conditions are most likely. As the concentration of oxygen falls below a certain threshold, fish die off. If we assume that the oxygen consumption increases as detritus concentration grows, then perhaps we may get away without an additional variable to track oxygen and simply assume that the fish die-off is triggered by high detritus concentrations. All these processes are described by the following equations in Madonna:

```
{Reservoirs}
    d/dt (Fish) = +Growth – Mortality
        INIT Fish = 0.1
    d/dt (Feed) = –Growth + Feeding – Loss
        INIT Feed = 0
    d/dt (Detritus) = +Accum – Decomp
        INIT Detritus = 0.1
{Flows}
    Growth = if Feed > 0 then C_growth*Feed*Fish/(Feed + C_Hs) else 0
        {We use the Mono function with saturation for fish growth}
    Feeding = if (C_feed > 0) then C_feed else 0
        {There may be many ways we plan to feed the fish – let us make sure that the
        scenario never goes to negative values}
    Mortality = (C_mort + Detritus^4/(C_mort_d^4 + Detritus^4))*Fish
        {Mortality is made of two parts: first is the loss of biomass due to metabolism
        and respiration, second is die off due to anoxia described by a step function that
        kick in when concentration of detritus becomes more than a certain threshold.}
    Loss = C_loss*Feed + Growth*0
        {A part of feed that is not consumed by fish turns into detritus. A trick to make
        sure that this flow is calculated AFTER fish Growth is taken care off. In order to
        calculate this flow we enforce that fish growth should be already calculated.}
    Accum = Loss + Mortality
    Decomp = C_decomp*Detritus
        {Natural decomposition of detritus due to bacterial and chemical processes}
{Functions}
    C_growth = 0.2
    C_mort = 0.02
    C_loss = 0.6
    C_feed = A*(TIME + B)^2 + C
    C_mort_d = 20
    C_decomp = 0.1
    A = 0.001
    B = –20
    C = 0.1
    C_Hs = 0.3
```

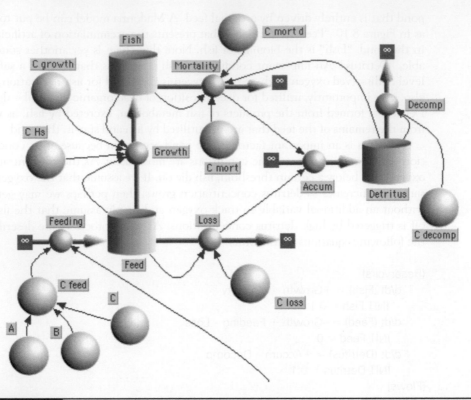

Figure 8.10   A fishpond model formulated in Madonna.

Note that in a way similar to the previous model, we have described the scenario of feeding as a second-order polynomial – except that this time we have formulated the polynomial in a somewhat different way. Instead of $A * T^2 + B * T + C$, we use the formula $A * (T + B)^2 + C$. By rearranging the coefficients, we get a much better handle on how we control the form of the feeding curve (Figure 8.11). While in the original form of the polynomial the role of parameters A and B was somewhat unclear, in the new formula they have a distinct impact on what kind of feeding strategy we generate. This is one of the examples of how by using the right approximation we get a better control of the changes that we try to introduce to our system.

Of course, we could always use the graphic function to input the feeding scenario. Like Stella, Madonna allows input of a graphic of an independent variable (Figure 8.12). We just draw the line by dragging the cursor or by inserting values into the table. This will certainly give the ultimate flexibility in terms of the form of the function we can create; however, every time we wish to modify this function, we will need to do it manually. This involves opening the graphic, changing the function, closing the graphic, running the model, seeing the results – then repeating the whole thing again. This is nice for manual operations, but there will be no chance to use any of the optimization algorithms available. By describing the input in a mathematical form as a function, we have several parameters that control the form of the input, and that can be changed automatically when running optimization. This is clearly an advantage, which comes at a cost: we will need to stay within a class of curves that will be allowed as input to the model. For example, no matter how we

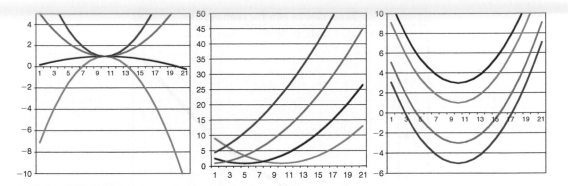

Figure 8.11 Changes in the feeding strategy resulting from variations of the three parameters in the equation. The role of each coefficient is clearly seen: A makes the curve either convex or concave, B shifts the graphic either horizontally, to the right or left, C shifts it vertically, up or down.

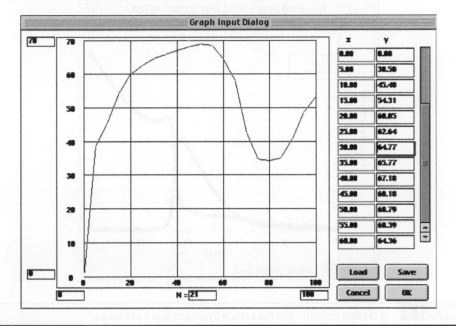

Figure 8.12   The graphic tool in Madonna – it looks a little rugged, but functions as well as in Stella.

change the parameters in Figure 8.11 we will not be able to generate the input as in Figure 8.12. This does not mean that there are no other functions out there that can be used to reproduce the curve in Figure 8.12. For example, by going to a polynomial of higher order – say, three or above – we will be able to get pretty close to the form of the input that is in Figure 8.12. However, this will now come at a cost of more parameters, more complexity, longer computation times, etc. There are always trade-offs.

So let us first choose some feeding strategy and make the model produce some reasonable results. With A = 0.001, B = –10, C = 0.2, and the rest of the parameters defined above in the equations, we get the feeding scenario and the fish dynamics

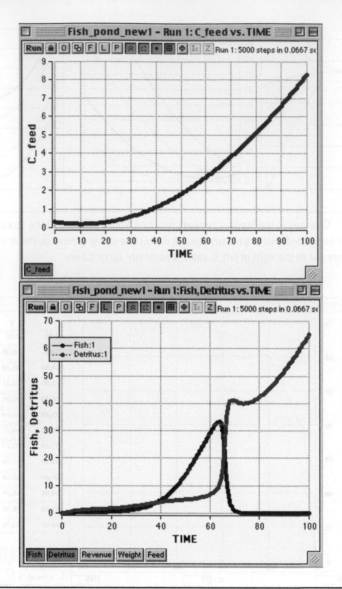

as shown in Figure 8.13. We see that the fish population gradually grows until a certain point where the amount of detritus that is excreted and produced by the decomposition of feed exceeds a threshold and causes massive die-off of fish. The fish population crashes, further adding to the detritus pool. Clearly, this is a condition we want to avoid. So we should use caution when supplying the feed into the pond: there is always a risk of a fish-kill if we let it grow too fast. With the existing scenario we can see that there is no huge accumulation of unused feed while the fish are still present, so we can conclude that in this case the fish-kill is really caused by products of fish metabolism, not overfeeding.

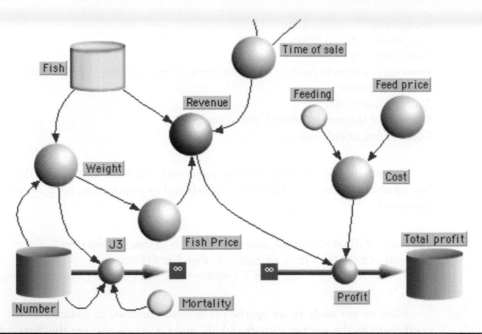

Figure 8.14   The economic part of the model that calculates the profits from fish sales and the expenses of fish feeding.

If the growth coefficient for fish were smaller, more feed would be channeled to detritus and its accumulation could occur even faster. The feeding process is clearly an important factor in this ecosystem performance. If we were to optimize this system we would seek a feeding scenario that would produce the highest fish biomass at a certain point, so that fish could be harvested at that time and sold for a profit. However, it could be a little more complex than that, since feed is also not cheap and has to be purchased at a cost. So it is more likely that we would be optimizing for the net profit rather than just the total fish biomass.

Let us build an economic submodel that will take care of all these additional processes and flows of money. The Madonna diagram is shown in Figure 8.14. Note that in addition to the ghost state variable for fish, we have two state variables: one to track the number of fish in the pond and the other one to calculate the total profit from the pond operation. The number of fish is needed to keep track of the average weight of the fish. We want to take into account the fact that larger fish with higher weight are more likely to cost more on the market. Another important parameter that we can introduce and use as a control is the Time_of_sale parameter, which tells us when exactly we will harvest the fish and sell them. The new Madonna equations are:

{Reservoirs}
d/dt (Total_profit) = +Profit
    INIT Total_profit = 0
d/dt (Number) = −J3
    INIT Number = 100

{Flows}
Profit = Revenue-Cost
J3 = if Weight > 0 then Mortality/Weight else Number/DT
    [Loss in number of fish when they die]

{Functions}
Fish_Price = 10 + 0.2*Weight
Feed_price = 2
Revenue = if Time > Time_of_sale AND Time < Time_of_sale + 2 then Fish_
Price*Fish else 0
Time_of_sale = 100
Weight = if Number > 1 then Fish/Number else 0
Cost = Feed_price*Feeding

By introducing the Time_of_sale parameter, we have also modified two earlier equations to describe the harvest of fish and to stop feeding after all the fish is harvested:

Feeding = if (Time < Time_of_sale and C_feed > 0) then C_feed else 0
Mortality = if (TIME > Time_of_sale + 1) then Fish/DT else
                    (C_mort + Detritus^4/(C_mort_d^4 + Detritus^4))*Fish

Now we are ready to set up the optimization process. In addition to the three coefficients in the feeding scenario (A, B, and C), let us also use the Time_of_sale parameter as a control. We have already realized that adding much feed is hardly a good strategy, so probably A is not going to be large – otherwise over the 100-day time period we may get quite high concentrations of feed, which will be damaging to the system. Let us set the limits for A as $0 < A < 0.001$. Negative numbers are also excluded, since we already understand that it makes little sense to add more feed at first, when the fish biomass is low, than later on, when there is more fish to consume the feed. So most likely the winning strategy will start low and then increase to match the demands of the growing fish. Let us leave some room for B: $-30 < B < 20$. As we remember, B places the minimum point on the curve relative to TIME = 0. We do not need a very large interval for C, which designates the minimal value (if A is positive) or the maximum (otherwise). Let us set $0 < C < 4$.

Now, by choosing the guess values somewhere within these ranges, setting the goal function to be minimized to –Total_profit, and pressing the OK button, we can start the optimization algorithm. This will return a value of Total_profit = 246542 after some 405 iterations of the model. The optimized control parameter values are Time_of_sale = 80.2895, A = 0.001, B = –26.5, and C = 0.122. The dynamics of Total_profit is are shown in Figure 8.15. We see that at first we get a loss, because we only spend money on feed purchases, but then at the end, when we finally sell the fish, we end up with a profit of 264. Using sliders, we can explore the vicinity of the optimized control parameters and see that apparently, indeed, the values identified are delivering a minimum to the objective function, so there is no reason to expect that we can find a better solution.

In some cases it is worth while exploring some very different areas of the control domain, just to make sure that the optimum we are dealing with is indeed a global and not a local one. It does look, for this given combination of parameters, like the optimum described above is global.

Let us check out how the weight factor in fish price affects the optimization results. Will we get significantly different results if there is a huge preference for really big fish and the price of such fish is considerably larger than the price for small fish?

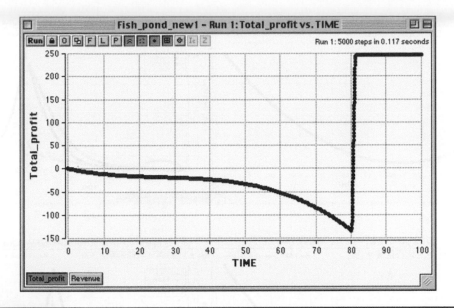

Figure 8.15   Dynamics of Total_profit for optimized model parameters with Time_of_sale as a control parameter.

Clearly there is some difference. Let us change the impact that weight has on price, using the formula: Fish_Price = 10 + 2 * Weight. Here we have increased the effect of Weight by an order of magnitude, changing it from 0.2 to 2. The optimized parameters are somewhat different: A = 0.0007, B = −30, C = 1.075, and Time_of_sale = 67. The Total_profit with these parameters is 579 (Figures 8.16A, 8.16B). Note, however, that the optimal value reported for B is on the lower limit chosen for this control parameter. This should cause some concern, since very likely it means that the algorithm would rather use a yet smaller value for B, but was not allowed to go there. Let us release this constraint and set B: −50 < B < 20.

Rerunning the optimization procedure, we find a different set of controls: A = 0.000347, B = −44, C = 1, and Time_of_sale = 83. The Total_profit with these parameters is 1537.77 − an almost threefold increase in comparison with the previous experiment (Figure 8.16C).

If we compare the performance of this system with what we can get from a system where Fish_Price = 10 + 0.2 * Weight, we will see that indeed the new feeding strategy results in fish that are fewer but larger, so we can take advantage of the higher market prices for bigger fish. Also note that if we rerun that 0.2 * Weight model with the larger parameter interval for B, −50 < B < 20, we will generate a slightly

> *If the optimum is found at the boundary of a parameter's domain or close to it, make sure that this constraint is real and important. You may be able to release it and find a much better optimal solution.*

higher Total_profit = 246.574. The control parameters we end up with will be A = 0.001, B = −27, C = 0.12, and Time_of_sale = 81. The differences are small, but noteworthy. Apparently the value of B = −26.5 was still a little too close to the boundary for the algorithm to move further below to −27, which gives a better result.

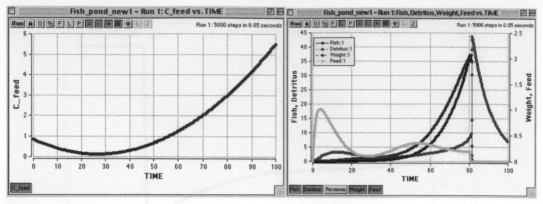

A

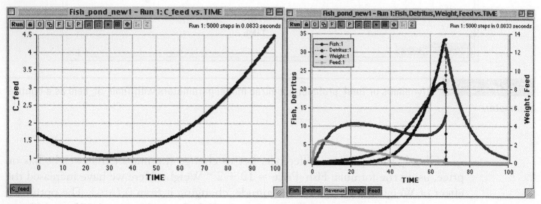

B

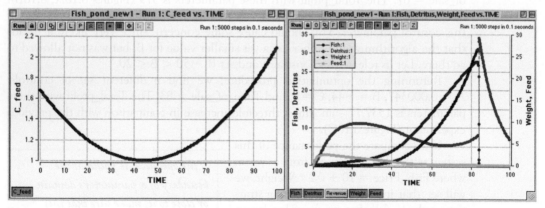

C

Figure 8.16   Experiments with the fish weight as a factor in the fish price and therefore the total profit.

A. Optimized results for low importance of weight in fish pricing: Fish_Price = 10 + 0.2 * Weight; B. Results for a higher preference for larger fish: Fish_Price = 10 + 2 * Weight. For optimal results we get higher fish weight, while total biomass is actually lower. The optimization process hits a constraint for B, which is set to B > −30; C. Removing the constraint allows a better optimal solution. We get a smaller number of fish but with much higher weight, which is rewarded by the objective function.

So by releasing the constraints we can end up with B = −27 and slightly different values for the other parameters, with an overall gain in the objective function − Total_profit.

All this would make sense, assuming that the model is correct. Unfortunately, if we take a closer look at the way we presented the fish numbers and weight in this model, we may very well start wondering. The number of fish should be an integer; otherwise it does not make sense. A fish is either alive or dead; we cannot have 85.6 fish in the pond. So far we have ignored that. At the same time, the weight is calculated as the total fish biomass divided by the number of fish. This makes sense at the beginning, when we are stocking the pond, but later on as one fish dies it certainly does not mean that the rest of the fish are gaining weight. The fact that the number of fish decreases does not imply that the remaining fish grow fatter! Does this mean that the whole model should be trashed, or some parts of it at least can be salvaged?

First, we should realize that actually if the number of fish decreases there is still some potential for weight increase, because there will be less competition for feed and therefore each individual fish will be eating more and growing faster. One quick fix that we can incorporate into the model equations is to make sure that Numbers are integers. This can be achieved by using a built-in function INT. INT(x) returns the largest integer that is less or equal than x. So if we write

$$J3 = if \ \ Weight > 0 \ \ then \ \ INT \ \frac{Mortality}{Weight} \ else \ \frac{Number}{DT}$$

we will be really subtracting something from the variable Number only when (Mortality/Weight) is larger than 1, and in this case we will be subtracting 1. If it is larger than 2, we will be subtracting 2 – and so on. Making this change we do not see a very big difference in model performance, but at least we can feel good that we do not have any half-fishes swimming around in our pond.

Second, we may also note that actually things are not so bad with the weight–number controversy. Indeed, the numbers in our model decline only when the total fish biomass also declines: Mortality is calculated as an outflow for the Fish variable. So the situation described above, when Weight is to increase with Number decreasing, is hardly possible; Fish will have to decline first, so the remaining Fish will be divided by the remaining Number, producing the same reasonable estimate for Weight. The only problem is when the fish population is losing weight but not dying. This situation is not tracked by our model, and can cause us some trouble. Indeed, decreasing weight of the population, say due to malnutrition, in our formalism will result in the decline of the Number instead of Weight.

Let us take a closer look at the results of the recent optimization. One of the reasons that B needed to be made smaller and smaller was to push the feeding curve further to the right, so that at first we had a pretty long period with almost no feed added to the pond and the fish population gradually starving and, under the chosen formalism, decreasing in numbers. If we plot the Number we will see that over the first 50 days or so it was gradually decreasing from 100 to about 50. Only after that feeding was started. So apparently the optimal strategy that was found was relying on a smaller number of fish in the pond. Let us test this directly and add the initial fish number INIT Number to the list of control variables that we optimize for. Now we will be optimizing for the number of fish that we stock in the pond (INIT Number), the feeding strategy (A, B, and C) and the time of harvest (Time_of_sale).

The more control parameters we have, the longer the algorithm runs. However, it still converges with a somewhat astonishing result: INIT Number = 1. If we keep only one fish in the pond, bearing in mind the extremely high value that we attribute to fish weight, we will be growing this one individual to some gigantic sizes and reaping a huge profit of 2420.

We have certainly learned some things about the system and about optimization. We have also identified some areas where the model can use some improvements. There is a potential problem with how we model fish weight. If we want a more realistic model of this system, we probably need to do it on an individual basis, describing the lifecycle of an individual fish and then looking at the whole pond as an aggregate of these individuals. Otherwise, when total fish biomass goes down it will be always difficult to attribute this either to the death of one or more individuals in the stock (in which case the weight of other individuals does not change) or to a gradual leaning of the whole population (when obviously the average weight of all individuals declines).

The simplest way to fix the model will be to use the Fish variable as the mean weight of fish in the pond, and then to have the Number variable representing the total number of fish. Then we will be doing the reverse calculation to get the total fish biomass: we will take the Fish and multiply it by Number. Since now some of the variables will be defined in units different than concentrations, we also need to make certain assumptions about the size of the pond. Suppose we are dealing with $10\,m \times 10\,m$ pond, $1\,m$ deep, so the total volume is $100\,m^3$. Let us see what the model will look like in this case.

The new model equations with comments are as follows:

{Reservoirs}

  d/dt (Fish_W) = +Growth − Metabolism

    {Fish_W is now the biomass of an individual fish in kg}

    INIT Fish_W = 0.01

      {We stock the pond with fishes, 10 g each.}

  d/dt (Feed) = − Growth + Feeding − Loss

    {The Feed is the concentration of feed in the pond, kg/m^3}

    INIT Feed = 0

  d/dt (Detritus) = +Accum − Decomp

    {Detritus is also the total concentration in the pond, kg/m^3}

    INIT Detritus = 0.01

      {Let us assume that at first the pond is really clean, so we have only 10 g of detritus in each m^3}

  d/dt (Total_profit) = +Profit

    INIT Total_profit = 0

  d/dt (Number) = −Mort

    {Number is the number of fish stocked in the pond. As fish may die, their number may decrease. We assume that dead fish are picked up and do not add to the Detritus pool.}

    INIT Number = 100

{Flows}

Growth = if   Number > 0.5   then   (1-Fish_W/10)*C_growth*Feed*Fish_W/(Feed + C_Hs) else 0

    {There is a limit to how big a fish can grow. We assume that this species does not get bigger than 10 kg. The condition on Number is to make sure that if all fish died at least they do not continue to grow in size.}

Feeding = if (Time < Time_of_sale and C_feed > 0) then C_feed else 0
{Same clamping on the feeding scenario to make sure it never starts to extract feed from pond.}
Methabolism = C_mort*Fish_W
Loss = C_loss*Feed + Growth*0
Accum = Loss + Metabolism*Number/100
{The fish metabolism produces detritus. There are "Number" of fish, so we multiply by Number. The size of the pond is 100 m3, so we divide by 100 to get concentration.}
Decomp = C_decomp*Detritus
Profit = Revenue-Cost
Mort = INT(if (TIME > Time_of_sale + 1) then Number/DT else (Detritus^4/(C_mort_d^4 + Detritus^4))*Number)

{Functions}
C_growth = 0.5
C_m = 0.02
C_loss = 0.1
C_feed = A*(TIME+B)^2 + C
C_mort_d = 2
C_decomp = 0.2
Fish_Price = 10 + 2*Fish_W
Feed_price = 2
Revenue = if   Time > Time_of_sale   AND   Time < Time_of_sale + 2   then   Fish_Price*Number else 0
Time_of_sale = 100
A = 0
B = 0.04
C = −1
C_Hs = 0.3
Cost = Feed_price*Feeding

Actually, this model turns out to be much better behaved and seems to produce even more reasonable results. You may notice in the future, when building many more models of your own, that the better your model gets, the more reasonable behavior it produces. In a way, the first indicator that most likely there is something wrong either with the logic or the formalism in your model is when you start getting something totally unexpected and hard to interpret.

Running optimization in this model produces the maximum Total_profit = 2572 with A = 0.00032, B = −19.43, C = 0.00034, and Time_of_sale = 67 (see Figure 8.17). The time of harvest is picked carefully to catch the moment when detritus approaches the threshold and starts to put the fish population at risk of extinction. A possible gain of a few grams in fish body

The better the model you build, the more reasonable behavior you will find in it. Many weird dynamics occur simply because the model is not quite correct.

weight is offset by more and more fish dying, and the size of the stock rapidly decreasing.

The feeding scenario is quite sensitive to the metabolism rate used in the model, the C_m parameter. If we change C_m from 0.02 to 0.01, the optimization results

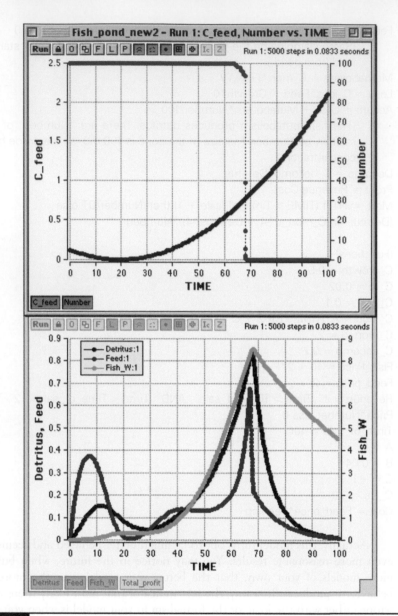

Figure 8.17   Result for a modified model that tracks the biomass of an individual fish.

change quite dramatically (Figure 8.18). Now, with a lower rate of metabolic loss, the accumulation of detritus occurs more slowly and it never reaches the critical conditions that may cause a fish die-off. Therefore, the optimization works only to try to get the fish weight to as high a value as possible, spending the least on feed. Notice that the feeding strategy now is significantly different from what we have been getting before. We end up with a higher Total_profit = 2947 with A = −0.000064, B = −32.4, C = 0.248, and Time_of_sale = 74.

Certainly, if we were to apply these modeling and optimization tools to some real-life system we would be constrained by actual monitoring data, and the model parameters would be measured in some experiments. The main purpose of this exercise

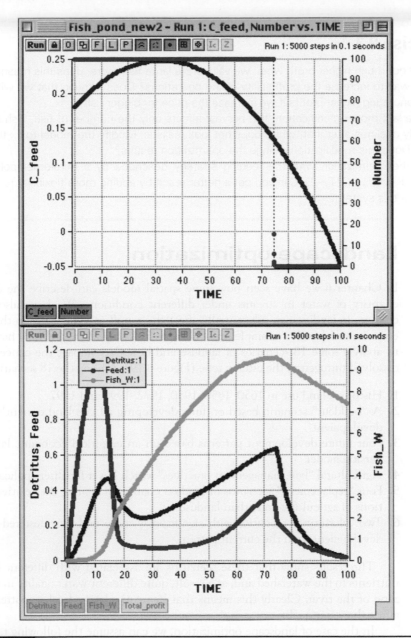

Figure 8.18   A very different feeding strategy works better when the metabolic rate is low. The optimum is reached when we get the fish weight to as high a value as possible, spending the least on feed.

has been to show how optimization works and how it can be used to derive possibly the best strategies for managing systems. There is hardly any other way in which, by means of reasoning or experiment, we could match the efficiency of the optimization magic, when in a matter of seconds or minutes hundreds and thousands of scenarios are compared and the best ones are chosen. We have also seen that there are always caveats and uncertainties that need to be carefully analyzed and realized when making the real management decisions.

Exercise 8.3

1. If we got much higher profit when we valued the large fish more, does this mean this can be a way to increase the profitability of the operations? Does it mean that we will always be generating higher profits if we increase the price for bigger fish?
2. In the last model modification, fish harvest affects only the number of fish. Fish biomass slowly declines after that, which does not look realistic. Modify the model to fix this problem. Does this modification change the optimization results?
3. Run the optimization with the feeding strategy described by a third-order polynomial: $A*T^3 + B*T^2 - C*T + D$. Can you get a better result by adding more flexibility to the controls in this system?

8.4   Landscape optimization

In Chapter 6 we have seen how some spatial models can describe the quality and quantity of water in streams under different conditions. We have also seen that changes in landuse can substantially affect the runoff and the water in the receiving estuaries or lakes. For example, the Patuxent Landscape Model (PLM) has been used to analyze some 18 scenarios of landuse and how they impact the concentration of dissolved nitrogen in the estuary zone (Figure 8.19). Among the 18 scenarios, we find:

1. Historical land use in 1650, 1850, 1950, 1972, 1990 and 1997
2. A "buildout" scenario based on fully developing all the land currently zoned for development
3. Four future development patterns based on an empirical economic landuse conversion model
4. Agricultural "best management practices" which lower fertilizer application
5. Four "replacement" scenarios of landuse change to analyze the relative contributions of agriculture and urban landuses
6. Two "clustering" scenarios with significantly more and less clustered residential development than the current pattern.

The results show that these scenarios are associated with different loadings of nutrients to the watershed and, as a result, quite different water quality in the estuary zone of the river. Clearly this means that by manipulating landuse patterns we can affect the water quality.

In the case of landscape optimization, we can assume the following task: find an optimal distribution of landuse (say, agricultural land) that would deliver the same or maximal amount of revenue, while at the same time minimizing the nutrient pollution of the river. Note that the problem now is quite different from that we have seen above. So far we have had the control parameters taking on numeric values. We changed various coefficients in the model to maximize or minimize a certain objective function. Now, we wish to optimize for maps. We want to feed various landuse maps into the model and find out the one that will produce the lowest concentration of nitrogen in the river.

This is a big change. We can modify a parameter incrementally, little by little, and it is quite obvious how to measure the change and how to direct the change in a parameter. How do we modify a map "little by little"? Where do we choose the next

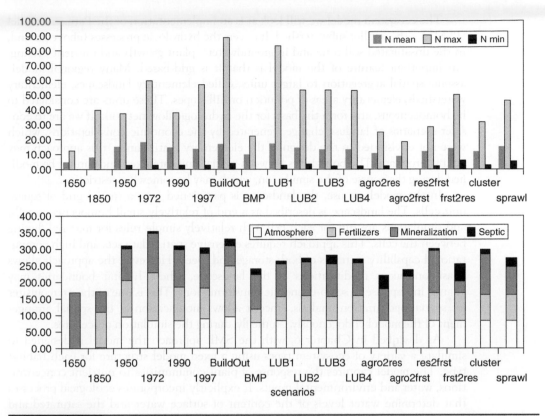

Figure 8.19   Nitrogen loading and concentration of nitrogen in the Patuxent River under different scenarios of landuse.
The different patterns of landuse result in different loading factors for nitrogen and, as a result, produce different levels of dissolved nitrogen in the estuary.

cell to make the change? If it is just one cell that we modify, where should this cell be located?

Besides, while parameters can be changed continuously, maps – especially landuse maps – are categorical. It means that numbers on a landuse map stand for different landuse types – for example, 1 is forest, 2 is corn, 3 is wheat, etc. But it could be also any other way round – 3 is forest, 1 is corn, 2 is wheat, etc. It really does not matter how we code the different landuse types. Therefore, a "little" change on the map does not mean much. We can change 1 to 2 or to 3 or to 99, but in effect we will be changing only one landuse type on the map. Instead of continuous changes, we get discrete variations of our control map.

There are no good optimization methods for this kind of task. Let us look at the example of a model built for the Hunting Creek watershed, which is located within Calvert County in Maryland, USA. The 22.5-km^2 watershed belongs to the drainage basin of the Patuxent River (2,356 km^2), which is one of the major tributaries of Chesapeake Bay. Soil types are well-drained, mostly severely eroded soils that have a dominantly sandy-clay loam to fine sandy loam subsoil. The annual rainfall varies between 400 and 600 mm. Main landuses of the watershed are forest and agricultural habitats. Rapid population growth, development and change in landuse and land cover have become obvious features of the landscape.

The ecosystem model we will look at is an implementation of the Patuxent model (PLM) for the smaller subwatershed. It covers the hydrologic processes (above ground, in the unsaturated soil zone and in groundwater), plant growth and nutrient cycling. An important feature of the model is that it is grid-based. Many regional models assume spatial aggregation to larger units, called elementary landscapes, elementary watersheds, elementary areas of pollution or hill-slopes. These units are considered to be homogeneous, and form the basis for the hydrologic flow network. If we are to consider scenarios of landuse change, generated by the economic considerations, which were not envisioned in the design of the elementary spatial units, this approach may be inappropriate. The boundaries between spatial units are fixed and cannot be modified during the course of the simulation, which may be somewhat restrictive.

In the model we use, the landscape is partitioned into a spatial grid of square unit cells. The landscape is described as a grid of relatively small homogeneous cells, and simulations are run for each cell with relatively simple rules for material fluxing between the cells. This approach requires extensive spatial data sets and high computational capability in terms of both storage and speed. However, the approach allows quasi-continuous modifications of the landscape, where habitat boundaries may change in response to socio-economic transformations. This is one of the prerequisites for spatial optimization analysis, since it allows modification of the spatial arrangement of the model endogenously, on the fly, during the simulation procedures.

As described in Chapter 6, with the SME approach the model is designed to simulate a variety of ecosystem types using a fixed model structure for each habitat type. The model captures the response of plant communities to nutrient concentrations, water and environmental inputs. It explicitly incorporates ecological processes that determine water levels or the content of surface water and the saturated and unsaturated soil zone, plant production, nutrient cycling associated with organic matter decomposition, and consumer dynamics. Therefore, the simulation model for a habitat consists of a system of coupled non-linear ordinary differential equations, solved with a 1-day time-step.

Let us now formulate the optimization task. The study area can be described as a set of discrete grid points $R = \{(i, j), 0 < n_i < i < N_i < N; 0 < m_j < j < M_j < M\}$ (Figure 8.20). N is the number of cells in the row, and M is the number of cells in the column. Not all of these cells are in the study area. A cell that belongs to the study area is denoted by $z \in R$. Six different landuse types are encountered in the study area: soybeans, winter wheat, corn, fallow, forest, and residential. We will assume that the residential areas are fixed, but otherwise landowners are free to decide what type of crop to grow in a cell, or whether to keep it forested or in fallow. Let $c(z)$ be the landuse (or habitat type) in cell z. The control parameters in our case are the landuse types that are chosen for each cell. The set of landuse types will be $L = \{$soybeans, winter wheat, corn, fallow, forest$\}$. Then $R_c = \{z \in R \mid c(z) \in L\}$ stands for the set of grid points that can be controlled with controls chosen from L. Let $H(c,z)$ be the yield of crop c (if any) harvested from cell z, and $N(z,t)$ be the amount of nitrogen that escapes from cell z at time t. The other control decision that farmers can make is the amount of fertilizer to apply: let $F(c,t)$ be the amount of fertilizer applied for the habitat type c at time t. The time of fertilizer application could be another important control parameter, but let us not further complicate the problem, and assume that fertilizers are timed according to the existing best management practices and the only factor we can control is the total amount applied.

Qualitatively, our goal is to find the optimum landuse allocation and fertilizer application to reduce nutrient outflow out of the watershed while increasing total

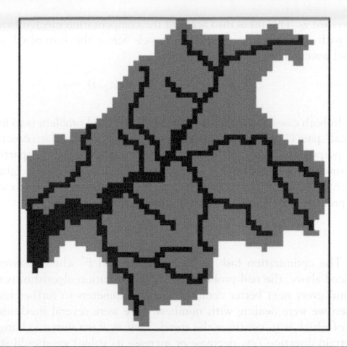

Figure 8.20   The study area for the Hunting Creek model.
Only the cells that are on the map will take part in the optimization. The cells in the water category will not change and therefore can also be excluded.

yield. So the objective function (performance criterion) will need to account for crop yield, fertilizer application and nutrient outflow. The first two factors are easier to compare, since we can operate in terms of prices. The revenue from the yield over the whole study area is

$$A = \sum_{z \in R} p_H(c) H(c, z)$$

where $p_H(c)$ is the current market price of crop c. The price of fertilizers applied is then

$$B = p_F \sum_{z \in R} \sum_{1 < t < T} F(z, t)$$

where p_F is the unit price of nitrogen fertilizer. Obviously, A is to be maximized while B is to be minimized, which means that $(A - B)$ is to be maximized. $A - B$ is the "economical" part of the goal function.

There are different ways of modeling the "ecological" part of the performance criterion. One possibility is to take into account the total amount of nutrients generated by all the cells in the study area,

$$C = \sum_{z \in R} \sum_{1 < t < T} N(z, t)$$

This is the distributed nutrient leaching. More realistic, and comparable with measurements at gauging stations, is the amount of nitrogen in the outlet cell of the watershed z_0. This takes into account the compensation mechanisms of uptake along the pathways of nitrogen while it travels across the watershed and estimates the actual water quality in the river estuary:

$$C = \sum_{1 < t < T} N(z_0, t)$$

In both cases, C is to be minimized. The crucial problem is to integrate the "ecological" part C and the economic part $A - B$ into a scalar objective function. For this purpose, C has to be expressed in units that can be compared with the dollar measure that we have in $A - B$. Let us assume that there is a weighting coefficient λ, which can convert our C measured in gN/m^2 into dollars, which we use to measure the profit $A - B$. Then we can formulate the goal function as

$$J = A - B - \lambda C \qquad (8.4)$$

The optimization task is: Find maps c^* and F^* which maximize $J \rightarrow$ max. As noticed above, the real problem for the optimization algorithm is to figure out how to find every next better combination of parameters to further improve our result. When we were dealing with numbers there were several methods, the most obvious of which is to continue the trend. That is, if we start to change a parameter in a certain direction (say decrease or increase its value) we should stay on this course as long as the results continue to improve. Or we could follow the gradient. That is, check a parameter change in one direction (increase), then the other (decrease) and see where the objective function performs better (say, has the minimal value). That will be the parameter value that we take as our next approximation.

But how do we do that in case of maps, especially categorical maps? If we changed from 1 to 2 going from soybeans to winter wheat, we could continue to 3, which is corn. But that would have little sense, since 3 may have also been forest or fallow. We just chose that value of 3 to represent corn. There is no real reason that a 3 and not a 4 should represent corn. There is no such thing as an increase or decrease of a category value: we are switching to a different landuse only, the number itself has no meaning. We may have easily used letters instead of numbers on the map.

We end up with a so-called combinatorial optimization problem. To get to the solution, we really need to sort through all the possible combinations of the five possible landuse types over the study area. The number of possible combinations for the task in (8.4) depends on the size of the study area. For example, for the Hunting Creek watershed, which is represented by $|R_c| = 1681$ controllable cells of $200 \times 200 \, m^2$ with five possible landuse types, we get $I_1 = 5^{1681}$ different patterns of landuse allocation. Remember that for each of these landuse maps we will need to run our model for at least 550 days to cover the vegetation season, including winter to accommodate for winter wheat, which is planted in the fall but grows in the spring. On a high-end workstation, the model takes about 3 minutes to run. On top of that we also want to test for various fertilizer application rates, but even without that it is clearly much longer than the time required to finish reading this book. Actually, the age of Earth is about $4.5 \cdot 10^9$ years, and we are asking for something around $6 \cdot 10^{1153}$ years. Even the best supercomputer will not help us. There should be a better way to solve the problem.

Generally, when mathematicians end up with a problem that they cannot solve they start simplifying it by making certain additional assumptions about the system. Let us do the same for our system by taking into account the following considerations.

After all, the landscape operates as a combination of grid cells, and perhaps we can assume that the connections between these cells are not that important. This means that perhaps we can get something if we solve the optimization problem for each individual cell, and then produce the overall landscape by combining the landuses that we find optimal for these cells.

In this case we will need to define a local objective function for each grid cell. This is structurally different, because it aims to map the regional goal function onto the processes in a grid cell. The basic idea is to try to split our global optimization problem, which is spatial, and which has cells spatially connected, into a combination of local optimization problems, ignoring the spatial connectivity between the cells.

For every grid cell, z, we define the objective function as a function of z. Let $A(z) = p_H(c) H(c,z)$ be the local profit from crop yield; $B(z) = p_F \Sigma_t F(z,t)$ be the local cost of fertilizers applied, and $C(z) = \Sigma_t N(z,t)$ be the amount of nitrogen leached locally. $A(z)$, $B(z)$ and $C(z)$ are now calculated for a specific cell. They do not require integration over the entire study area. Based on this, the local goal function for every cell is then:

$$J(z) = A(z) - B(z) - \lambda C(z) \qquad (8.5)$$

and the optimization task is: For each cell $z \in R_c$ find $c^*(z) \in L$ and $F^*(z)$ which maximize $J \to (z)$ max. Once we find the landuse and the fertilizer application that is optimal for each individual cell, we can then produce the global solution as a map made of these local optimal solutions (actually two maps: one for landuse, the other one for fertilizer application).

The problem is now reduced to optimization of landuse and fertilizer application for every grid cell – but now this becomes feasible. Indeed, assuming homogeneous landuse and several discrete stages of possible total fertilizer input, say six stages $F \in \{0, 25, 50, 75, 100, 150\,kg/ha\}$, our task $I_2 < |F| |L| = 36$ combinations. Considering that no fertilization takes place for $c \in \{forest, fallow\}$, we get $I_2 = 26$ combinations. Yes, this approach neglects any neighborhood effects. We have also implicitly introduced another assumption – that is, that the effect of fertilizer is smooth and continuous, with no significant thresholds. Otherwise it would be incorrect to use the six-step scale of fertilizer application that we described above. But making these assumptions we reduced the task to something we can easily solve. Indeed we need to run the model only 26 times and then produce the global solution by simply choosing the optimal landuse and fertilizer rate for each cell.

Actually, the local task gives us a worst-case scenario. In terms of nutrient outflow, the global approach could take into account the retention capability of the landscape, when the next cell downstream captures nutrients leached from one cell. The local approach no longer allows that, and therefore gives us a worst-case upper estimate of the net nutrient outflow.

The solution of the local task performs a grid search through the entire control space, assuming a homogeneous landuse and identical fertilizer amounts for each cell. So what we need is to run the Hunting Creek model assuming that the whole area is covered by one of the agricultural crops, and do it six times for each crop changing the fertilizer application rate. That will be 4 (landuses) $\times$ 6 (fertilizer rates) = 24 model runs. Then, in addition, we run the model using an all-forested and all-fallow landuse. These are not fertilized. Those are the 26 model runs estimated above. These runs are sufficient to give us the $A(z)$, $B(z)$, $C(z)$ of the local objective function as maps. Using these, we can calculate $J(z)$ for all cells and choose the maximal value for each cell. This solution corresponds to a certain landuse and fertilizer rate in each cell. Putting these

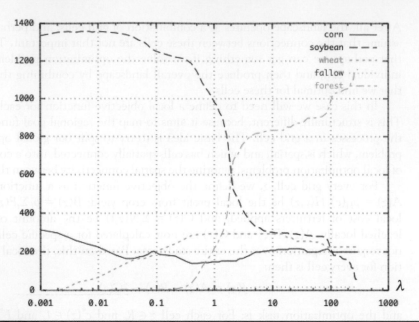

Figure 8.21   Distribution of landuse in the optimal pattern as a function of the environmental awareness parameter λ.
We start with a fully ploughed watershed all covered by soybeans, the most profitable crop. As λ grows, there are fewer crops and more forest in the watershed.

into another map, we get a solution to the global task. This pair of maps can be then fed into a spatial simulation to calculate the value of the global performance criterion.

The estimation of local optimum landuse maps does not require any computational effort. Once we have all possible combinations in the maps $A(z)$, $B(z)$ and $C(z)$, we can study how weighting parameter λ affects the results. As you may recall, this λ parameter represents the relative importance or weight of the environmental concerns, in our case measured in terms of nitrogen content in the estuary. Figure 8.21 shows the results of a parameter study for the Hunting Creek watershed plotting the number of different landuse types as a function of λ. The corresponding results for the total fertilizer application are presented in Figure 8.22. As we can see, while the rate of fertilizer application quickly drops as environmental awareness grows, there is also a significant change in the composition of landuse in the watershed. These graphs do not tell us much about the spatial distribution of landuse. Let us take a look at some of the spatial output.

Figure 8.23 shows maps of optimum land for several λ values. We start with a zero value for λ, which is the "why would I care about the environment!" scenario. In this case, we get the monoculture solution: plant the most valuable crop (in this case soybeans) in the entire study area, wherever possible. The only other cells that remain are the residential and open-water ones, since those are non-controllable cells. As we start increasing λ, some forest appears. The more we get concerned with nutrient outflow (and push λ up), the more forest will appear in the study area. At the same time, agricultural cells change to crops with a better nutrient-uptake/yield efficiency. This succession of crops also depends on the market prices of the crop. If we were to run these calculations today the results would most likely be different, because of the jump in price of corn, instigated by growing demand for corn-based ethanol.

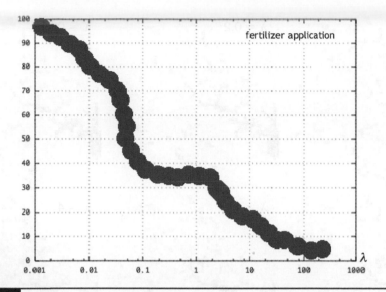

| **Figure 8.22** | Application of fertilizer as a function of λ. |

Application of fertilizers plummets as environmental concerns about water quality start to dominate.

What is also remarkable is that we can actually see how, with growing λ, forests first appear along the stream network and then gradually spread out. There is nothing in the objective function that would be directly responsible for that, yet there is a clear pattern. Could we interpret this as yet another evidence of the important role of forest buffers? Clearly, we get the most "bang for bucks" when forests are located along the streams.

Of course, once the global solution is produced, based on the local one, it makes sense to check whether it is really optimal in the global sense. The most obvious way to do this is to disturb the solution and see if the results we get are consistently "worse" than what the optimal result delivers. We can use the Monte Carlo method and randomly choose some cells and randomly change the landuse in them. Then we can run this new disturbed map through the model and check if the result gets any better than that found above. In the case of the objective function that uses water quality as an indicator of environmental quality, we do not find any better solutions than that identified by the local algorithm. Apparently for this task our model simplification was not damaging in any way, and we achieve a solution to the global optimization task. For nutrient content, the neighborhood connections seem to be negligible.

Unfortunately, the local method does not work so well in all cases. For example, another way to account for environmental conditions is to look at watershed hydrology. As we have seen, changing landuse types also changes the infiltration and evaporation patterns, which in turn affect how much water ends up in surface runoff. Very often as a result of deforestation we see an increase in peak flow (the maximum flows after rainfalls are elevated), while the baseflow plummets (the flow between rainfalls under dry conditions). If we try to incorporate this concern into our objective function by, say, maximizing the baseflow, we get another optimization task. If we try to apply the same set of assumptions, and find an optimum using the local algorithm, we may be quite disappointed to find that the corresponding global solution does not seem to be optimal. Running some Monte Carlo tests, we easily

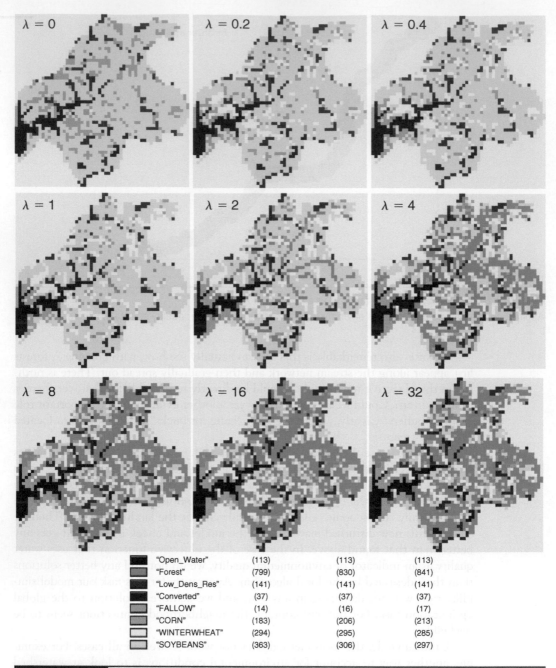

"Open_Water"	(113)	(113)	(113)
"Forest"	(799)	(830)	(841)
"Low_Dens_Res"	(141)	(141)	(141)
"Converted"	(37)	(37)	(37)
"FALLOW"	(14)	(16)	(17)
"CORN"	(183)	(206)	(213)
"WINTERWHEAT"	(294)	(295)	(285)
"SOYBEANS"	(363)	(306)	(297)

Figure 8.23   Change in landuse pattern as a function of λ.

find that the local solution can be improved by changing some cell categories on the map. So the method is not universal and does not work for all systems and objective functions. However, when it does work, it produces a very fast and efficient way to find the optima. For example, in the same system if we were to optimize for NPP (net primary production – another important proxy used in ecosystem services analysis), we would find the method working very nicely.

Exercise 8.4

1. Think of other objective functions that can be used in the context of this same problem. Suppose you have a certain amount of money that you can spend on Best Management Practices (BMP), and you wish to improve the water quality as much as possible while still generating the same profit from agriculture. How can you describe this in the objective function?

2. The other ecosystem services that you are most likely to want to consider in this context are the CO_2 retention capacity (an important factor for climate change mitigation), Net Primary Production, phosphorus retention capacity, and flood control. Can you formulate an objective function that would optimize the landuse for all these factors at the same time?

8.5   Optimality principles

There are several reasons why optimization can be useful. Instead of running numerous scenarios through the model and then comparing the results to find the best outcome, we may formulate a certain goal that we want the system to reach, and then let the computer sort through the numerous parameter and pattern combinations to reach that goal. In this case, the variations of parameters and functions in the model input (control) are performed automatically, as well as the processing of the output. The core of this process is the algorithm of numerical optimization, which makes the decision on how to define the next scenario to be analyzed based on the available information about the results of previous model runs. This optimization procedure connects the scenarios, the simulation process and the objective function or performance criterion. As we have seen above, optimization algorithms perform a systematic search in the space of control variables to find an input vector, which controls the systems in the desired way, specified by the objective function.

In most cases, to assess the outcome of a scenario in terms of environmental, economic, toxic or social aspects, more than one output variable needs to be considered. To compare different simulation scenarios (or to analyze a multidimensional decision problem) we integrate output variables into a scalar value. This function that we choose to integrate or aggregate the several output variables is the goal function or performance criterion, and is a mathematical formalization of the state of the system that should be maximized or minimized to reach the desired state.

Yet another way we can look at optimization is when we treat it as part of the model. Indeed, some systems seem to be operating as if there are certain optimal principles embedded in them – or at least we may assume the existence of these principles. For example, suppose we are building a model of plant growth. We know that there is the photosynthesis process that produces new "juice", or fuel, for growth. We also know that there are different parts of the plant, and that in some cases the plant grows leaves or roots or branches. Most sophisticated plant models use a lot of experiments to determine how this new fuel is distributed among the different plant parts, yet still there is much uncertainty about it; the rate coefficients seem to vary from one species to another, and change under different climatic and ambient conditions.

This could be expected, since clearly this distribution is not always the same. When we have drought conditions, and there is not enough water, plants tend to grow roots to get more water from the ground. They may even shed some leaves to cut on evapotranspiration. When conditions are favorable, plants will grow leaves. If there is not enough light, they will try to grow the trunk and the branches, to get higher up towards the sunshine. All these mechanisms can be pretty hard to describe and model. Racsko and Svirezhev came up with an interesting idea to model this based on optimality principles (Racsko, 1978).

Suppose the plant has a goal, which is to grow as much as possible under the existing conditions. If that is the case, then on each time-step the new fuel should be distributed among the different parts of plant in a way that will ensure maximal production of new fuel over the next time-step. This means that for every time-step we will solve an optimization problem:

$$\max_{p} F(t, c, q)$$

where F is the newly produced fuel at time t that is to be distributed among leaves, branches and roots according to the proportions $p = (p_1, p_2, p_3)$, respectively, $p_1 + p_2 + p_3 = 1$. $c = (c_1, c_2, \ldots)$ is the vector of ambient conditions (temperature, soil moisture, etc.), and q is the vector of model parameters (photosynthetic rate, respiration rate, etc.). So at each time-step we assume that the ambient conditions will be the same as during the previous time-step, and then optimize for the best distribution of the fuel available to produce the most fuel during the next time-step.

Note that by making this assumption about some optimality principle involved in the process of plant growth, we have eliminated a lot of unknown parameters that otherwise would have to be either measured or calibrated. The only problem is that in most cases we do not really know whether these optimality principles really exist and we are reproducing some real process (plants *deciding* what to do depending on environmental conditions!), or whether the optimality that seems to be in place is just an artifact of a combination of many other processes such as the ones brought by the evolutionary process and natural selection in living systems, many of which we do not really know or understand. Here is another example.

Above, we have seen how different levels of environmental awareness result in different patterns of landuse distribution. Each value of the awareness coefficient λ creates a landuse distribution that is optimal in a certain sense. That is, depending upon how high we value environmental quality in comparison with economic profit, we get different patterns of landuse (Figure 8.23). This begs for a reverse problem statement. We know what the existing landuse distribution is. Is there a λ that will describe it? Or, in other words, can we judge the environmental awareness by the landuse pattern observed in an area?

To answer this question, we will need to compare maps from the set generated during our optimization process with a map of landuse for, say, 1990, which is available. We know how to compare numbers. We can figure out how to compare many numbers at the same time. That is what we are doing when calibrating a model and using an error model. This error model is our way of wrapping up several numbers into one to make the comparisons needed. However, in the case of map comparisons the task becomes more complicated. It is not just the total number of cells in different categories that we are interested in; it is also their spatial arrangement.

For example, in Figure 8.24 we see that the map on the left has the same number of black cells as the map on the right. The number of gray cells is also the same. However, the maps obviously look very different. On the other hand, maps in Figure 8.25 also have the same number of cells in different categories and do look alike,

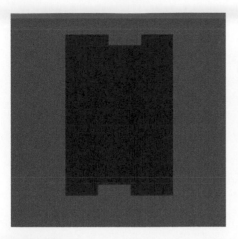

Figure 8.24   While the number of cells in different categories can be the same, the maps will look quite different.

Figure 8.25   In other cases the number of cells in different categories can be the same, but there will be no single cell that exactly matches the corresponding cell in the other map – yet the maps will look similar.

even though not a single pair of corresponding cells on the two maps has matching colors. So it is not just the total number of cells that matters, but also the pattern, the spatial arrangement, of the cells.

The human eye is a pretty powerful tool for spatial map comparisons. We are quite good at distinguishing patterns and finding similar maps, as long as we have an agreement on a criterion for comparisons. Figure 8.26 shows some maps that were offered as part of a survey to compare some machine algorithms with human identification. Most of the algorithms of map comparison that try to account for pattern are based on the idea of a moving window where, in addition to a cell-by-cell comparison, we start looking at an increasingly expanding vicinity of cells and search for similarities in these neighborhoods, not just at the cell-by-cell comparison (Figure 8.27). Some of these methods get quite close to visual comparisons, and can be used for objective automated map comparisons.

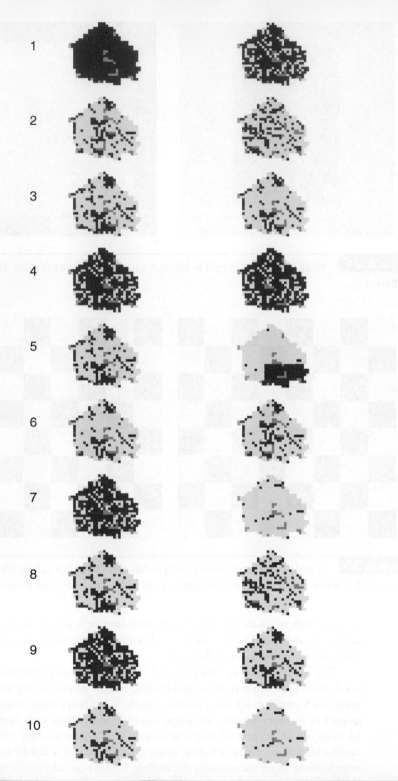

Figure 8.26 Pairs of maps offered for comparison in the survey. Most of the participants said that pair 4 is the closest – would you agree?

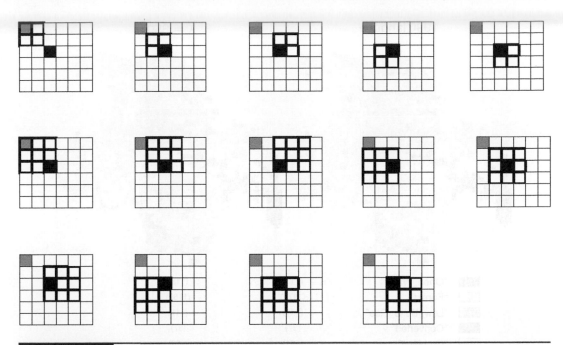

Figure 8.27   How a moving window algorithm works.
We start with a cell-by-cell comparison. Then we start including the vicinity of a cell and see if there are any matches in the neighborhood. Then we can gradually expand the window to see if there are more matches. The matches in the wider windows get a lower ranking in the overall comparison index. Do you think you were doing something like this when you where visually comparing the maps in Figure 8.26?

This is exactly what we need to solve the problem: find an optimum landscape that will match a real landuse map. Using the local method described above, we can easily generate the whole series of maps for the various values of λ. Before we get into the complex methods of map comparison, let us first find the λ for which the number of cells in different categories of the optimal solution match those of the 1990 landuse map.

That is where a surprise is waiting for us. As we change the value of λ, we can see how the number of forested cells gradually increases. At $\lambda = 350$, we find that the numbers of cells in all categories (forest and aggregated agriculture – we did not have any information about the actual crop allocations) in both the optimal map and the landuse data match. What is really surprising, is that, looking at the map that corresponds to this $\lambda = 350$, we find that it is not just the number of cells that we have matching; it is also the pattern that looks amazingly alike. See for yourself, looking at the two maps in Figure 8.28. Even without any map comparison algorithms, it is pretty clear that the maps have a lot in common.

Does this mean that we have inadvertently found another optimality principle that, in this case, governs landuse change? Are the landuse patterns that we currently have indeed results of some optimization? It is clearly too premature to jump to this kind of conclusion. Many more case studies should be considered and a variety of objective functions should be tested to find out if there is really something meaningful in this result. However, it does make sense to assume that indeed humans apply

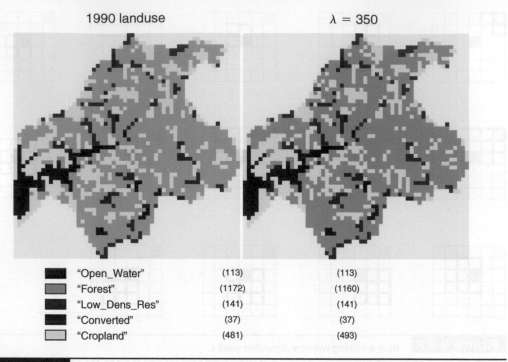

1990 landuse		λ = 350

		1990 landuse	λ = 350
■	"Open_Water"	(113)	(113)
▨	"Forest"	(1172)	(1160)
■	"Low_Dens_Res"	(141)	(141)
■	"Converted"	(37)	(37)
▢	"Cropland"	(481)	(493)

Figure 8.28   Comparison of the real 1990 landuse map with the result of optimal landuse allocation with λ = 350. The patterns seem to be remarkably close. Does this mean that the existing landuse is a result of some optimization process? If we find this process, would it help us figure out what can be the future landuse maps?

some optimality principles in their landuse allocation decisions. As a matter of fact, it should not be surprising at all that agricultural land is allocated in areas where the yields and profits are maximized. What is surprising is that the λ factor actually does play a role. But again, chances are that it is not the environmental concerns that stop further expansion of agriculture but something entirely different, and it is just that the agriculture stays in areas where it is most profitable while other cells get transferred to other landuse.

Still, this comparison gives us a numeric value for λ that suddenly becomes a meaningful index that can be used to value certain ecosystem services. As noted above, the units for λ are $/(g N/m^2)$, so every g N/m^2 that is allowed to escape from the land and travel to the estuary of the watershed has this dollar value ($350) under current landuse conditions. We can now run the model with no forest, compare the amount of nitrogen that will be released in that case to what we have now, and derive the value of the nitrogen-retention service provided by the forest ecosystem on this watershed. Some back-of-the-envelope calculations can tell us that if, according to Figure 8.19, the difference in nitrogen runoff between an all-forested watershed in 1650 and an all-agricultural watershed can be an order of magnitude or more, then each square meter of forest will be producing $3,500 worth of ecosystem service. If we multiply that by the total area of forest in Hunting Creek watershed, we will get … well, a lot of dollars.

Anyway, optimization is an exciting tool to explore. It can help us to understand many of the features of the systems that we are studying. By running the model so many times under various conditions, it reveals behavioral patterns that otherwise would remain obscure and unexplored.

Further reading

A good primer on optimization is a book by Diwekar, U. (2003). *Introduction to Applied Optimization.* Kluwer, 333 pp.

For more details on the Chernobyl accident see the Nuclear Energy Agency website at http://www. nea.fr/html/rp/chernobyl/allchernobyl.html, *where you can download the report* "Chernobyl: Assessment of Radiological and Health Impacts. 2002 Update of Chernobyl: Ten Years On". *A good brief account is given at* http://www.uic.com.au/nip22.htm. *More links can be found at* http://www. chernobyllegacy.com/.

There is a large body of literature on discounting. A brief analysis of discounting and how it applies to sustainability, can be found in Voinov, A. and Farley, J. (2007). Reconciling Sustainability, Systems Theory and Discounting. *Ecological Economics,* 63:104–113. *Another interesting paper is* Sumaila, U., Walters, C. (2005). Intergenerational discounting: a new intuitive approach. *Ecol. Econ.* 52, 135–142.

The IPCC – The Intergovernmental Panel on Climate Change – is probably one of the most respected bodies on global warming and the future of climate on this planet (http://www.ipcc. ch/). *Despite all the criticism that it gets from the climate change nay-sayers, they have produced a series of reports that present the existing consensus among a vast majority of scientists. There are currently four reports available, in which the climate change forecast has been getting increasingly grim and alarming, while more efforts were requested from the governments and international organizations. The* "IPCC Second Assessment: Climate Change 1995" *is available at* http://www.ipcc. ch/pdf/climate-changes-1995/ipcc-2nd-assessment/2nd-assessment-en.pdf.

A model of a fishpond that can be considered as a more complex version of the one presented here appears in Svirezhev, Yu.M., Krysanova, V.P., Voinov, A.A. (1984). Mathematical modelling of a fish pond ecosystem. *Ecol.Modelling,* 21: 315–337.

A general description of the Patuxent model that we used here can be found in Costanza, R., Voinov, A., Boumans, R., Maxwell, T., Villa, F., Voinov, H. and Wainger, L. (2002). Integrated ecological economic modeling of the Patuxent river watershed, Maryland. *Ecological Monographs,* 72 (2): 203–231. *For more details on the processes and model structure see* Voinov, A., Fitz, C., Boumans, R., Costanza, R. (2004). Modular ecosystem modeling. *Environmental Modelling and Software.*19, 3: pp. 285–304.

The spatial optimization technique was developed in Seppelt R., Voinov, A., 2002. Optimization Methodology for Land Use Patterns Using Spatial Explicit Landscape Models. *Ecological Modeling,* 151/2–3 pp. 125–142. *Further details are in* Seppelt R., Voinov, A., 2003. Optimization Methodology for Land Use Patterns – Evaluation based on Multiscale Habitat Pattern Comparison. *Ecological Modeling,* vol 168(3): 217–231.

The application of optimality principles for modeling plant growth was the topic of the PhD dissertation of Peter Racsko, with Yuri Svirezhev as his advisor. The PhD thesis is available only in Russian Racsko, P. (1979). Imitacionnaja model dereva kak elementa lesnogo biogenocenoza (Simulation model of the tree growth, as the element of the forest biogeocenosis). *Vopr. Kibern.* 52, USSR Academy of Sciences, Moscow, pp. 73–110 *(In Russian with English summary), and was never published in any international journals. The principle was further applied to agricultural crops and modified to include the reproductive organs (seeds) that plants grow and that after a certain biological time become the main and only recipient of the newly produced material in the plant. This is described in* Racsko, P. and Semenov, M. (1989). Analysis of Mathematical Principles in Crop Growth Simulation Models, *Ecological Modelling,* 47: 291–302.

A good review of model comparison techniques is offered by Kuhnert, M., Voinov, A., Seppelt, R. (2006). Comparing Raster Map Comparison Algorithms for Spatial Modeling and Analysis. *Photogrammetric Engineering & Remote Sensing,* Vol. 71, No. 8: 975–984.

9. The Practice of Modeling

9.1 Why models don't work
9.2 Participatory and adaptive modeling
9.3 Open-source, web technologies and decision support
9.4 Conclusions

SUMMARY

There are no formulas and almost no figures in this chapter, and this is because now we are figuring out what happens to models when they are brought out for human use and action. This is when we need to speak, interact and communicate more to explain why and how we built our models and what use they are. We will look at a brief history of global climate-change modeling as a terrible example of failure to communicate the scientific results in a timely fashion, to make people understand the possible disasters and make them act accordingly. Participatory modeling will be then considered as one possible tool of stakeholder interaction that has the potential of overcoming these disconnects between the modelers and the society at large. Participatory modeling uses the modeling process as a way of joint learning and understanding to build consensus and help make better decisions. The open-source paradigm offers a promising framework to support participatory modeling and other open and shared decision support tools.

Keywords

Global climate change, failures of governance, uncertainties, IPCC, Kyoto Protocol, Shared Vision Planning, mediated modeling, Companion modeling, stakeholders, scenarios, model transparency, modularity, open-source software, Linux, General Public License, gift economy, web tools, intellectual property, collaborative research, community modeling, open data, open access publication, watershed management.

9.1 Why models don't work

So now we know how to build a model, how to make it run, how to analyze it and produce results. Does this mean that we are ready for success stories? Unfortunately, there is still one element missing. How do we make people listen and act according to the findings from our model? After all, in most cases we were building the model to find out something and to make the right decisions based on our findings. The gist of modeling is to simplify the reality to improve the understanding of real-world processes. But we do it for a purpose: we want to find solutions for the real world problems and to make better decisions to improve life and avoid disaster. Otherwise, why bother modeling?

> *The thoughtless and selfish, indeed, who fear any interference with the enjoyment of the present, will be apt to stigmatize all reasoning about the future as absurd and chimerical. But the opinions of such are closely guided by their wishes.*
>
> W.S. Jevons, 1865

In most cases, people have preconceived notions about the problem. They come to the table with some ideas about the solution. In many cases, they are so entrenched in their opinions that it becomes almost impossible to find a common ground. We all build models. But our models are different.

In engineering or physics, after all, we have the "hard" science, the experiment and, ultimately, the system that will either work or not. If we want to model a bridge, we know that there will be the ultimate test that will tell us whether the model was right or wrong. If the bridge collapses, then the model was wrong – and no one wants that to happen. If we model weather and forecast a rainfall, we will soon know whether we were right or wrong; we can then adjust our model and, most importantly, again the society at large wants our model to be correct.

In ecology or ecological economics, things get messy. First, there is the additional uncertainty that comes from humans being part of the system. Human behavior may be very complex, unpredictable and heterogeneous. There is no single law (like Newton's law or the law of gravity in physics) that can be directly applied to every human and will hold. The science of psychology is trying to come up with some general rules of human behavior, but we are clearly not there yet. Different people have different preferences for goods and services, have different goals and aspirations, and different levels of ecological awareness. All these factors cause different patterns of human behavior in an ecological–economic system, and will steer the system to different outcomes. Economics has tried to explain economic behavior. Some 20 years ago, the assumption that individuals are purely rational, fully informed maximizers (they maximize either their utility or profit) was a shared understanding among economists. However, the evidence from experimental economics has altered these views. Apparently, people are sometimes irrational, do not possess all the information to make a decision in a complex ecological–economic environment, have different levels of risk awareness, and so on. This actually means that people may respond differently to the state of an ecological–economic system and exhibit different strategies of resource consumption. For example, some American Indians certainly had very different views on natural resources (remember the seventh-generation principle we mentioned in Chapter 2, page 34) than the white people who came to their land. Which behavior model should we choose, if we go beyond one individual and need to model the economic behavior for a whole region or country? If the rational maximizing individual model is not valid, then which one is? Now, economists talk about heterogeneous consumer models, but those are far more complicated, with many more uncertainties. Thus, although in "hard" science we know for sure that water will flow downhill (or that an object with a certain mass will fall down to the ground with a certain speed), we can never be 100 percent sure about human behavior. It is as if we are modeling a bridge, not knowing for sure whether the right amount of concrete will be poured or whether the bridge will be actually used for travel or for a rock-'n'-roll concert.

Second, humans are also users of the model. They may have the desire and power to ignore, twist and distort the results of the model. They have their own priorities and vested interests, and may "like" some results and "dislike" others. There are likely to be parties that do not want our model to produce certain results. For example, if you are predicting rainfall and I make a living betting on drought and selling sunglasses, I will want your model to be wrong – or at least to make sure that nobody believes in your forecast.

There was a well-known dispute between Paul Ehrlich and Julian Simon. Ehrlich, an ecologist and professor of population studies, forecast that some of the main resources would gain in price over the next several years. Simon, a mainstream economist, claimed that their price would drop. His theory was that it is not the natural resources that may be a limiting factor, but rather the number of human brains that are there to solve problems. So the higher the world population, the merrier it will be. Eventually, Simon bet that the price of any set of raw materials would be lower 10 years from now than it is today. Ehrlich and his supporters took up the challenge and, in October 1980, chose five metals: chrome, copper, nickel, tin and tungsten. Simon won the bet as, by October 1990, the composite price index of these five metals had fallen by more than 40 percent. Note, however, that this was not a fair game, since actually Ehrlich (as well as other environmentalists) was working hard during those years to try to lower demand for natural resources. So actually he was betting against himself.

This is just to illustrate that it makes little sense to predict the behavior of open systems, where humans themselves are likely to change those systems. If you think that global warming is happening, you are more likely to do something about it. So don't bet on it, since it is likely, thanks to the efforts of yourself and people who think your way, that the process may be slowed down. You may be betting against yourself. Instead, keep on with the good work.

The recent global climate change saga provides a spectacular example of how this happens. It dates back almost 200 years, to when Edme Mariotte, Horace Benedict de Saussure, Fourier and Poulliet did their experiments, collected data and laid the foundation for some theoretical generalizations. By the 1850s, Tyndall was measuring various gases' absorption–emission behavior. Arrhenius wrapped it all up in his 1895 talk to the Swedish Royal Academy and in a subsequent April 1896 paper, "On the Influence of Carbonic Acid in the Air upon the Temperature of the Ground" (*The London, Edinburgh, and Dublin Philosophical Magazine and Journal of Science*). That CO_2 can change the absorption right in the middle of Earth's outgoing blackbody spectrum has been understood for a long time.

In 1938, Guy Stewart Callendar discovered that global warming could be brought about by increases in the concentration of atmospheric carbon dioxide due to human activities, primarily through burning fossil fuels. This is the untold story, just recently discovered by a historian, James Fleming, of the remarkable scientist who established the carbon-dioxide theory of climate change. These findings were based on some simplified theoretical models. Then, during the twentieth century, these models were consistently improved, incorporating more detail and mechanisms, culminating in a dozen or more Global Circulation Models (GCMs) designed to understand global climate in as much details as possible and to predict its possible future. Currently, there are at least some 20 models used for climate prediction around the world, with the same acronym, they are now called Global Climatic Models.

By the 1990s, scientists had started to raise red flags and blow all sorts of whistles and horns, trying to focus the attention of the public on the simple fact that increased CO_2 and other GHG concentrations leads to global warming, and that we are rapidly increasing the amount of CO_2 in the atmosphere by burning huge amounts of fossil fuels. It seems simple – but there is one important caveat. If this is indeed a problem, then the most obvious solution is that we need to burn less fossil fuels; however, this also means that we will need to consume less gasoline, drive less and, consequently, most likely deliver less profit to the oil corporations. We have already seen in Chapter 7 how corporations and their lobbyists can rule the world. Apparently, this is exactly what is happening. The Bush Administration, which is notorious for its links with Big Oil, has diligently been doing its job of slandering climate change research and preventing any meaningful mitigation efforts. This might be compared with an ostrich sticking its head into the sand just to ignore the fact that there is imminent danger – except now we know that actually no ostrich would really do that. When ostriches feed, they sometimes do lay their heads flat on the ground to swallow sand and pebbles, which helps them to grind the food that they eat; although from a distance it may indeed look as though the bird is burying its head in the sand, it is clearly smart enough not to do that when there is something dangerous coming up. What about the politicians?

The official US stance for the past 6 years has been that scientific findings are "unconvincing," and "too uncertain" to call for any action. There was always something missing from the models. That is not surprising. As we know, models are always designed to simplify, to explain. The climatic system is so complex that there will always be certain things that the models will not cover. Besides, as Marika Holland, of the National Center for Atmospheric Research, says, there are some processes that "are just not well understood, and because of that have not been incorporated into climate models" (http://www.cgd.ucar.edu/oce/mholland/). However, it does not mean that with those processes included the results will turn around. In fact, according to Dr Holland, the sea ice is melting faster than models have predicted. There are many reasons for the underestimates. For example, models do not fully capture heat transport between ocean and atmosphere, or faster warming as reflective ice gives way to darker, heat-absorbing waters. Actually, it has been consistently observed that modelers tend to be conservative in their predictions, filtering out models that clearly overestimate the changes seen so far, but accepting the results where everything is too well-behaved and stable.

For any modeler, it is obvious that there is something not included in the model, and that there is always uncertainty in the results. Does this mean that models are useless? Certainly not! If several models, especially built independently, point in the same direction, then that is a huge reason for concern. If these models are scrupulously tested by third parties, and if there is a scientific consensus that the models are correct, then we had better start to act. However, that is where we find a gap between modeling and real life. It turns out that no matter how good a model is, whether it will be used for the betterment of humanity or not may be decided by forces that have nothing to do with science or modeling.

Twenty years of denial: a brief history of climate research

June 23, 1988: James Hansen (NASA) testifies to the Senate committee about the greenhouse effect. He says he is 99 percent sure that "the greenhouse effect has been detected, and it is changing our climate now."

Companies and industry associations representing petroleum, steel, autos and utilities form lobbying groups with names like the Global Climate Coalition (GCC) and the Information Council on the Environment (ICE). The goal is to "reposition global warming as theory rather than fact," and to sow doubt about climate research." ICE ads ask, "If the Earth is getting warmer, why is Minneapolis [or Kentucky, or some other site] getting colder?"

1992: The United Nations "Earth Summit" is held in Rio de Janeiro, with climate change high on the agenda. The ICE and GCC lobby hard against a global treaty to curb greenhouse gases, and are joined by the George C. Marshall Institute, a conservative thinktank. Just before Rio, it releases a study concluding that models of the greenhouse effect have "substantially exaggerated its importance." The small amount of global warming is because the Sun is putting out more energy.

US President George H.W. Bush is undecided. The Head of his Environmental Protection Agency (EPA), William Reilly, supports binding cuts in greenhouse emissions. Political advisers insist on nothing more than voluntary cuts. The Rio treaty calls for countries voluntarily to stabilize their greenhouse emissions by returning them to 1990 levels by 2000. (As it turns out, US emissions in 2000 are 14 percent higher than in 1990.) Avoiding mandatory cuts is a huge victory for industry.

The press does not take sides; it qualifies "every mention of human influence on climate change with 'some scientists believe'." In fact, the vast majority of scientific opinion already accepts that human-caused GHG emissions are contributing to warming. Talk radio host Rush Limbaugh tells listeners "more carbon dioxide in the atmosphere is not likely to significantly contribute to the greenhouse effect. It's just all part of the hoax." In the Newsweek Poll, 42 percent say the press "exaggerates the threat of climate change."

1996: William O'Keefe, Vice President of the American Petroleum Institute and Leader of the GCC, suggests that there is too much "scientific uncertainty" to justify curbs on greenhouse emissions. The "Leipzig Declaration on Global Climate Change" is released, where over 100 scientists and others, including TV weathermen, say they "cannot subscribe to the politically inspired world view that envisages climate catastrophes." Few of the Leipzig signers had actually carried out climate research.

1997: Kyoto, Japan, over 100 nations negotiate a treaty on making Rio's voluntary and largely ignored greenhouse curbs mandatory. The worried coal and oil industries ramp up their message that there is too much scientific uncertainty to justify any such cuts.

The Intergovernmental Panel on Climate Change (IPCC) – the international body that periodically assesses climate research – issues its second report. Its 2,500 scientists conclude that, although both natural swings and changes in the Sun's output might be contributing to climate change, "the balance of evidence suggests a discernible human influence on climate."

US President Clinton, while a strong supporter of GHG cuts, does not even try to get the Senate to ratify the Kyoto treaty. The Republican Party has a majority in both houses and is in denial. Republicans have also received significantly more campaign cash from the energy and other industries that dispute climate science.

April 1998: A dozen people from the Marshall Institute, Fred Singer's group and Exxon meet at the American Petroleum Institute's Washington headquarters. They propose a $5 million campaign to convince the public that the science of global warming is riddled with controversy and uncertainty.

January 2000: The National Academy of Sciences announces that, contrary to the claim that satellites finding no warming are right and ground stations showing warming are wrong, it turns out that the satellites are off. The planet is indeed warming, and at a rate, since 1980, much greater than in the past.

2001: Inauguration of President George W. Bush. As a candidate, he had pledged to cap carbon dioxide emissions. He was expected to reiterate that pledge in his speech after inauguration. The line was never said.

Bush disavows his campaign pledge, and in March withdraws from the Kyoto treaty.

The IPCC releases its third assessment of the studies of climate change. Its conclusion: the 1990s were very likely the warmest decade on record, and recent climate change is partly "attributable to human activities." The early years of the new millennium are setting heat records. In the summer of 2003, a heatwave in Europe kills tens of thousands of people. Consultant Frank Luntz writes: "you need to continue to make the lack of scientific certainty a primary issue … you should challenge the science by recruiting experts who are sympathetic to your view."

2003: The Senate calls for a national strategy to cut greenhouse gases. The White House leadership staff indicate that this is unacceptable. The whole thing disappears without much notice.

2003: James Inhofe of Oklahoma takes over as Chairman of the Environment Committee. In a 2-hour speech, he disputes the claim of scientific consensus on climate change. Despite the discovery that satellite data show warming, he argues that "satellites, widely considered the most accurate measure of global temperatures, have confirmed" the absence of atmospheric warming. Might global warming, he asks, be "the greatest hoax ever perpetrated on the American people?" Another denial study is partly underwritten with $53,000 from the API.

16 February 2005: The Kyoto Protocol comes into force after Russia finally ratifies it.

There is careful management of what federal scientists and officials write and say. Former coal and oil lobbyists are appointed to key jobs overseeing climate policy. Officials make sure that every report and speech cast climate science as dodgy, uncertain, controversial, and therefore no basis for making policy. An ex-oil lobbyist, working for the White House Council on Environmental Quality, edits a 2002 report on climate science by adding "lack of understanding" and "considerable uncertainty" throughout the text.

2007: The Democrats take the Congress and the Senate. Al Gore testifies to both chambers on climate change. Inhofe tells allies he will filibuster any climate bill that mandates greenhouse cuts.

February 2007: The IPCC releases its Fourth Assessment and reports that it is "very likely" (>90 percent) that heat-trapping emissions from human activities have caused "most of the observed increase in globally averaged temperatures since the mid-20th century."

In the Newsweek Poll, 38 percent of those surveyed identify climate change as the nation's gravest environmental threat – three times the number in 2000. ExxonMobil is blamed by senators for giving $19 million over the years to the CEI and others who are "producing very questionable data" on climate change. Senator Jay Rockefeller says the company has to cut back its support for such groups.

Bush announces that he will convene a global-warming summit next month, with a 2008 goal of cutting greenhouse emissions.

The Newsweek Poll finds less than half in favor of requiring high-mileage cars or energy-efficient appliances and buildings. While majorities in Europe and Japan recognize a broad consensus among climate experts that greenhouse gases are altering climate, the influence of the denial machine in USA remains strong. Although the figure is less than in earlier polls, 39 percent (down from 64 percent last year) say there is "a lot of disagreement among climate scientists" on the basic question of whether the planet is warming; 42 percent say there is a lot of disagreement that human activities are a major cause of global warming. Only 46 percent say the greenhouse effect is being felt today.

As of writing this text, there is still no agreement on GHG control in USA.

This account is largely based on a Newsweek article, "The Truth About Denial," by Sharon Begley, August 13, 2007 (http://www.msnbc.msn.com/id/20122975/site/newsweek/page/0/). Looking at readers' comments regarding this article and posted on the web, it is shocking how many people still refuse to accept the facts about climate change.

The problem is really not with the models; the problem is that models sometimes produce results that people do not want to know. So what do we do then? How can we make people listen, think, believe?

We saw in Chapter 7 that as the rate of flows in a system increases, we need to make faster decisions to keep it under control. A delay in taking appropriate action may result in system collapse, in a transition to an entirely different state from which return is extremely expensive or impossible. The decision-making time-step should decrease, or otherwise we do not have time to adapt, to restructure – especially when we are dealing with a combination of positive feedback effects, as in the case of global climate change. There is certainly much hope for new technologies, and so far humanity has indeed been successful in finding technological solutions – except, of course, for those cases when we failed … Jared Diamond, in his *Collapse: How Societies Choose to Fail or Succeed*, gives some pretty vivid examples of situations where the right technological solutions might have come too late, leading to vast socio-economic and ecological crises. Note that, unlike those ancient civilizations that disappeared on local and regional scales, today we are really talking about global processes, global systems and, therefore, if they are to happen – global collapses. It appears that we are entering the phase of "inconvenient truths" and unpopular decisions. We see how this is the case with climate; we have also seen above how this is going to be the case with "peak oil".

So, on the one hand we have systems that are becoming increasingly complex with very fast flows of material and energy; on the other, we have a growing human population with diverse values and priorities and growing access to information. On the one hand we have a need for urgent and well-informed decisions; on the other, we have better organized groups of stakeholders who have vested interests and want to play a role in the decision-making process. As a result, in many cases standard governance and decision-making become inadequate and start to fail. Many more regulatory decisions end up in court in lengthy and costly litigation procedures, creating social discontent, aggravation and frustration among people.

To deal with this increasing complexity of decision-making, we can delegate the decision-making process to experts, who have the knowledge and skills to make the right decisions in a timely manner. By doing this, we gradually become more dependent on the selected few and divorce the governance from the public, putting the whole democratic process at risk. At the same time we may indeed gain in efficiency of decision-making – as, for example, in China, which manages to make some very important (though unpopular) decisions quickly and has the means to enforce them. (For example, the one-child family planning regulation has managed to curb population growth; another example is the recent decision of the Chinese Government to stop all further conversion of agricultural lands for biofuel production. While the USA continues to subsidize biofuel, unable to overcome the lobbying power of agricultural corporations, China has made some very swift and timely decisions in this regard.) Certainly, this type of decision-making may be efficient – decisions are made and implemented quickly. However, the downside is that, as mentioned above, we are betting on one Wise King. Everything may work well as long as he is indeed wise – but if he goes crazy, we have little power to replace him. Also note that centralized decisions that are unpopular are hard to implement, require much enforcement, and usually fail.

Alternatively, we need to invest heavily in educating the public and in creating means and methods for public participation in the decision-making process. Recognizing the need to reinforce the process with local knowledge and iterative participatory interactions in order to derive politically feasible and scientifically sound solutions, governments and international organizations have embraced concepts of public involvement, and devolution of decision-making to lower and lower levels. For example, the Shared Vision Planning process that has been developed in the Army Corps over the past 30 years is a promising way to find understanding and acceptance among the various stakeholders that may be interested in the outcomes of a project and knowledge produced by models. The new web technologies and services provide new means of interaction and dissemination of data and knowledge.

As human domination over the environment grows and as the complexity of natural systems is further elevated by the complex human socio-economic systems built on them, decision-making processes become more constrained by feasible options and time horizons, while the consequences of wrong decisions become more dramatic and affect larger geographic areas. Under such circumstances, standard scientific activities are inadequate if we wish to continue on the democratic path of development. They must be reinforced with local knowledge and iterative participatory interactions in order to derive solutions which are well understood, politically feasible and scientifically sound. We need new ways to understand and embrace the inconvenient truths of today.

9.2   Participatory and adaptive modeling

(This section was written in collaboration with Erica Brown Gaddis.)

As argued by Oreskes *et al.* (1994), and as we discussed in Chapter 4, models do not tell us the "truth" about the system. They should be rather viewed as a process of striving towards the truth. The best model is a process in which we learn about the system and understand how best to manipulate and manage it. As we start administering this management, or as something starts to change in the environment, the system also changes and the model is no longer valid. We can succeed only if *the model is viewed as a process* that is designed to accommodate these changes and adapt to them. A good model should evolve with the system; it should be able to change both quantitatively and qualitatively as the system changes and as our understanding about the system improves.

In recent years, there has been a shift from top-down prescriptive management of ecological resources towards policy-making and planning processes that require ongoing active engagement and collaboration between stakeholders, scientists and decision-makers. Participatory modeling (PM) is the process of incorporating stakeholders (often including the public) and decision-makers into an otherwise purely analytic modeling process to support decisions involving complex ecological questions. It is recognized as an important means by which non-scientists are engaged in the scientific process, and is becoming an important part of environmental planning, restoration and management. Previously science was conducted outside of the policy-making process, allowing scientists to develop ecological models derived from analysis and observation of the natural world, thereby contributing an objective opinion to the policy-making process without accounting for the values, knowledge or priorities of the human system that affects and is affected by ecological systems. The shift towards more open and integrated planning processes has required the adaptation of the scientific modeling process to incorporate community knowledge, perspective and values.

Participatory modeling is particularly compatible with the rising focus on ecosystem-based management, integrated water resources management and adaptive management, all of which incorporate systems theory and aim to protect and improve ecological resources while considering economic and social concerns in the community. These approaches have been adopted by, among others, the Water Framework Directive of the European Commission, and supported by the National Research Council in the United States. The latter recommends that the processes of analysis and deliberation be integrated in such a way that systematic analysis is combined with community values critical to decision-making. PM provides a platform for integrating scientific knowledge with local knowledge and, when executed, provides an objective, value-neutral place for a diverse group of stakeholders to contribute information regarding an ecosystem of interest. Recognition that effective environmental management requires input from both scientific and social processes is key to developing effective partnerships between scientists and stakeholders that live and work within an ecosystem.

PM (of which clones are also known as "mediated modeling", "companion modeling" or "shared vision modeling") draws on the theory of post-normal science, which dictates that in problems characteristic of highly complex systems, there is no one correct, value-neutral solution. Stakeholder participation in environmental research and management has been justified for multiple reasons. PM supports democratic principles, is educational, integrates social and natural processes, can legitimize a local decision-making process, and can lead participants to be instrumental in pushing forward an agreed agenda. The extent to which the public or representative stakeholder group can effectively participate in ecological research and management is determined by the methods employed in engaging stakeholders, the inclusion of diverse groups, group size, incorporation of local knowledge and expertise, and the time available for the process to develop. The development of unique, practical and affordable solutions to ecological problems is often best accomplished by engaging stakeholders and decision-makers in the research process.

The idea stems from the feeling that any modeler develops while working on the model. You may have experienced it yourself when working with some of the models earlier presented in this book. The feeling is that as you go through all the essential steps of model building you get a really good understanding of all the processes and interactions involved and develop a certain intimacy with the system, learning what is more important and what can be approximated, getting a handle on the inputs and understanding how they may affect the outputs. You also learn to appreciate the uncertainties embedded in the system, and realize that even with these uncertainties there is certain level of confidence, or a comfort zone, that may be large enough to make a decision. And that decision will probably be the best-informed one for the current state of knowledge.

It seems as though you start thinking that if only everyone could share your understanding of what is going on in the system, then it should not be a problem to communicate the results and make the right decisions. So that is exactly what you do when you open up the modeling process and invite everybody potentially interested in the system and the decisions to participate in this collaborative group study. If it is recognized that during the modeling process the modeler gains much understanding about the system workings, about what is most essential and what controls the system behavior, then this rich and exciting experience that comes from the modeling process should be shared, and the whole decision-making process designed around the modeling process. The modeling process itself becomes the decision-making tool, and the decision-making becomes part of the modeling process.

Models are used to formalize concepts of ecological and socio-economic processes and, as such, explore existing dynamics and characteristics. Models can also be predictive or used to compare proposed management plans and explore their effects on other processes. Modeling tools are especially useful in communicating complex processes, spatial patterns and data in a visual format that is clear and compelling and, when appropriately applied, can empower stakeholders to move forward with concerted efforts to address an environmental or socio-economic problem. Both monitoring and modeling are scientific tools that can support good decision-making in ecosystem-based management, and are often most powerful when used together. Monitoring data collected at varying scales can be used as inputs to models, to calibrate and validate the accuracy of a model, or to address specific research questions using statistical models. Development of ecological models often indicates the types of information that are important in understanding dynamics but for which no data are available. Whereas selective monitoring can give a good description of patterns and linkages within a system, it may be more difficult and expensive to determine the driving forces of these patterns. Simulation models help to determine the mechanisms and underlying driving forces of patterns otherwise described statistically. In many cases, the monitoring efforts that go along with modeling can serve as a good vehicle to engage the local stakeholders in the process. When stakeholders see how samples are taken or, ideally, take part in some of the monitoring programs, they bond with the researchers and become better partners in the future decision support efforts.

The modeling of physical, biological and socio-economic dynamics in a system requires attention to both temporal dynamics and spatial relationships. There are many modeling tools that focus on one or the other. To be useful in a participatory framework, models need to be transparent and flexible enough to change in response to the needs of the group. Simulation (process) models may be formalized in software such as Stella, Simile or Madonna, which we have considered in this book. These and other software packages have user-friendly Graphic User Interfaces (GUI) which make them especially helpful when models are demonstrated to stakeholders or when they are formulated in their presence and with their input. In this context, complex simulation models or programming directly in C++ or other languages may be less effective, no matter how powerful the resulting models are. In some cases, tools as generic and simple as Excel turn out to be even more useful in engaging the stakeholders in a meaningful collaborative work than the far more powerful and accurate complex models.

To make these state-of-the-art complex models useful for the decision-making process, additional efforts are essential to build interfaces or wrappers that will allow them to be presented to the stakeholders, or embedded into other models (modularity). In general, process models may be very helpful to explain and understand the systems to be analyzed; however, they are not practical for exploring the role of the spatial structure of an ecosystem. Alternatively, Geographic Information Systems (GIS) explicitly model the spatial connectivity and landscape patterns present in a watershed, but are weak in their ability to simulate a system's behavior over time. Ecosystem-based management demands the coupling of these approaches such that spatial relationships, linkages and temporal dynamics can be captured simultaneously. There are many specific models developed to analyze the spatio-temporal dynamics of specific systems or processes. So far, there are not many generic tools that combine temporal and spatial modeling. One is the Spatial Modeling Environment (SME), which we have seen above. Simile, too, offers some powerful linkages to spatial data and processing. There are also modules programmed as components of GISs, say using the scripting language or Avenue in ArcINFO.

Agent-based models provide yet another modeling technique that is useful in participatory workshops. They offer some powerful techniques to engage the stakeholders in a dialogue, with some role-playing games leading to more clearly defined rules of behavior for agents. Again, for the participatory context a GUI is essential. NetLogo or StarLogo are two modeling frameworks that offer very user-friendly interfaces and have a relatively simple learning curve. NetLogo also has a module called HubNet (see, for example, http://ccl.northwestern.edu/netlogo/models/CompHubNet-TragedyoftheCommonsHubNet), which allows several people to work on the model while sitting behind different computers at different places. This can be an excellent environment to work on participatory modeling projects.

Forms of participation

Stakeholder participants engage in the decision-making process in the form of model selection and development, data collection and integration, scenario development, interpretation of results, and development of policy alternatives. It is generally recognized that engaging participants in as many of these phases as possible and as early as possible, beginning with setting the goals for the project, drastically improves the value of the resulting model in terms of its usefulness to decision-makers, its educational potential for the public, and its credibility within the community.

Model selection and development

Selecting the correct modeling tool is one of the most important phases of a PM exercise, and should be determined based on the goals of the participants, the availability of data, the project deadlines and funding limitations, rather than being determined by scientists' preferred modeling platform and methodology.

In terms of model development, stakeholders are very helpful in identifying whether there are processes or ecological phenomena that have been neglected in the modeling process. Stakeholders can also be called upon to verify basic assumptions about the dynamics, history and patterns of the ecosystem. In addition, community stakeholders can frequently validate assumptions about typical human behavior in the system. This often anecdotal evidence may be the only source of model assumptions about human behavior in a system. When combined with technical knowledge of ecological processes, such evidence may be key to identifying new and more appropriate management solutions. The PM approach is based on the assumption that those who live and work in a system may be well informed about its processes and perhaps have observed phenomena that would not be captured by scientists. This two-way flow of information is a key characteristic of successful PM.

Data collection and availability

Stakeholders often play a key role in research activities by contributing existing data to a research process or by actively participating in the collection of new data. Some stakeholders, particularly from governmental agencies, may have access to data that are otherwise unavailable due to privacy restrictions or confidentiality agreements. These data can often be provided to researchers if aggregated to protect privacy concerns, or if permission is granted from private citizens. In addition, some stakeholders are aware of data sources that are more specific to a particular ecosystem or locale, such as climatic data and biological surveys.

Stakeholders can also engage in ecological sampling and monitoring. This can be a particularly effective entry point to a community that is ready to "act" on a perceived problem and is not satisfied with more meetings and discussions of a problem. Monitoring by citizen stakeholders, in particular, provides other benefits to the research process. In many cases, they live close to monitoring sites or have access to private property such that more frequent and/or more complete monitoring can take place at significantly less cost than one individual researcher could complete independently. Citizens also gain benefits by becoming more familiar with their ecosystem – an educational opportunity that may be shared with other community members.

Scenario development

Stakeholders are best placed to pose solution scenarios to a problem. Many of them have decision-making power and/or influence in the community, and understand the relative feasibility and cost-effectiveness of proposed solutions. In addition, engaging local decision-makers in the scenario-modeling stage of the research process can lead to development of more innovative solutions.

Interpreting results and developing policy alternatives

A primary goal of a PM exercise is to resolve the difference between perceived and actual sources of an ecological problem. Whereas stakeholders might have proposed scenarios based on their perception of the problem or system, they may be particularly adept at proposing new policy alternatives following initial model results from a scenario-modeling exercise. The PM process can further facilitate development of new policies through development of a collaborative network between stakeholders and their respective agencies or constituents throughout the research process. Stakeholders are important communication agents to deliver the findings and the decision alternatives to the decision-makers in the federal, state or local governments. They are the more likely to be listened to than the scientists, who may be perceived as foreign to the problem or the locality. Governments certainly have a better ear for the electorate.

Criteria for Successful Participatory Modeling

PM is a relatively new activity, and as such the field is just beginning to define itself and the criteria that qualify a project as a good or successful PM exercise. Below are some of the key criteria that may be useful.

1. *Representative involvement, openness.* Regardless of the method used to solicit stakeholder involvement, every attempt should be made to involve a diverse group of stakeholders that represent a variety of interests regarding the question at hand.

While key stakeholders should be carefully identified and invited to the process, there should be also an open invitation to all interested parties to join. This will add to the public acceptance of and respect for the results of the analysis. If a process is perceived to be exclusive, key members of the stakeholder and decision-making community may reject model results.

2. *Scientific credibility*. Although PM incorporates values, the scientific components of the model must adhere to standard scientific practice and objectivity. This criterion is essential in order for the model to maintain credibility among decision-makers, scientists, stakeholders and the public. Thus, while participants may determine the questions that the model should answer and may supply key model parameters, the structure of the model must be scientifically sound. It does not mean that the model should be all encompassing and complex; to the contrary, it should be as simple as possible. It is crucial, however, to be extremely clear and honest about all the assumptions and simplifications made.

3. *Objectivity*. Facilitators of a PM project must be trusted by the stakeholder community as being objective and impartial, and therefore should not themselves be direct stakeholders. In this regard, facilitation by university researchers or outside consultants often reduces the incorporation of stakeholder biases into the scientific components of the model. On the other hand, it is essential that stakeholders trust the facilitators and scientists, and a certain track record in the local area and perhaps even recognition of researchers by the local stakeholders, based on passed research or involvement, can be helpful.

4. *Transparency*. Key to effective stakeholder engagement in PM is a process that is transparent. Transparency is not only critical to gaining trust among stakeholders and establishing model credibility with decision-makers, but also key to the educational goals often associated with PM.

5. *Understanding uncertainty*. Many ecological and socio-economic questions require analysis of complex systems. As problem complexity increases, model results become less certain. Understanding scientific uncertainty is critically linked to the expectations of real-world results associated with decisions made as a result of the modeling process. This issue is best communicated through direct participation in the modeling process itself.

6. *Flexibility*. The modeling process should be flexible and adjustable to accommodate the new knowledge and understanding that comes from the stakeholder workshops. Stakeholders might come up with ideas and factors that modelers had not anticipated, but modelers should be ready to incorporate these into the model.

7. *Model adaptability*. The model developed should be relatively easy to use and update after the researchers have moved on. This requires excellent documentation and a good user interface. If non-scientists cannot understand or use the model, it will not be applied by local decision-makers to solve real problems.

8. *Incorporation of stakeholder knowledge*. Key to success with any participatory approach is that the community participating in the research be consulted from the initiation of the project, and help to set the goals for the project and the specific issues to be studied.

9. *Influence on decision-making*. Results from the modeling exercise should have an effect, through some mechanism, on decisions made about the system under study.

Is there anything special about models that would be most appropriate for PM? Indeed, there are certain features that would make a model better suited for use with stakeholders.

Choosing a tool

The problem of choosing an appropriate tool is difficult, because learning each one requires some time and effort, which can be quite considerable. Therefore, it is often the case that once modelers have mastered a particular modeling language or system, they are inclined to use the same acquired skills next time they need to analyze a different system – even when this other system is quite unlike the first one, and even when the modeling goals are different. As Bernard Baruch (or, according to alternative sources, Abraham Maslow) is supposed to have said, "If all you have is a hammer, everything looks like a nail." Anyway, it is quite natural for people to try to do what they already know how to do. As a result, modelers who are equally proficient in a variety of modeling techniques are quite rare, and good comparisons of modeling tools are also hard to find.

In choosing a tool, the following should be considered:

1. *Inclusiveness.* PM cannot rely on several particular models. The modeling engine supporting PM should be able to incorporate a variety of models, presented as modules. These modules should be interchangeable to serve particular needs of a project, and to present state-of-the-art modeling and data analysis. The modeling interface serving these needs should operate as a middleware product, or coupler, that can take various modules and make them work in concert. Modules in this context present both software objects for simulation and data objects. The challenge is to make these modules talk to each other and perform across a variety of temporal and spatial scales and resolutions.

2. *Modularity.* In the modular approach, we do not intend to design a unique general model. Instead, the goal is to offer a framework that can be easily extended and is flexible regarding modification. A module that performs best in one case may not be adequate in another. The goals and scale of a particular study may require a completely different set of modules that will be invoked and further translated into a working model. There is a certain disparity between the software developer's and the researcher's views upon models and modules. For a software developer, a module is an entity, a black box, which should be as independent as possible, and as easy as possible to combine with other modules. This is especially true for the federation approach to modular modeling, and is well demonstrated by web-based modeling systems. The utility of such applications may be marginal from the research viewpoint.

   For a researcher, a model is predominantly a tool for understanding the system. By plugging together a number of black boxes, for which specifics and behavior is obscure and hardly understood, we do not significantly increase our knowledge about the system. The results generated are difficult to interpret when there is not enough understanding of the processes that are actually modeled. The decomposition of such systems requires careful analysis of spatial and temporal scales of processes considered, and is very closely related to specific goals of the model built.

   In this context, the modular approach can be useful if the focus is shifted from reusability and "plug-and-play" to transparency, analysis and hierarchical description of various processes and system components. With the modules being transparent and open for experiment and analysis, the researcher can better understand the specifics of the model formalism that is inherited. It is then easier to decide whether a module is suitable, or whether it should be modified and tuned to the specific goals of a particular study.

3. *Transparency.* In PM, the models are used to explain rather than to predict. It is important therefore to be able to dive into the model structure and be clear about the processes that are included and the assumptions made. This immediately adds value to simpler models and modeling tools. In some cases, the benefits of gaining stakeholder "buy-in" into the model and process by working together on simple models that they understand outweigh the lack of detail and lower accuracy that we get from such models in comparison with the more sophisticated but less comprehensible models. A simple model that can be well communicated and explained may be more useful than a complex model that may be taking more features into account but with narrow applicability, high costs of model and data, and much uncertainty.

It is also important to make sure that we clearly draw the boundaries of the system that is researched and modeled, and realize that we are not supposed to be modeling the whole world in all its complexities. For example, if a study is concerned with scenarios driven by global warming, it should not be our goal to reproduce, understand and defend the extremely complex Global Circulation Models that are used to generate future climates. We will be much better off clearly describing the output from those models, with the associated range of predicted change, as a forcing function that is out of the scope of our analysis and should be used as a given for our purposes. Otherwise, we are at risk of getting ourselves involved in the highly contentious debate about the "truths" of climate change instead of analyzing the risks and outcomes that we face within our system.

4. *Visualization.* Models should be impressive on the output side; they must present results in an appealing and easy-to-understand form. Interfaces must allow multiple levels of complexity and interactivity to serve different stakeholder groups.

5. *Affordability.* The models used in the PM process should be affordable for the stakeholders in different levels of governance. This means that either the modeling tools should be made available over the web and run on the server side, so that users will not need to purchase expensive licenses, or the models themselves should be freeware or shareware.

6. *Flexibility, extendibility.* When something is missing in models, there should be a way to add it to the existing structure rather than rebuilding the whole model again from scratch simply because one element is missing. This is especially crucial in the PM process, when models should be developed quickly in response to the concerns and new information coming from the stakeholders.

It is really important to be inventive on the visualization side. For example, one very popular way to present the level of a certain impact is to use a color code ranging from green (safe and good) to yellow (moderate but bearable) and then to red (bad, unsafe and unhealthy). This color code is widely adopted in some of the EPA reports and web pages (see, for example, http://www.epa.gov/reg3artd/airquality/airquality.htm).

Chris Jordan, a graphic designer and photographic artist, uses an ingenious way to show the scale of various process and stocks. He starts to picture simply certain items (say, plastic bottles or aluminum cans) and then zooms out, getting more and more items into the picture. Showing, say, 2.5 million plastic bottles, which is how many are used in the US every hour, creates a powerful message. Or the 11,000 jet trails, equal to the number of commercial flights in the US every 8 hours, or the 2.3 million folded prison uniforms, equal to the number of Americans incarcerated in 2005. See http://www.chirsjordan.com/current_set2.php, or check out the PBS website at http://www.pbs.org/moyers/journal/09212007/profile4.html.

Some of these considerations clearly point us in the direction of open-source (OS) development. The OS paradigm delivers ultimate transparency and flexibility in the products developed. These products are also free for the user. It only makes sense that taxpayers' money be spent on products that will be available for the taxpayers, stakeholders, at no additional cost. Federal agencies should promote and support open-source software for a variety of reasons, such as transparency, extendibility, security, low cost, etc.

There are numerous implementations of the method that vary in their level of success and achievement. Let us mention a few.

Solomons Harbor Watershed, Maryland

Excessive nutrient loads to the Chesapeake Bay from surrounding cities and rural counties has led to eutrophication, especially in small harbors and inlets. The Maryland Tributary Strategies, Chesapeake Bay 2000 Agreement and Calvert County Comprehensive Plan call for reductions in nutrients entering the Bay in order to reduce impacts on aquatic natural resources. Though the goal set for phosphorus appears to be achievable, reductions in nitrogen lag well behind the target. Most sewage in rural residential areas of Maryland, such as Calvert County, is treated by on-site sewage disposal systems (septic systems). Almost all of the nitrogen pollution that enters local waters from Calvert County comes from non-point sources, of which the Maryland Department of Planning estimates 25 percent comes from septic systems. In this project we initiated a PM effort to focus on the most densely populated watershed in Calvert County that drains to Solomons Harbor. Despite high population densities, only a small portion of the watershed is serviced by sewer. There are no major point sources of nitrogen in the watershed.

Two different modeling tools were used to analyze and visualize the fate of nitrogen from three anthropogenic sources: septic tanks, atmospheric deposition, and fertilizer. The first is a simple dynamic model of a septic tank and leach-field system using Stella™ software, which allows the user to evaluate alternative septic technologies. The second modeling tool is the spatially explicit Landscape Modeling Framework (LMF), developed by the Gund Institute for Ecological Economics and discussed in Chapter 6.

Participation in the study was solicited from community stakeholders who were instrumental in understanding how models could be applied to local decision-making, in making appropriate model assumptions and in developing politically feasible scenarios. The model results found that septic tanks may be a less significant contributor to surface water nitrogen pollution in the short term, whereas fertilizer used at the home scale is a more significant source than previously thought. Stakeholders used the model results to develop recommendations for the Calvert County Board of Commissioners. Recommendations include mandating nitrogen removal septic tanks for some homes, but primarily focus on intensive citizen education about fertilizer usage, local regulation of fertilizer sales, reduction in automobile traffic, and cooperation with regional regulatory agencies working to reduce regional NOx emissions.

St Albans Bay Watershed, Vermont

Lake Champlain has received excess nutrient runoff for the past 50 years due to changes in agricultural practices and rapid development of open space for residential use. The effect of excess nutrients has been most dramatically witnessed in bays such as St Albans Bay, which exhibits eutrophic algal blooms every August. The watershed feeding St Albans Bay is dominated by agriculture at the same time that the urban area is growing. In the 1980s, urban point sources of pollution were reduced by upgrading the St Albans sewage treatment plant. At the same time, agricultural non-point sources were addressed through the implementation of "Best Management Practices" (BMPs) on 60 percent of the farms in the watershed, at a cost of $2.2 million (USDA, 1991). Despite the considerable amount of money and attention paid to phosphorus loading in St Alban Bay, it remains a problem today. The focus has remained primarily on agricultural landuses in the watershed, and as a result has caused considerable tension between farmers, city dwellers, and landowners with lake-front property.

Recently, the Lake Champlain TMDL allocated a phosphorus load to the St Albans Bay watershed that would require a 33 percent reduction of total phosphorus to the bay. We initiated a PM effort to apportion the total load of phosphorus from all sources, including diffuse transport pathways, and identify the most cost-effective interventions to achieve target reductions.

A group of stakeholders was invited to participate in the 2-year research process and members were engaged in the research at multiple levels, including water quality monitoring, soil phosphorus sampling, model development, scenario analysis, and future policy development. Statistical, mass-balance and dynamic landscape simulation models were used to assess the state of the watershed and the long-term accumulation of phosphorus in it, and to describe the distribution of the average annual phosphorus load to streams in terms of space, time and transport process. Watershed interventions, matched to the most significant phosphorus sources and transport processes, were developed with stakeholders and evaluated using the framework.

Modeling results suggest that the St Albans Bay watershed has a long-term net accumulation of phosphorus, most of which accumulates in agricultural soils. Dissolved phosphorus in surface runoff from the agricultural landscape, driven by high soil phosphorus concentrations, accounts for 41 percent of the total load to watershed streams. Direct discharge from farmsteads and stormwater loads, primarily from road sand washoff, were also found to be significant sources.

The PM approach employed in this study led to identification of different solutions than stakeholders had previously assumed would be required to reduce the phosphorus load to receiving waters. The approach led to greater community acceptance and utility of model results, as evidenced by local decision-makers now moving forward to implement the solutions identified to be most cost-effective.

Redesigning the American Neighborhood, South Burlington, Vermont

Urban sprawl and its associated often poorly-treated stormwater have a big impact on water quality and quantity in Vermont. Converting agricultural and forested land to residential and commercial use has significantly changed the capacity of the watersheds to retain water and assimilate nutrients and other materials. Currently, as some studies suggest, storm discharges may be 200 to 400 times greater than historical levels (Apfelbaum, 1995).

As mentioned in Chapter 6, Redesigning the American Neighborhood (RAN) is a project conducted by the University of Vermont to find cost-effective solutions to the existing residential stormwater problems at the scale of small, high-density residential neighborhoods (http://www.uvm.edu/~ran/ran). The project is focusing on a case study of the Butler Farms/Oak Creek Village communities in South Burlington,

VT, to address the issue of targeting and prioritizing best management practices (BMPs). The idea was to engage local homeowners in a participatory study that would show them how they contribute to the stormwater problem and introduce them to existing alternative methods of stormwater mitigation through low-impact distributed structural and non-structural techniques.

The project started slowly, with only a few homeowners willing to participate in the process. However, soon the neighborhood learned that their homes were subject to long-expired State stormwater discharge permits, and that their neighborhood's stormwater system did not meet stringent new standards. As is often the case, problems with home sales, frustration with localized flooding, and confusion about the relationship between the City's stormwater utility and the State permit impasse led to frustration and even outright anger on the part of residents. The tension increased after the homeowners realized that in order for the City to take over the existing detention ponds and other stormwater structures, they had to be upgraded to currently active 2002 standards. Since then, the interest and involvement of residents in Stormwater Study Group has been heightened, but stakeholder meetings have become forums for conflict between homeowners and local municipalities.

The modeling component was mostly based on spatial analysis using the ESRI ArcGIS 9.2 capabilities for hydrologic modeling. As high-resolution LIDAR data became available, it became possible to generate clear visualization and substantial understanding about the movement of water through the neighborhoods, and to develop new approaches to resolve the stormwater management conundrum. The Micro Stormwater Network has helped to visualize rain flowpaths at a scale where residents have been able to make the connection with processes in their backyard. The Micro Stormwater Drainage Density (MSDD) index was instrumental in optimizing the location of BMPs of small and mid-scale management practices, and had an important educational and trust-building value.

At present, the homeowners seem to prefer decentralized medium and small-scale interventions (such as rain gardens) to centralized alternatives such as large detention ponds.

Cutler Reservoir TMDL process, Utah

Cutler Reservoir, in the Cache Valley of Northern Utah, has impounded the Bear, Logan and Little Bear Rivers since 1927. Cutler Dam is operated by PacifiCorp–Utah Power and Light to provide water for agricultural use and power generation. Cutler Reservoir supports recreational uses and a warm water fishery while providing a habitat for waterfowl and a water supply for agricultural uses. Cutler Reservoir has been identified as water-quality limited due to low dissolved oxygen and excess phosphorus loading. The Utah Division of Water Quality initiated the process of developing a Total Maximum Daily Load (TMDL) for the Cutler Reservoir in 2004, with the goal of restoring and maintaining water quality to a level that protects the beneficial uses described above.

Participation from local stakeholders is encouraged throughout the TMDL process, and has been formalized in the development of the Bear River/Cutler Reservoir Advisory Committee, which has representation from all the major sectors and interests of the local community. The advisory committee has been meeting monthly since August 2005, and has informed the TMDL process by contributing data and knowledge of physical and social processes in the watershed, and identifying solutions to help reduce pollution sources.

Watershed-loading models and a reservoir-response model (Bathtub) are in preliminary development stages at the time of writing, and will benefit from feedback from the advisory committee. It is expected that committee members will continue to provide feedback to the TMDL process while working with their respective constituents to provide direction to UDEQ in developing and implementing a watershed management plan. They will also be helpful in identifying funding needs and sources of support for specific projects that may be implemented.

James River Shared Vision Planning, Virginia

The James River in Virginia will potentially face significant water supply development pressures over the next several years due to growing population and development pressure. The Corps' Norfolk District has already received one application for a Clean Water Act Section 404 permit for Cobb Creek Reservoir, and initial inquiries by the Virginia Department of Environmental Quality indicate the potential for more applications in the near future. USEPA Region III has formally requested that Norfolk District prepare a basin-wide assessment that considers all the proposed water supply projects on the James River and make permitting decisions based on a cumulative impacts analysis.

These factors point to the need for a comprehensive planning process, involving all the key agencies and stakeholders, in order to identify broadly acceptable and sustainable solutions for water management within the basin. Due to historic water conflicts in the state, the Shared Vision Planning (SVP) process (http://www.sharedvisionplanning.us) has been proposed as the method for conducting this comprehensive process. The Army Corps of Engineers has pioneered participatory decision-making since the 1970s (Wagner and Ortolando, 1975, 1976). The Shared Vision Planning process is a PM approach in which stakeholders are involved in creating a model of the system that can be used to run scenarios and find optimal solutions to a problem. Shared Vision Planning relies on a structured planning process firmly rooted in the federal Principles and Guidelines, and in the circles of influence approach to structuring participation (Palmer et al., 2007).

The James River Study (JRS) began with a general workshop in the winter of 2006, entitled "Finding and Creating Common Ground in Water Management." The purpose of this open meeting was to start a continuing dialogue among the various stakeholders involved, including those with divergent interests. A major objective of the workshop was to describe and introduce the use of collaborative modeling to facilitate learning and decision-making across various governmental and non-governmental groups. While there was good participation in the workshop, the process stalled when working groups were to be formed. Only a few stakeholders signed up to continue with the PM effort, and during the following months the process almost stopped. It took some time to realize that in fact the project got stuck amidst some major controversy between two key stakeholders. In addition, there was some internal opposition to the project within the Army Corps. Under these conditions, not surprisingly, stakeholders who knew about these conflicts were skeptical about the project and reluctant to participate. As of today, a consensus seems to be emerging between the stakeholders regarding the goals of the project, and a fresh start is planned in the near future.

To a certain extent, these and other PM projects tend to follow the flow chart for a generic PM process presented in Figure 9.1. Note that there may be a lot of variations of and deviations from this rather idealized sequence. When dealing with

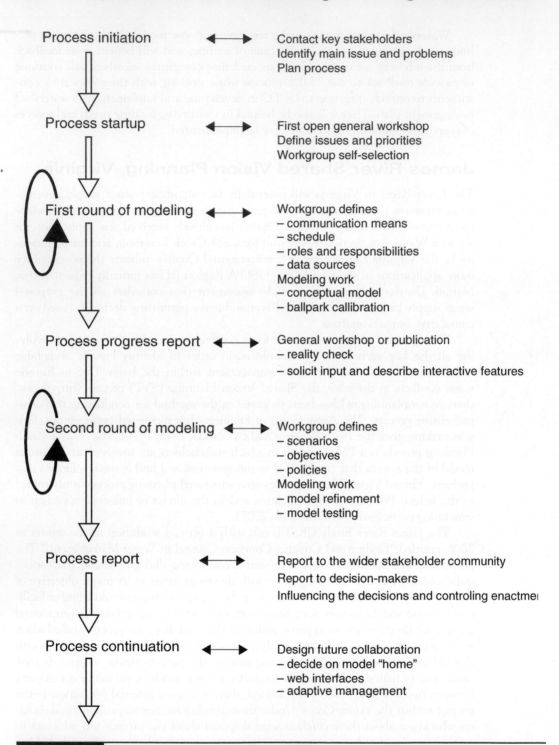

Process initiation	◄——►	Contact key stakeholders Identify main issue and problems Plan process
Process startup	◄——►	First open general workshop Define issues and priorities Workgroup self-selection
First round of modeling	◄——►	Workgroup defines – communication means – schedule – roles and responsibilities – data sources Modeling work – conceptual model – ballpark callibration
Process progress report	◄——►	General workshop or publication – reality check – solicit input and describe interactive features
Second round of modeling	◄——►	Workgroup defines – scenarios – objectives – policies Modeling work – model refinement – model testing
Process report	◄——►	Report to the wider stakeholder community Report to decision-makers Influencing the decisions and controling enactmer
Process continuation	◄——►	Design future collaboration – decide on model "home" – web interfaces – adaptive management

Figure 9.1   A flow chart for a generic PM process.

Note that each particular project will most likely develop in its own way, driven by the stakeholders involved.
That is perfectly fine; however, it is good to keep some of the keystones in mind.

people, we have to be ready for surprises and need to adapt the whole process to specific needs of particular projects and stakeholder groups. However, this diagram may be something to keep in mind when planning a PM process.

Some lessons learned, or a guide to success

1. Identify a clear problem and lead stakeholders

Although most watershed management decisions benefit from stakeholder input and involvement, some issues might not raise the interest of a wide group of stakeholders. If the problem is not understood or considered to be important by stakeholders, then it will be very difficult to solicit involvement in a participatory exercise. For example, the Virginia project had a very difficult startup because there was clear disagreement between stakeholders regarding the importance of the study. While it was quite clear to all that there would be growing problems with water supply in the area, the situation did not look bad enough to get local people really involved, while agencies had their own agendas and were not exactly clear on the purposes of the study.

Education of the community about water resource issues and the impact of decisions on the community is often a good first step. This can often be accomplished through the media, town hall meetings, or volunteer and community-oriented programs.

In some cases, it is helpful when there is a strong governmental lead in the process. The Calvert group sprouted from an open meeting where all citizens residing in the watershed were invited to comment on proposed regulation of septic systems by the County Planning and Zoning commission. The possibility of new regulation caught the attention of the public, and interested parties were willing to participate in the study. In other cases, interest from some stakeholders may only arise after a policy change that directly impacts them. The RAN project started with several stakeholder workshops, where homeowners were addressed about the looming problems associated with untreated stormwater. The reception was lukewarm, with very low attendance. Things changed quite dramatically when the city of South Burlington approved legislation that created a stormwater utility, which would take over stormwater treatment from the homeowners, but only after they brought their runoff up to certain standards. It turned out that their titles were no longer valid, since all their permits related to stormwater had expired a while ago. The interest in the RAN project immediately jumped, but even then for some homeowners the involvement of university researchers was seen as an impediment.

Never underestimate the "luck factor." Working with people, it takes just one or two stakeholders who choose to take an obstructionist position to damage the process. Similarly, one stakeholder that "gets it" and is interested and actively participating can significantly enhance the effort.

2. Engage stakeholders as early and often as possible

Establishment of a community-based monitoring effort can be a particularly effective entry point to a community that is ready to "act" on a perceived problem and is not satisfied with more meetings and discussion. Monitoring by citizens, in particular, provides other benefits to the research process. In many cases, they live close to monitoring sites or have access to private property such that more frequent and/or more complete monitoring can take place at significantly less cost than one individual researcher could complete independently. Citizens also gain benefits by becoming more familiar with their watershed – an educational opportunity that may be shared with other community members. When stakeholders see how samples are taken or, ideally, take part in some of the monitoring programs, they bond with the researchers and become better partners in future research and decision-support efforts.

In the St Albans Bay watershed, there was a lack of recent data regarding the general state of the watershed, including water quality, discharge, and soil phosphorus concentrations. At the same time, there was a highly motivated group of citizens organized through the St Albans Area Watershed Association eager to begin "doing" something in the watershed immediately. In partnership with this group and the Vermont Agency of Natural Resources, a citizens' volunteer monitoring program was established with 25 monitoring sites around the St Albans Bay watershed. Most of the 500+ water-quality samples and stage-height data were collected by a group of 15 volunteers drawn from the community over 2 years. The resulting data would not have been available otherwise, and the process engaged a group of local citizens in the research process. This early engagement proved valuable during the latter stages of the project, when a stakeholder group was assembled for the PM exercise. The partnership that grew from the monitoring effort also built trust between the researchers and watershed activists working in the community.

A key to success with any participatory approach is that the community participating in the research be consulted from the initiation of the project and help to set the goals for the project and specific issues to be studied (Beirele and Cayford, 2002). Stakeholder participants engage in the decision-making process in the form of model selection and development, data collection and integration, scenario development, interpretation of results, and development of policy alternatives. It is generally recognized that engaging participants in as many of these phases as possible and as early as possible, beginning with setting the goals for the project, drastically improves the value of the resulting model in terms of its usefulness to decision-makers, its educational potential for the public, and its credibility within the community (Korfmacher, 2001).

3. Create an appropriately representative working group

Participatory modeling may be initiated by local decision-makers, governmental bodies, citizen activists or scientific researchers. In the United States, most PM activities are initiated by governmental bodies (Duram and Brown, 1999). Depending upon the type of participation and the goals and time restrictions of the project, stakeholders may be enlisted to participate in a variety of ways. In some projects stakeholders are sought out for their known "stake" in a problem or decision, and are invited to join a working group. In other cases, involvement in the working group may be open to any member of the public.

Regardless of the method used to solicit stakeholder involvement, every attempt should be made to involve a diverse group of stakeholders that represent a variety of interests regarding the question at hand. When less well-organized stakeholder groups do not actively participate, watershed managers can obtain information about their opinions through other means such as public meetings, education, or surveys (Korfmacher, 2001).

In this sense, the St Albans Bay watershed modeling process may have failed somewhat in that the stakeholder group formed rather organically from those that currently work on issues or are directly affected by watershed management, including local, state and federal natural resources, planning and agricultural agencies, as well as farmers and watershed activists. A deliberate attempt was made to involve members of the business and residential community without success, due to a lack of interest in the process – perhaps because they perceived themselves to have no stake in the outcome.

4. Gain trust and establish neutrality as a scientist

This can be achieved by adhering to the second and third criteria on scientific credibility and objectivity, as presented above. It is helpful when we can refer to past examples and success stories, or refer to existing models that are known to stakeholders and perhaps published in peer-reviewed literature. However, it is even more important to keep the model clear to all participants, to have a good hand on all assumptions and formalizations used in the model. For example, the models developed for use in the St Albans Bay and Solomons Harbor watersheds have been peer-reviewed and accepted by the scientific community (Gaddis, 2007; Gaddis *et al.*, 2007). Model development is still underway for the James River and Cutler Reservoir.

5. Know the stakeholders and acknowledge conflict

In some cases, stakeholders may have historical disagreements with one another. One purpose of the PM method is to provide a neutral platform upon which disputing parties can contribute and gain information. However, it is important to watch for such historic conflicts and external issues that may overshadow the whole process. In addition, we have found that when the outcome of a modeling exercise is binding, such as in the development of a TMDL, parties may be both more engaged but also defensive if they perceive that the process will result in a negative impact on them or their constituents. For example, point-source polluters may look for ways to hold up a TMDL process in order to prolong a load-reduction decision. These sources of contention may be masked as scientific dissent when they are actually political. When conflict within the group becomes unmanageable, it is important to set out rules for discussion and, in some cases, to hire a professional facilitator.

In the James River project, there had been a long history of tension between some stakeholders on issues of water planning. The Shared Vision Planning process got caught in this controversy, and could move nowhere further until some consensus was reached between stakeholders. In theory, the modeling process was supposed to be open to all stakeholders, should be truly democratic and transparent, and should not depend upon local misunderstanding between some stakeholders. In practice, the historic network of connections (both professional and personal) between stakeholders is evident and can come to dominate the participatory process.

6. Select appropriate modeling tools to answer questions that are clearly identified

A critical step early in the PM process is the development of research questions and goals of the process. The questions identified should be answerable, given the time and funding available to the process. In addition, it is important that all stakeholders agree on the goals of the process such that a clear research direction is embraced by the entire group before detailed modeling begins.

Selecting the correct modeling tool is one of the most important phases of a PM exercise, and should be determined based on the goals of the participants, the availability of data, the project deadlines and funding limitations, rather than being determined by scientists' preferred modeling platform and methodology. Models are used to formalize concepts of watershed, stream and receiving-water processes, and as such explore existing dynamics and characteristics. Models can also be predictive, or used to compare proposed management plans and explore their effects on other processes. Modeling tools can be especially useful in communicating complex processes, spatial patterns and data in a visual format that is clear and compelling and, when appropriately applied, can empower stakeholders to move forward with concerted efforts to address an ecological problem.

It is important to maintain "model neutrality." It is common for modelers to turn to the models and modeling platforms that are most familiar to them. It is important, however, always to survey the available tools and select one that is most appropriate to the points of interest of the stakeholders. The Solomons Harbor watershed project was initially geared towards a fairly sophisticated spatial modeling effort based on our experience in integrating dynamic spatial models. While this modeling was still being performed, the project focus turned to some fairly simple balance calculations that helped move the decision-making process in the right direction.

To be useful in a participatory framework, models need to be transparent and flexible enough to change in response to the needs of the group. As we noted above, in some cases tools as simple as Excel can be the right choice. Major benefits of Excel are that it is readily available in most cases, and many stakeholders are already intimately familiar with it. Simulation (process) models help determine the mechanisms and underlying driving forces of patterns otherwise described statistically; however, they are not practical for exploring the role of the spatial structure of an ecosystem. Alternatively, Geographic Information Systems (GIS) explicitly model the spatial connectivity and landscape patterns present in a watershed, but are weak in their ability to simulate a system's behavior over time. Model complexity must be dictated by the questions posed by the stakeholder group. Models that are too simple are less precise and explanatory; however, a model that is too complex can lose transparency among the stakeholder group. In many cases, a simple model that can be well communicated and explained is more useful than a complex model that has narrow applicability, high costs of data, and much uncertainty.

In addition to a Stella implementation of the simple TR-55 routing model, the RAN project has been using GIS analysis. The spatial visualization of streamflows at the fine scale that was allowed by the LIDAR data was a turning point in the discussions, when stakeholders could actually see how their local decisions could make a difference.

7. Incorporate all forms of stakeholder knowledge

The knowledge, data and priorities of stakeholders should have a real, not just cursory, impact on model development, both in terms of selecting a modeling platform and in setting model assumptions and parameters. Stakeholders often contribute existing data to a research process or actively participate in the collection of new data. Some stakeholders, particularly from governmental agencies, may have access to data that are otherwise unavailable to the public due to privacy restrictions or confidentiality agreements. These data can often be provided to researchers if aggregated to protect privacy concerns, or if permission is granted from private citizens. In addition, some stakeholders are aware of data sources that are more specific to the watershed, such as locally collected climatic data.

The PM approach is based on the assumption that those who live and work in a system may be well informed about its processes and may have observed phenomena that would not be captured by scientists. Stakeholders can also be very helpful in identifying whether there are hydrologic, ecological or human-dominated processes that have been neglected in the model structure. Stakeholders can also verify basic assumptions about the dynamics, history and patterns of both the natural and socio-economic systems. Farmers and homeowners possess important local and lay knowledge about the biophysical and socio-economic system being researched. Anecdotal evidence may be the only source of assumptions about human behavior in a watershed, many of which are important inputs to a simulation model (i.e. frequency of fertilizer application). This type of knowledge, when combined with technical knowledge of watershed processes, is key to identifying new and more appropriate solutions to environmental problems.

The modeling process should be flexible and adjustable to accommodate the new knowledge and understanding that comes from stakeholder workshops. This requires that models be modular, robust and hierarchical to make sure that changes in components do not crash the whole system.

8. Gain acceptance of modeling methodology before presenting model results

Giving stakeholders the opportunity to contribute to and challenge model assumptions before results are reported also creates a sense of ownership of the process that makes it more difficult to reject results in the future. This can only occur, however, if the models developed are transparent and well understood by the public or stakeholder group (Korfmacher, 2001). In some cases, it can reduce conflict between stakeholders in the watershed, since model assumptions are often less controversial than model results.

The development of the modeling tools used in the St Albans Bay watershed was very transparent. Stakeholders were repeatedly given the opportunity to comment on model assumptions and parameters selected, and were even consulted on alternative modeling frameworks when appropriate. However, the model is not "user-friendly" due to the architecture of modeling framework selected.

9. Engage stakeholders in conversations regarding uncertainty

Stakeholders that have participated in all the stages of the model building activities develop trust in the model and are less likely to question the reliability of the results. Primarily, this is because they know all the model assumptions, the extent of model reliability, and that the model incorporated the best available knowledge and data; they also understand that there will always be some uncertainty in the model results.

10. Develop scenarios that are both feasible and ideal

Stakeholders are in a better position to judge what the more realistic and effective interventions are, and what the most feasible decisions might be (Carr and Halvorsen, 2001).

In the Solomons Harbor watershed, Maryland, an interesting question emerged from the discussion of scenarios that could reduce nitrogen to Solomons Harbor. Given limited resources for modeling, is it better to focus on the scenarios that the research team suspect will have the greatest impact on water quality, or those that are easiest and therefore likely to be implemented politically? Scenarios are very different for each perspective. A consensus was reached here through discussion to test both sets of scenarios. By testing feasible scenarios, we get a sense of what can reasonably be achieved in the short term, given current funding and political realities. Ideal scenarios push stakeholders to think beyond conventional solutions and to recognize the boundaries and time lag involved with what they aim to accomplish. Besides, another most cost-efficient and productive scenario emerged from the participatory fact-finding exercise: to focus on reduction of residential fertilizer application and other airborne sources of nitrogen in the area.

11. Interpret results in conjunction with stakeholder group; facilitate development of new policy and management ideas that arise from modeling results

Stakeholders can help to interpret the results and present them in the way that will be better understood by decision-makers at various levels of governance. They can advise on how best to visualize the results in order to deliver a compelling and clear message.

In the St Albans Bay watershed, many of the modeling results were not expected by the stakeholder group. Some of the most important sources and pathways of phosphorus movement to receiving waters (dissolved phosphorus from agricultural fields, road-sand washoff and tile drainage) were not addressed by most of the proposed scenarios. Some processes had previously been considered significant by the stakeholder group. However, several stakeholders have indicated that they intend to use the information gleaned from the project to direct existing funding sources and adapt policies to the extent possible to address the most significant phosphorus transport processes and sources in the watershed. The municipalities in the watershed have agreed to investigate alternatives to road sand for winter deicing of roads.

The TMDL process currently underway for Cutler Reservoir, Utah, requires that the results of the PM study be included in the prescribed management changes included in the TMDL document submitted for approval to the USEPA. These decisions include required nutrient-load reductions according to load allocations for various point and non-point sources throughout the watershed, as well as a Project Implementation Plan designed to achieve these reductions.

In the Solomons Harbor watershed, Maryland, unexpected results led the working group to adapt management goals and policies for Calvert County. Fertilizer and atmospheric deposition were found to have a significantly larger effect than the community had thought on nitrogen loads in Solomons Harbor, although none of the proposed septic management scenarios are likely to have a real effect on the trophic status of the harbor in the short-term. Nonetheless, upgrading septic tanks is still a good environmental decision, since it will improve groundwater quality and, in the long term, affect surface-water quality. Furthermore, it is the only regulation that can be easily and immediately implemented at the local level. The model results were first presented to the smaller working group over two meetings and were a severe test of participant confidence, since new results were somewhat contrary to previous estimates. The working group took a very positive and constructive approach. While acknowledging the inherent uncertainties in the modeling process, they began to explore new solutions and policy recommendations. Rather than abandoning the proposed policies to reduce nitrogen from septic tanks, the working group chose to expand its policy recommendations to include all sources of nitrogen to the watershed. The research team found this to be a distinctly positive outcome of the PM exercise. The working group came up with the following conclusions about the types of policy options that are realistic and available to the Solomons Harbor community:

- Atmospheric deposition cannot be directly influenced by local citizens, except through reduction of local traffic and lobbying regional officials to reduce NOx emissions from coal-fired power plants
- Fertilizer usage can be most easily influenced through educational initiatives, since policy changes will require involvement of other governmental and citizen groups beyond the Department of Planning and Zoning, which is currently leading the initiative to reduce nitrogen to the harbor.

12. Involve members of the stakeholder group in presenting results to decision-makers, the public and the press

An important final step in the PM method is the dissemination of results and conclusions to the wider community. Presentations to larger stakeholder groups, decision-makers, and the press should be made by a member of the stakeholder working group. This solidifies the acceptance of the model results and cooperation between stakeholders that were established during the PM exercise. In addition, members of the community are often more respected and have a better handle on the impact of policy decisions on the local community's issues.

In the Solomons Harbor watershed, two members of the working group presented their recommendations to the larger stakeholder group following a presentation of the modeling results by one member of our research team. During this meeting, the Director of Planning and Zoning for Calvert County solicited feedback on proposed policy recommendations and later refined them for a presentation to the Calvert County Board of Commissioners. We emphasize here that the role of the research team in this process was to support the discussion rather than to recommend our own policy ideas.

In the St Albans Bay watershed, several stakeholders participated in the presentation of model results to the local press and general public in May 2006. Several interagency partnerships appear to have been strengthened and trust developed in previously opposing groups as a result of the PM exercises.

13. Treat the model as a process

There are always concerns about the future of participatory efforts. What happens when the researchers go away? If we look at how collaborative model projects are developed, there is a clear similarity with the open-source paradigm, where software is a product of joint efforts of a distributed group of players. Ideally, the process should live on the web and continue beyond a particular project. It is a valuable asset for future decision-making and conflict resolution. It can be kept alive with incremental funding or even donations, with stakeholders able to chip in their expertise and knowledge to keep it going between peaks of activity when bigger projects surface. There are examples of web and modeling tools that can provide this kind of functionality and interoperability, so there is real promise that this might actually happen.

This last lesson brings up a whole new issue of how to use and reuse models. Where and how do models "live," and how can we make the most of them? It appears that the new web technologies and the new dispersed way of collective thinking, research and development have the potential to become the new standard of modeling and decision-making.

9.3   Open-source, web technologies and decision support

(Parts of this section stem from discussions at the International Environmental Modeling and Software Society workshop on Collaborative Modeling, and the resulting position paper with Raleigh Hood, John Daues, Hamed Assaf and Robert Stuart.)

Computer programming in the 1960s and 1970s was dominated by the free exchange of software (Levy, 1984). This started to change in the 1980s, when the Massachusetts Institute of Technology (MIT) licensed some of the code created by its employees to a commercial firm and also when software companies began to impose copyrights (and later software patents) to protect their software from being copied (Drahos and Braithwaite, 2002).

Probably in protest against these developments, the open-source concept started to gain ground in the 1980s. The growing dominance of Windows and the annoyingly secretive policies of Microsoft certainly added fuel to the fire. The open-source concept stems from the so-called hacker culture. Hackers are not what we usually think they are – software pirates, vicious producers of viruses, worms and other nuisances for our computers. Hackers will insist that those people should be called "crackers." Hackers are the real computer gurus, who are addicted to problem-solving and building things. They believe in freedom and voluntary mutual help. It is almost a moral duty for them to share information, solve problems, and then give the solutions away just so other hackers can solve new problems instead of having to re-address old ones. Boredom and drudgery are not just unpleasant but actually evil. Hackers have an instinctive hostility to censorship, secrecy, and the use of force or deception.

The idea of software source code shared for free is probably best known in connection with the Linux operating system. After Linus Torvalds developed its core and released it to software developers worldwide, Linux became a product of joint efforts of many people, who contributed code, bug reports, fixes, enhancements and plug-ins. The idea really took off when Netscape released the source code of its Navigator, the popular Internet browser program, in 1998. That is when the term "open source" was coined and when the open-source definition was derived. Both Linux and Navigator (the latter now developed as the Firefox browser under mozilla. org) have since developed into major software products with worldwide distribution, applications and input from software developers.

> *The basic idea behind open source is very simple: when a programmer can read, redistribute, and modify the source code for a piece of software, the software evolves. People improve it, people adapt it, people fix bugs. And this can happen at a speed that, if one is used to the slow pace of conventional software development, seems astonishing.*
>
> Raymond, 2000a

Motivated by the spirit of traditional scientific collaboration, Richard Stallman, then a programmer at MIT's Artificial Intelligence Laboratory, founded the Free Software Foundation (FSF) in 1985 (http://www.fsf.org/). The FSF is dedicated to promoting computer users' rights to use, study, copy, modify and redistribute computer programs. Bruce Perens and Eric Raymond created the Open Source Definition in 1998 (Perens, 1998). The General Public License (GPL), Richard Stallman's innovation, is sometimes known as "copyleft" – a form of copyright protection achieved through contract law. As Stallman describes it:

> *To copyleft a program, first we copyright it; then we add distribution terms, which are a legal instrument that gives everyone the rights to use, modify, and redistribute the program's code or any program derived from it, but only if the distribution terms are unchanged.*

The GPL creates a commons in software development "to which anyone may add, but from which no one may subtract."

One of the crucial parts of the open-source license is that it allows modifications and derivative works, but all of them must be then distributed under the same terms as the license of the original software. Therefore, unlike simply free code, that can be borrowed and then used in copyrighted, commercial distributions, the open-source definition and licensing effectively makes sure that the derivatives stay in the open-source domain, extending and enhancing it. The GPL prevents enclosure of the free software commons, and creates a legally protected space for it to flourish. Because no one can seize the surplus value created within the commons, software developers are willing to contribute their time and energy to improving it. The commons is protected and stays protected.

The GPL is the chief reason that Linux and dozens of other programs have been able to flourish without being privatized. The Open Source Software (OSS) paradigm can produce innovative, high-quality software that meets the needs of research scientists with respect to performance, scalability, security, and total cost of ownership (TCO). OSS dominates the Internet, with software such as Sendmail, BIND (DNS), PHP, OpenSSL, TCP/IP, and HTTP/HTML. Many excellent applications also exist, including Apache web server, Mozilla Firefox web browser and Thunderbird email client, the OpenOffice suite, and many others.

OSS users have fundamental control and flexibility advantages. For example, if we were to write a model using ANSI standard C++ (as opposed Microsoft C++), we could easily move the code from one platform to another. This may be convenient for a number of reasons – from simply a preference for one developer to another, to moving from a desktop PC environment to a high-performance computing environment. Open Standards, which are publicly available specifications, offer control and flexibility as well. Examples in science include Environmental Markup Language (EML) and Virtual Reality Markup Language (VRML). If these were proprietary, use would be likely limited to one propriety application to interface with one proprietary format or numerous applications, each with its own format. We need only imagine the limitations on innovation if commonly used protocols like ASCII, HTTP or HTML were proprietary. To organize this growing community, the Open Source Development Network (OSDN) (http://www.osdn.com) was created. Like many previous open-source spin-offs, it is based on the Internet and provides the teams of software developers distributed around the world with a virtual workspace where they can discuss their ideas and progress, any bugs, share updates and new releases. The open-source paradigm has become the only viable alternative to the copyrighted, closed and restricted corporate software.

What underlies the OSS approach is the so-called "gift culture" and "gift economy" that is based on this culture. Under gift culture, you gain status and reputation within it not by dominating other people or by being special or by possessing things other people want, but rather by giving things away – specifically, by giving away your time, creativity, and the results of your skill. We can find this in some of the primitive hunter–gatherer societies, where a hunter's status was not determined by how much of the kill he ate, but by what he brought back for others. One example of a gift economy is the potlatch, which is part of the pre-European culture of the Pacific Northwest of North America. In the potlatch ceremony, the host demonstrates his wealth and prominence by giving away possessions, which prompts participants to reciprocate when they hold their own potlatch. There are many other examples of this phenomenon. What is characteristic of most is that they are based on abundance economies. There is usually a surplus of something that it is easier to share than to keep for yourself. There is also the understanding of reciprocity – that by doing this, people can lower their individual risks and increase their survival (Raymond, 2000).

In hunter–gatherer societies, freshly killed game called for a gift economy because it was perishable and there was too much for any one person to eat. Information also loses value over time and has the capacity to satisfy more than one. In many cases, information gains rather than loses value through sharing. Unlike material or energy, there are no conservation laws for information. On the contrary, when divided and shared, the value of information only grows – the teacher does not know less when he shares his knowledge with his students. While the exchange economy may have been appropriate for the industrial age, the gift economy is coming back as we enter the information age.

It should be noted that the community of scientists, in a way, follows the rules of a gift economy. The scientists with highest status are not those who possess the most knowledge; they are the ones who have contributed the most to their fields. A scientist of great knowledge but with no students and followers is almost a loser – his or her career is seen as a waste of talent. However, in science the gift culture has not yet fully penetrated to the level of data and source-code sharing. This culture has been inhibited by an antiquated academic model for promotion and tenure that is still prevalent today. This culture encourages delaying the release of data and source code to ensure that credit and recognition are bestowed upon the scientist who collected the data and/or developed the code. This model (which was developed when data were much more difficult to collect and analyze, and long before computers and programming existed) no longer applies in the modern scientific world, where new sensor technologies and observing systems generate massive volumes of data, and where computer programs and numerical models have become so complex that they cannot be fully analyzed or comprehended by one scientist or even small teams.

Knowledge-sharing and intellectual property rights

For centuries, nobody cared about "owning knowledge." Either people freely shared ideas, or they were kept secret. The idea of giving knowledge out yet retaining some sort of connection to it, rights for it, was hard to comprehend. Actually, it is still a pretty fluid concept, regardless of the numerous laws and theories that have been created since the British Statute of Anne, from 1710, which was the first copyright act in the world. Victor Hugo struggled with the concept back in 1870:

> *Before the publication, the author has an undeniable and unlimited right. Think of a man like Dante, Molière, Shakespeare. Imagine him at the time when he has just finished a great work. His manuscript is there, in front of him; suppose that he gets the idea to throw it into the fire; nobody can stop him. Shakespeare can destroy Hamlet, Molière Tartufe, Dante the Hell.*
>
> *But as soon as the work is published, the author is not any more the master. It is then that other persons seize it: call them what you will: human spirit, public domain, society. It is such persons who say: I am here; I take this work, I do with it what I believe I have to do, [...] I possess it, it is with me from now on....*

An Act for the Encouragement of Learning, by Vesting the Copies of Printed Books in the Authors or Purchasers of such Copies, during the Times therein mentioned.

Whereas Printers, Booksellers, and other Persons, have of late frequently taken the Liberty of Printing, Reprinting, and Publishing without the Consent of the Authors or Proprietors of such Books and Writings, to their very great Detriment, and too often to the Ruin of them and their Families: For Preventing therefore such Practices for the future, and for the Encouragement of Learned Men to Compose and Write useful Books; May it please Your Majesty, that it may be Enacted, ... That from and after the Tenth Day of April, One thousand seven hundred and ten, the Author of any Book or Books already Printed, ... or ... other Person or Persons, who hath or have Purchased or Acquired the Copy or Copies of any Book or Books, in order to Print or Reprint the same, shall have the sole Right and Liberty of Printing such Book and Books for the Term of One and twenty Years, to Commence from the said Tenth Day of April, and no longer.
(http://www.copyrighthistory.com/anne.html)

Formally, an intellectual property (IP) is a knowledge product, which might be an idea, a concept, a method, an insight or a fact, that is manifested explicitly in a patent, copyrighted material or some other form, where ownership can be defined, documented, and assigned to an individual or corporate entity (Howard, 2005). It turned out that in most cases it was the corporations, companies, producers and publishers who ended up owning the intellectual property rights and being way more concerned about them than authors, even though originally the idea was for the "Encouragement of Learned Men to Compose and Write useful Books."

Although the concept of public domain was implicitly considered by the Statute of Anne, it was clearly articulated by Denis Diderot, who was retained by the Paris Book Guild to draft a treatise on literary rights. In his *Encyclopedie*, Diderot advocated the systemic presentation and publication of knowledge of all the mechanical arts and manufacturing secrets for the purpose of reaching the public at large, promotion of research, and weakening the grip of craft guild on knowledge (Tuomi, 2004). With these pioneering ideas, Diderot set the stage for the evolvement of public domain, which includes non-exclusive IP that is freely, openly available and accessible to any member of the society.

At the same time, Diderot was part of a debate with another French Enlightenment prominent mathematician, philosopher, and political thinker, the Marquis de Condorcet (1743–1794), who was voicing even more radical ideas about intellectual property rights. Diderot argued that ideas sprang directly from the individual mind, and thus were a unique creation in and of themselves. Indeed, they were, in his words, "the very substance of a man" and "the most precious part of himself." Ideas had nothing to do with the physical, natural world; they were subjective, individual and uniquely constituted, and thus were the most inviolable form of property. For Diderot, putting ideas in public domain did not encroach on the property rights for these ideas. For him, copyright should be recognized as a perpetual property right, bestowed upon an author and inherited by his or her offspring.

Condorcet went much further. In sharp contrast to Diderot, he argued that ideas did not spring directly from the mind but originated in nature, and were thus open to all. Condorcet saw literary works as the expression of ideas that already existed. The form of a work might be unique, but the ideas were objective and particular, and could not be claimed as the property of anyone. Unlike land, which could only be settled by an individual or a family, and passed down by lineage to offspring, ideas could be discovered, used and cultivated by an infinite number of people at the same time.

For Condorcet, individuals could not claim any special right or privilege to ideas. In fact, his ideal world would contain no authors at all. Instead, people would manipulate and disseminate ideas freely for the common good and the benefit of all. Copyright would not exist, since no individual or institution could claim to have a monopoly on an idea. There go our patents!

Public domain and exclusive IP rights represent the two extremes in IP regimes, with the former providing a free sharing of knowledge and the latter emphasizing the rights of owners in limiting access to their knowledge products. Since the inception of the concept of intellectual property rights, it has been argued that protecting these rights provides adequate compensations for owners and encourages innovations and technological development. However, historical evidence and published research do not support these claims, and point to lack of concrete evidence that confirms them (National Academy of Engineering, 2003). Also, and increasingly, many technological innovations are the result of collaborative efforts in an environment that promotes non-exclusive intellectual rights. Although most of these efforts are in the software development domain (e.g. development of Linux), it is interesting to note that the tremendous growth and development in the semi-conductor industry are mainly attributed to the highly dynamic and connected social networks of the Silicon Valley in the 1960s, which was regarded as a public domain region, since information and know-how were freely shared among its members.

In the world of business, preservation of exclusive IP rights is seen as a necessity to maintain competitive edge and protect expensively obtained technology. Patents that were designed to stimulate innovation are now having the opposite effect, especially in the software industry. As Perens describes: "Plagued by an exponential growth in software patents, many of which are not valid, software vendors and developers must navigate a potential minefield to avoid patent infringement and future lawsuits" (Perens, 2006a). The big corporations seem to solve the problem by operating in a *détente* mode: by accumulating huge numbers of patents themselves, they become invulnerable to claims from rivals – competitors don't sue out of fear of reciprocity. However, now we see that whole companies are created with the sole purpose of generating profit from patents. These "patent parasites" make no products, and derive all of their income from patent litigation. Since they make no products, the parasites themselves are invulnerable to patent infringement lawsuits, and can attack even very large companies without any fear that those companies will retaliate. One of the most extreme and ugly methods is known as patent farming – influencing a standards organization to use a particular principle covered by a patent. In the worst and most deceptive form of patent farming, the patent holder encourages the standards organization to make use of a principle without revealing the existence of a patent covering that principle. Then, later on, the patent holder demands royalties from all implementers of the standard (Perens, 2006b).

Certainly, these patent games are detrimental for small businesses. According to the American Intellectual Property Law Association, software patent lawsuits come with a defense cost of about $3 million per annum. A single patent suit could bankrupt a typical small or medium-sized applications developer (let alone an open-source developer) even before the case were fully heard (NewsCom, 2005). The smaller patent holder simply cannot sustain the expense of defense, even when justified, and is forced to settle and license patents to the larger company. The open-source community is also constantly under the threat of major attacks from large corporations. There is good reason to expect that Microsoft will soon be launching a patent-based legal offensive against Linux and other free software projects (NewsForge, 2004).

Unfortunately, universities are increasingly seeking to capitalize on knowledge in the form of IP rights. However, only a few of these universities are generating significant revenues from licensing such rights (Howard, 2005). This applies equally to individual researchers who may seek to protect and profit from their findings.

Software development and collaborative research

Just as public domain and exclusive IP rights represent the two extremes in IP regimes, the software development process can occur in one of two ways – either the "cathedral" or the "bazaar" (Raymond, 2000a). The approach of most producers of commercial, proprietary software is that of the cathedral, carefully crafted by a small number of people working in isolation. This is the traditional approach we also find in scientific research. Diametrically opposed to this is the bazaar, the approach taken by open-source projects. Open source encourages people to tinker freely with the code, thus permitting new ideas to be easily introduced and exchanged. As the best of those new ideas gain acceptance, it essentially establishes a cycle of building upon and improving the work of the original coders (frequently in ways they didn't anticipate). The release process can be described as release early and often, delegate everything you can, be open. Leadership is essential in the OSS world – i.e., most projects have a lead that has the final word on what goes in and what does not. For example, Linus Torvalds has the final say on what is included in the kernel of Linux. In the cathedral-builder view of programming, bugs and development problems are tricky, insidious, deep phenomena. It takes months to weed them all out – thus the long release intervals, and the disappointment when long-awaited releases are not perfect. In the bazaar view, most bugs become shallow when exposed to a thousand co-developers. Accordingly, frequent release leads to more corrections, and, as a beneficial side-effect, you have less to lose if a bug gets through the door.

It is clear that the bazaar approach can work in general scientific projects, and in modeling applications in particular. Numerous successful examples, especially in Earth system modeling, attest to this fact. However, we must also recognize that there is a difference between software development and science, and that software engineers and scientists have different attitudes regarding software development. For a software engineer, the exponential growth of computer performance offers unlimited resources for the development of new modeling systems. Models are therefore viewed by engineers as just pieces of software that can be built from blocks or objects, almost automatically, and then connected over the web and distributed over a network of computers. It is simply a matter of choosing the right architecture and writing the appropriate code. The code is either correct or not; either it works or it crashes. Not so with a scientific model. Rather, most scientists consider that a model is useful only as an eloquent simplification of reality that needs profound understanding of the system to be built. A model should tell us more about the system than simply the data available. Even the best model can be wrong and yet still quite useful if it enhances our understanding of the system. Moreover, it often takes a long time to develop and test a scientific model.

As a result of this difference in point of view and approach, we tend to see much more rapid development of new languages, software development tools and open-code and information-sharing approaches among software engineers. In contrast, we see relatively slow adoption of these tools and approaches by the research modeling community. This is in spite of the fact that they will undoubtedly catalyze more rapid scientific advancements. As web services empower researchers, it is becoming clear that the biggest obstacle to fulfilling this vision of free and open exchange among scientists is cultural. Competitiveness and conservative approaches will always be with us, but developing ways to give meaningful credit to those who share their data and their code will be essential in order to change attitudes and encourage the diversity of means by which researchers can contribute to the global academy (*Nature*, 2005). It is clear that a new academic model that promotes open exchange of data, software and information is needed. Fortunately, the success of the open-source approach in software development has encouraged researchers to start considering similar shared open approaches in scientific research. Numerous collaborative research projects are now based on Internet communications, and are led simultaneously at several institutions working on parts of a larger endeavor (Schweik *et al.*, 2005). Sometimes, such projects are open and allow new researchers to participate in the work. Results and credit are usually shared among all the participants. This trend is being fueled by the general trend of increasing funding for large collaborative research projects, particularly in the Earth sciences.

Open-source software vs community modeling

The recent emergence of open-source model development approaches in a variety of different Earth science modeling efforts (which we refer to here as community modeling) is an encouraging development. Although the basic approach is the same, we can also identify several aspects of research-oriented community modeling that distinguish it from an open-source software development. For example, there have been a number of successful community modeling efforts (Table 9.1). However, unlike most open-source software development projects, these have been blessed by substantial grant and contract support (usually from federal sources), and exist largely as umbrella projects for existing ongoing research. It is probably also fair to say that

Table 9.1	Some ongoing community modeling projects		
Name	**Website and players**	**Scope**	**Projects**
CMAS Community Modeling and Analysis System	http://www.cmascenter. org/Funding – US EPA, Lead – Carolina Environmental Program at the University of North Carolina at Chapel Hill	Development of Air Quality and Meteorological models, extensions of the Models-3/ CMAQ. Outreach, user-support	CMAS-Supported Products: Community Multiscale Air Quality (CMAQ) Modeling System, Meteorology Chemistry Interface Processor (MCIP), Sparse Matrix Operator Kernel Emissions (SMOKE), System Package for Analysis and Visualization for Environmental (PAVE), data Input/Output Applications Programming Interface (I/O API), MM5 Meteorology Coupler (MCPL), Multimedia Integrated Modeling System (MIMS)
ESMF Earth System Modeling Framework	http://www.esmf.ucar. edu/	High-performance, flexible software infrastructure for use in climate, numerical weather prediction, data assimilation, and other	Earth science applications
CCSM Community Climate System Model	http://www.ccsm.ucar. edu/ – NCAR	Global atmosphere model for use by the wider climate research community	Working Groups: Atmosphere Model, Land Model, Ocean Model, Polar Climate, Biogeochemistry, Paleoclimate, Climate Variability, Climate Change, Software Engineering
CSTM National Community Sediment-Transport Model	http://woodshole.er. usgs.gov/project-pages/ sediment-transport/ – Woods Hole	Deterministic models of sediment transport in coastal seas, estuaries, and rivers	CTSM modules implemented in ROMS and FVCOM hydrodynamic models. Regional applications: Massachusetts Bay, Hudson River, Adriatic Sea
CCMP Chesapeake Community Model Program	http://ccmp.chesapeake. org – Chesapeake Research Consortium	Estuary, river and watershed modeling for water quality in the Chesapeake Bay	Baywide Hydrodynamic models: Quoddy, ROMS, POM, Biogeochemical models: CH3D_biowp, Larvae tracking IBM: CBOLT, Watershed: CBP-HSPF and V5 data

(Continued)

Table 9.1	(Continued)		
Name	**Website and players**	**Scope**	**Projects**
WATer and Environmental Research Systems (WATERS) Network	http://www.cuahsi.org/ http://cleaner.ncsa.uiuc. edu/home/	Hydrologic sciences, complex, large-scale environmental systems, education, outreach, and technology transfer	CUAHSI Consortium of Universities for the Advancement of Hydrologic Science, CLEANER Collaborative Large-scale Engineering Analysis Network for Environmental Research

most of the existing Earth science community models are not truly "open source" – i.e. access to the codes and rules governing modification and redistribution are usually more restrictive than, for example, those under GPL.

In general, in community modeling there is usually a much smaller number of participants because the research community is much smaller and more specialized than the broad field of software development. Because the pool is smaller, it may be harder to find the right people, both in terms of their skills and their willingness to collaborate within an open modeling paradigm. Similarly, there is generally a much smaller number of users of open-source research-oriented models, which may be very specialized and usually require specific skills to use. This is mostly because scientific models are very often focused on simulating a specific phenomenon or addressing a specific scientific question or hypothesis, and also because the scientific community is very small compared with the public at large. Along these same lines, research-oriented models are generally more sophisticated and difficult to use than software products that are developed for the public. It is certainly much harder to run a meaningful scenario with a hydrodynamic simulation model than to aim a virtual gun at a virtual victim and press the "shoot" button in a computer game (though it might be argued that to a large extent this difference in difficulty of use has more to do with the primitive state of the user interface of most scientific codes). It is also generally true that scientific codes require more sophisticated documentation and a steeper learning curve. Documenting scientific models is a real problem – it is not what researchers normally enjoy doing, and the need for doing it is rarely appreciated and funded. On the other hand, documentation is a crucial part of the process if we anticipate others using and taking part in the development of our models.

Open research modeling is also much more than open programming. As mentioned above, software development has a clear goal, an outcome. The product specifications can be well established and designed. In contrast, research modeling is iterative and interactive. The goal often gets modified while the project evolves. It is much more a process than a product. It is usually harder to agree on the desired outcomes and the features of the product. In some respects, modeling is more like an art than a science. Following this analogy, how do you get several artists together to paint one picture? This is particularly true in ecological modeling, where there is no overarching theory to guide model structure and where a variety of different formulations can be used to represent a particular process. These aspects of scientific modeling actually make it highly amenable to open programming approaches, which naturally allow a high degree of flexibility.

A significant impediment to developing open research models is the lack of infrastructure – there are still very few good software tools to support community research and modeling projects. Once again, there is an obvious gap between software and application. There is software that potentially offers some exciting approaches and new paradigms to support modularity, data-sharing, web access or flexible organization – all the major components required for successful model integration and development. The most recent trends in software design are compared with the Lego constructor over the web (Markoff, 2006) – exactly what we need for modular models. However, this is yet to be developed and applied to the modeling process, and embedded into the modeling lexicon and practice. Yet another difference is that most research modeling projects takes years to develop. This is in contrast to some of the software hacks that can be invented and implemented in a matter of hours, quickly gaining recognition and respect in the software development community. Research is a much slower and tedious process, where small incremental ideas and successes may be very important, but are much harder to document, disseminate and appreciate.

Finally, returning to the central problem, we really need to change the traditional culture and attitudes of research scientists – that is, promote a shift in the mindset and psychology that drives scientific research. Historically, most science has been driven by individual efforts and talent. The talent and ingenuity of individuals will always be critical in scientific exploration, but with the growing amount of data, knowledge and information, most of the breakthrough achievements are now produced by team efforts, where teams and teamwork rather than individuals are key. This trend is being driven to a large extent by the increasing emphasis in scientific research on large projects aimed at solving complex interdisciplinary problems, such as simulating and predicting the Earth system response to global warming. It is becoming increasingly difficult to identify the sole individual who cries "Eureka!" and solves the problem. Even when this does occur, very often the recognition is biased by past success, hierarchy, and personalities. There is an obvious need for new award and credit systems that will stimulate sharing and teamwork rather than direct personal gain, credit and fame.

By sharing the data and concepts over the web, potential users are invited to join in collaborative research and analysis of the future trends of watershed development. Their feedback is solicited for further dissemination and improvement of knowledge about the watershed system. The management and decision-making are disclosed to the public, offering a broad spectrum of views and values, and inviting stakeholders to become participants in a truly democratic process of decision-making.

Beyond separate projects involving PM, we can envision them coming together in an integrated effort to support whole ecosystem and watershed management, which is a holistic and integral way of research, analysis and decision-making at a watershed scale. In the 1990s and even earlier, there was much hope for this approach. It certainly implies more than just the regional scale of analysis. The method stresses the need to integrate not only physical and biological factors, but also political and socio-economic ones. The major impetus for watershed management stemmed from the understanding that science needs to be linked to planning, and that decision-making should be based on broad citizen involvement. Thus it is important that the information is shared between the stakeholders and that it is processed into a format readily perceived by wide and diverse groups, institutions and individuals. Moreover, the watershed delineates a physical boundary and not a political one, creating the need for methods that will allow management and communication between many administrative entities such as towns, counties and states. One of the problems that watershed management immediately encountered was the mismatch between the existing administrative hierarchies and the physical and societal boundaries and groupings that represented the watershed dynamics. Appropriate institutions are required that can operate in a flexible manner over alternative regional divisions.

The fact that ecosystem management seeks alternative mechanisms to purely market forces based on the existing policy equilibrium seems to be very bothersome to traditional economists (Fitzsimmons, 1994). They argue that the ecosystem concept is inappropriate for use as a geographic guide for public policies. Mostly, though, they are concerned that the ecosystem approach will significantly expand federal and other non-market control of the use of privately-owned land, and lead to increased restrictions on the use of public lands for economic purposes.

Lackey (1998) identified five general characteristics for ecosystem management problems:

1. Public and private values and priorities are in dispute, resulting in mutually exclusive decision alternatives
2. There is political pressure to make rapid and significant changes in public policy
3. Private and public stakes are high, with substantial costs and risks (some irreversible) to some groups
4. The technical, ecological and sociological facts are highly uncertain
5. Policy decisions will have effects outside the scope of the problem.

These seem like exactly the type of difficulties that can be resolved with PM. He concludes that "solving these kinds of problems in a democracy has been likened to asking a pack of four hungry wolves and a sheep to apply democratic principles to deciding what to eat for lunch" (Lackey, 1998: 22). The outcome may seem quite obvious, except that with people there is always less certainty about how problems are resolved, and in the long run there is still a chance for the sheep to persuade the wolves to become vegetarians. The success of this endeavor becomes very much dependent on how efficiently the new technology is developed and used, since it is our scientific, cultural and social development that makes *Homo Sapiens* special and leaves certain space for optimism. In this context we do not view technology as a panacea that can cure all the problems of environmental degradation and resource depletion, but rather as a means of understanding, educating, and resolving conflict.

Regional management implies a close interaction and linkage between the numerous agents acting in the region. The efficacy of this interaction is a function of the information that is shared among and used by all the stakeholders. In many cases, it depends not so much on the quality and amount of the information available (what science has been mostly concerned with all this time) but rather on how well the information is disseminated, shared and used. And that is exactly the function that the PM techniques can offer, especially if they are enhanced by the Web technologies.

As with the advent of any new technology, it has taken some time to realize all the benefits and advantages that the Internet can deliver. Until 1992 it was the realm of a relatively small contingent of scientists and engineers, who were using it to communicate data among themselves, and both the sender and the recipient of information were usually personally defined. The Web opened a new page in the use and development of the Internet. Information was no longer personally targeted; once posted to the net it became open to any user who had the interest and time to view it. Basically, the Web to the Internet is the same as the radio is to postal services. Instead of mailing a letter to a definite addressee, information could be now aired as if being broadcast over a radio or television network, with the sender no longer knowing who the recipient is to be. In this way, the audiences expanded dramatically and are still growing rapidly. A major advantage of the Web, compared with other mass media, is that it is relatively cheap. As a result, in addition to the businesses that are eager to employ another opportunity for advertisement and sales, the Web offers a whole new way of outreach and communication to governmental, academic and non-profit organizations. Even individuals can afford to establish their presence in this mass media.

Another advantage of the Web is that it provides for direct feedback from the recipient, who can now interact with the information displayed. Instead of just passively viewing information, website visitors can change and modify it remotely. Users are offered search engines that can direct them to the most relevant information available; they can revisit sites and refer others to them. Unlike other mass media, the Web is more stable and persistent in the sense that, unlike other mass media such as radio, where once information has been aired it is no longer retractable, on the Web the information stays where it was and can be easily referenced and downloaded.

In spite of these novel features, most of the use of the Internet does not seem to be much different from that of the traditional mass media or archived information (libraries, data sets, etc.). Business is driving a vast majority of web applications towards advertisement and sales in a way very similar to that which may be observed on radio and TV, and in unsolicited mail and catalogs. There are just a few examples when the Web is used in an innovative way that employs some of its unique features.

The consensus building power of the "informational superhighway" created on the Web has not been used to "full speed." We argue that there are a number of features that make the Web an exceptionally important tool for watershed management in particular, and for decision support and management in general. The Web is:

- *Open*. The Internet is one of the most readily available and reliable media, providing information across geographical, administrative, social and economic boundaries. It is relatively cheap, and can be .accessed by all the stakeholders in a watershed and outside of it. The fact that it requires a computer (or advanced TV set – "Web-TV") and an Internet connection is becoming less and less restrictive as more Internet Service Providers (ISP) enter the market. For those who do not have Web access at home or at work, there are public providers (libraries, "web-cafés," etc.) that also have become more available. This direct access to all the necessary information and, reciprocally, the ability to disseminate the facts that are of concern to particular stakeholders is an important prerequisite of watershed management.
- *Interactive*. It is most important for management purposes that the user has the option of interacting with the provider of information and with other stakeholders. With the Internet, this can be accomplished either via e-mail or directly through forms, wikis or blogs that can be part of web pages and transmitted to

the server. These contributions can be further manually or automatically processed and posted back on the Web. In this case, information is not only passively perceived, as in case of the traditional media (radio, press, newsletter, etc.); it also stimulates direct feedback. Moreover, users can modify the content and format of the existing pages by ordering excerpts from data bases or providing scenarios for model runs, and thus creating their own output to be immediately viewed on the Web. They may also provide additional information to the Web in response to the published requests or as a representation of their own findings and concerns.

- *Fast.* Communications via the Internet are probably the fastest and the most economic, since they do not require any intermediate carriers (as in ordinary mail) and materials (paper). Once the information is updated on the server, it becomes immediately available for further use and processing. The feedback in many cases can be handled automatically and directly channeled to the appropriate web link or interest group.

- *Spatially distributed.* Internet access is offered over telephone lines and therefore covers almost the entire planet. The various nodes on the Internet can correspond and represent the spatially distributed data of different stakeholders in the watershed and outside it. The web tools allow information to be linked together; search engines are created to find the necessary information and data. In this way, concerns and awareness can be shared across different geographic localities. This gives a broader picture of the system at stake within the framework of external systems and concerns.

- *Hierarchical.* The hierarchical structure supported by the Web design allows organization of the data in logical and efficient ways when various branches on the Web may present specific fields, domains and interest groups. The links on web pages can stitch the whole structure together, offering cross-references and alternative views whenever necessary. For example, the watershed hierarchy of subwatersheds and sub-subwatersheds can be easily mirrored on the Web, with specific groups of pages representing each particular level. The hierarchical structure also offers levels of protection for the information, allowing certain domains to be completely open to all users, others only read-permitted, and still others accessible only to limited users and interest groups, providing the necessary extent of privacy and discretion.

- *Flexible.* Additional benefits that are offered by the interactive features allow the data to be processed by users according to their own goals and interests. This is especially important for modeling tools, because by employing the Web they can be made directly accessible to the user, and sufficiently flexible and user-friendly to be used meaningfully and efficiently. Currently, web applications are being used at the high-school level to teach science and ecology. The scope of potential uses ranges from running particular scenarios, which stakeholders can formulate based on their concerns, to adjustments in scale and structural detail of the model in response to special needs and projects.

All the important features and tools to augment and improve decision support and management seem to be present, and it then becomes a matter of using them efficiently. This is really handy for supporting the PM process and making it evolutionary and adaptive over the web, such that it can remain an ongoing activity even when the current project has reached its goals and a certain decision has been made.

No matter how good and appropriate a decision, an open system tends to change and evolve, and decisions will eventually need to be reassessed and adapted to new developments and new data. The web presence of stakeholders and their previous efforts as part of a PM project, together with modeling tools and data that have been developed and researched, should remain available for future applications. Future projects will then not need to start from scratch, as there will be access to all the previously collected information, and, even more importantly, the social capital of social networks and links developed as part of the previous PM adventure.

A PM project becomes a kind of open-source project with various stakeholders contributing to it in various roles. Some will be administering the process and guiding its progress, others will be contributing bits of data and knowledge, others will be developing models and analytical tools, while yet others will be writing documentation and disseminating results to other interested parties. This is very similar to the structure of many open-source software projects, thousands of which are administered by SourceForge at http://www.sourceforge.net – a powershop for open software development.

9.4 Conclusions

Much human creativity is geared towards moving energy and materials rather than information, even though information has become another crucial component of human welfare and livelihood. Information, unlike energy and materials, is not subject to conservation laws. By copying information from sources and distributing it to new destinations, we do not lose information at the source. This is what is known as non-rival goods in ecological economics (Daly and Farley, 2004). As with gravity, by using information we do not decrease the ability of others to use it. Nevertheless, exchange of information is restricted by patent law, as well as by institutional, cultural and traditional hurdles that create protective barriers hindering the free flow of this valuable commodity. In this way, we are making it excludable. It is not surprising that private companies are often reluctant to share data and software, because it can impact their profits in a competitive market. Unfortunately, barriers to information exchange are also significant in the academic community, where the long-standing emphasis on publication and (perhaps unwarranted) fear of misuse of released data and software have inhibited free and open exchange. Promotion and tenure at academic institutions is still largely dependent upon the volume of peer-reviewed publications and success in securing grant and contract funds. As a result, academic scientists have little or no incentive to spend the time and effort that is required to document and disseminate their data and/or their models and code for the greater good of the research community. This problem is exacerbated by the fact that grant and contract funding for research rarely provides direct support for documentation and dissemination activities. The issue is particularly acute when it comes to sharing the source code of models and data analysis software – even if a scientist or engineer is amenable to sharing the code, the effort required to provide documentation to make it useful is often viewed as an insurmountable obstacle.

Funding agencies worldwide seem to recognize clearly the pressing need to enhance communication and promote open exchange of data and information among scientists and between academic and private institutions via the Internet.

The National Science Foundation, for example, has initiated several new major research initiatives that are aimed at developing and/or explicitly requiring this enhanced communication. These initiatives include NEON (National Ecological Observatory Network), CLEANER (Collaborative Large-Scale Engineering Analysis Network for Environmental Research), CUAHSI (Consortium of Universities for the Advancement of Hydrological Science Inc.) and ORION (Ocean Research Interactive Observatory Network), to name just a few. The European Union has funded such open-source projects as Harmon-IT and Seamless. All of these initiatives embrace the idea that developing the infrastructure needed to allow free and open exchange of large volumes of data and information will be crucial for making rapid scientific advancements in the future. For example, the success of current efforts to develop Earth observatories in both terrestrial (e.g. NEON) and marine (e.g. ORION) environments will be critically dependent upon the successful development of this infrastructure, because these observatories will have to collect, process and disseminate large volumes of data and assimilate them into models in a timely manner.

The challenges we face in creating a new research paradigm are many. Substantial improvements in hardware (e.g. network and computing infrastructure) and software (e.g. database manipulation software and data-assimilating numerical models), and a much higher level of standardization of data formats, will be required. New means for carrying out real-time data processing and automated data quality control will also have to be developed. However, we believe that one of the greatest challenges we face in this endeavor is building the community-modeling and information-sharing culture that will be required for success. How do we get engineers and scientists to put aside their traditional modes of doing business? How do we provide the incentives that will be required to make these changes happen? How do we get our colleagues to see that the benefits of sharing resources far outweigh the costs? Timely sharing of data and information is in the best interests not only of the research community, but also of the scientist who is doing the sharing – substantial additional benefits will be derived through new contacts, collaborations and acknowledgement that are fostered by open exchange. Numerous examples attest to this fact. The real challenge we face is getting our colleagues to recognize the potential benefits that can be derived from adopting a community-modeling and information-sharing culture. In addition, we need to dispel the unwarranted fears that many scientists and engineers harbor: that they will be "scooped" if they release their data too soon or blamed if there is a bug in their code. Finally, we need to accept the fact that releasing undocumented or poorly documented software is preferable to not releasing it at all.

Further reading

The end of human civilizations is analyzed by Diamond, J. (2005), *in Collapse: How Societies Choose to Fail or Succeed*, Penguin, 592 pp.

There is a growing body of literature on human behavior. You can read more about complex economic behavior in W.B. Arthur, S.N. Durlauf, and D. Lane, 1997. The economy as an evolving complex system II. *Santa Fe Institute Studies in the Science of Complexity*, Vol. XXVII.: Addison-Wesley; K.L. Judd, and L. Tesfatsion, 2006. Handbook of Computational Economics Volume II: Agent-Based Computational Economics. Elsevier B.V.; Kirman, A.P., Whom or what does the representative individual represent? *Journal of Economic Perspectives*, 1992.

6(2): pp. 117–136. *There is a whole Journal of Economic Behavior & Organization published by Springer.*

An interesting analysis of human behavior from a psychologist's viewpoint is found in Whybrow, P.C. (2006). *American Mania: When More Is Not Enough.* W. W. Norton, 352 pp.

To read more about climate change you can start with the pages from the Union of Concerned Scientists at http://www.ucsusa.org/global_warming/. *Here you will find all the basic data. For more in-depth analysis, go to the* Real Climate *blog at* http://realclimate.org/. *It has articles for all levels, starting from very basic facts up to quite sophisticated and technical discussions of particular problems and issues.*

The history of climate change research is described by James Fleming, 2007. *Intimate Universality: Local and Global Themes in the History of Weather and Climate.* Science History Publications, 284 pp.

Some important issues related to model failures are discussed in a position paper by two dozen authors: B.S. McIntosh, C. Giupponi, A.A. Voinov, C. Smith, K.B. Matthews, M. Monticino, M.J. Kolkman, N. Crossman, M. van Ittersum, D. Haase, A. Haase, J. Mysiak, J.C.J. Groot, S. Sieber, P. Verweij, N. Quinn, P. Waeger, N. Gaber, D. Hepting, H. Scholtent A. Sulisu, H. van Delden, E. Gaddis, H. Assaf, 2008. Bridging the gaps between design and use: developing tools to support environmental management and policy. In: Jakeman, A., Chen, S., A. Rizzoli, A.A. Voinov, (Eds.). *State of the art and futures in Environmental Modelling and Software.* Elsevier. (in press)

The almost classic description of what post-normal science means for us and how it changes some of our paradigms, can be found in Funtowicz, S. and Ravetz, J.R. (1993). Science for the post-normal age. *Futures* 25 (7), 739–755.

Kasemir, B., Jäger, J., Jaeger, C.C. and Gardner, M.T. (eds.) (2003). *Public participation in sustainability science: A handbook.* Cambridge University Press, Cambridge. 316 pp. – *This is an important collection of papers on participatory research. There is not very much about modeling in there, but it does lay out some very important principles for stakeholder involvement and public participation.*

For more details about HubNet – the participatory modeling component of NetLogo see: http://ccl. northwestern.edu/netlogo/docs/hubnet.html. *There is also a link to a collection of participatory projects that use this tool:* http://ccl.northwestern.edu/partsims.html

Other relevant papers on participatory modeling are: Korfmacher, K. S., 2001. The politics of participation in watershed modeling. *Environmental Management* 27(2), 161–176.

Beirele, T.C. and Cayford, J. (2002). Democracy in practice: Public participation in environmental decisions. *Resources for the Future,* Washington, D.C..

Carr, D.S. and Halvorsen, K. (2001). An evaluation of three democratic, community-based approaches to citizen participation: Surveys, conservations with community groups, and community dinners. *Society and Natural Resources,* 14: 107–126.

Duram, L.A. and Brown, K.G. (1999). Assessing public participation in U.S. watershed initiatives. *Society and Natural Resources,* 12: 455–467.

Gough, C.E. and Darier, (2003). Contexts of citizen participation. Public participation in sustainability science. In: B. Kasemir, C.C. Jaeger, J. Jager and M.T. Gardener (eds.), *Public participation in sustainability science: A handbook.* Cambridge University Press, Cambridge, p. 316.

The Solomons Harbor project is described in Gaddis, E. J., Vladich, H., and Voinov, A. (2007). Participatory modeling and the dilemma of diffuse nitrogen management in a residential watershed. *Environ. Model. Software.* 22, 5, 619–629.

For more information about the St.Albans project see: USDA (1991). St. Albans Bay Rural Clean Water Program. United States Department of Agriculture, Vermont Water Resources Research Center; Hyde, K., Kamman, N., Smeltzer, E. (1994). History of phosphorus loading to St. Albans Bay, 1850–1990. Lake Champlain Basin Program. Techical Report No. 7B; Brown

Gaddis, E.J., Voinov, A. (2008). Participatory modeling of phosphorus reduction scenarios in a mixed use watershed in Vermont (*in preparation*).

Apfelbaum S.I. (1995). Role of Landscapes in Stormwater Management. In: *National Conference on Urban Runoff Management: Enhancing Urban Watershed Management at the Local, County, and State Levels*. Chicago, Illinois. EPA/625/R-95/003

Shared Vision Planning (SVP) ideas have been advocated inside the Army Corps of Engineers, but rarely got published in peer reviewed literature. Ironically some of their very first reports that are very relevant to ideas of participatory research go back more than 30 years: Wagner T.P., Ortolando L. (1976). Testing an iterative, open process for water resources planning. Fort Belvoir, Va.: U.S. Army Engineer Institute for Water Resources. 66 pp. (IWR contract report no. 76–2); Wagner T.P., Ortolando L. (1975). Analysis of New Techniques for Public Involvement in Water Planning. Water Resources Bulletin, V.11, N 2, pp.329–344. *An important recent paper on SVP –* Palmer, R.A., Cardwell, H.E., Lorie, M.A., Werick, W., (2008). Disciplined Planning, Structured Participation, and Collaborative Modeling – Applying Shared Vision Planning to Water Resources Decision Making, ASCE J Water Resources Management and Planning, *is still in review at this time. In the meanwhile, to find more about circles of influence, see* Voinov, A., W.E. Cox, and H. E. Cardwell (2007). Pilot Collaborative Modeling Study for Regulatory Issues on the James River. *World Environmental and Water Resources Congress 2007: Restoring Our Natural Habitat. ASCE.*

The importance of good visuals is hard to overestimate. To find some exciting ideas on how to prepare your visuals check out the book series by Edward Tufte: 1990 – *Envisioning Information*; 1997 – *Visual Explanations*; 2001 – *The Visual Display of Quantitative Information*; 2006 – *Beautiful Evidence*. Graphics Press.

Some key writers on open source and general public license ideas are Bruce Perens, Richard Stallman and Eric Raymond. *Much of their work can be found on the web.*

Perens, B., 1998. The open source definition. http://perens.com/Articles/OSD.html

Perens, B., 2006a. Software Patents vs. Free Software. http://perens.com/Articles/Patents.html

Perens, B., 2006b. The Problem of Software Patents in Standards. http://perens.com/Articles/PatentFarming.html#2

Raymond, E., 2000. A Brief History of Hackerdom. http://www.catb.org/~esr/writings/cathedral-bazaar/hacker-history/index.html

Raymond, E., 2000a. Homesteading the Noosphere. http://www.catb.org/~esr/writings/cathedral-bazaar/homesteading/

There is also a book by Eric Raymond, 1999. *The Cathedral and the Bazaar*. O'Reilly, 268 pp.

Levy, S., 1984. *Hackers*. Anchor/Doubleday, New York – *This digs into the history of the phenomenon of hackers and what they are. For more about information and property rights see:* Drahos, P. and Braithwaite J., 2002. *Information Feudalism: Who Owns the Knowledge Economy?* New York: The New Press.

The history of copyright and the Diderot–Condorcet controversy is described by Walker, D. (2000). *Heirs of the Enlightenment: Copyright During the French Revolution and Information Revolution.* http://quizzebox.quintessenz.at/pipermail/edri-forum/2002-December/000054.html

For more about this see Post, Robert, Editor, 1991. *Law and the Order of Culture*. Berkeley: University of California Press, http://ark.cdlib.org/ark:/13030/ft9q2nb693/

Other relevant publications are: Tuomi, Ilkka, 2004. *Knowledge sharing and the idea of the public domain*, UNESCO 21st Century Dialogues, *Building World Knowledge Societies*, Joint Research Centre, Institute for Prospective Technological Studies, Seoul, Korea; National Academy of Engineering, 2003, *The impact of academic research on industrial performance*, National Academies Press, Washington.

On patents see: Howard, J., 2005. *The emerging business of knowledge transfer: creating value from intellectual products and services*. Canberra: Australian Government Department of Education, Science and Training.

Newscom, 2005. The open-source patent conundrum. http://news.com.com/2102-1071_3-5557340.html?tag=st.util.print

Newsforge, 2004. HP memo forecasts MS patent attacks on free software. http://www.newsforge.com/article.pl?sid=04/07/19/2315200

To Conclude

There was once a time when humans were few, weak and vulnerable, on a large, hostile planet. They endeavored not to succumb, not to adapt to the environment, but instead to try something different on the evolutionary trail. They began to change the environment. The clear and obvious goal was to grow, to gain power, to take control. In the beginning, this was a battle with no clear winners. Sometimes humans succeeded, and would develop into mighty civilizations, and their numbers grew along with their power to harness the environment. But then something would go wrong, civilizations would collapse, human power would diminish, and they would have to start again somewhere else. In aggregate, it was a more or less equal battle until something really remarkable changed the world.

Humans learned to harness fossil energy. Suddenly they became masters of past worlds, of the energy that had accumulated over millennia in the past and was stored there, waiting for the right moment to come. Suddenly, the new evolutionary path became really fueled. Humans achieved the power and the luxury to allow some of their best minds just to think; they no longer needed to hunt, or to sow, or to build. With the power of concentrated old energy it was no problem to provide these minds with all they needed in terms of food, clothing or shelter. They could spend their entire lives thinking, inventing, designing, coming up with new, better solutions for the new alternative human evolution. That is when human evolution, 'advancement' really took off, and population began to advance in huge leaps. Local pockets of civilization became united on a global scale into one technocratic civilization, and the goal still remained the same – to expand, grow, empower.

And so the human population grew, both in terms of its numbers and in terms of its rates of consumption. Currently we really are at a turning point: a paradigm shift is badly needed. There are three reasons for this:

- Climate Change;
- Resource depletion and peak oil;
- Globalization.

Climate change is happening already and its change is likely to accelerate. We find numerous evidences for that. A recent study has shown that 150 years of records show trends toward fewer days of ice cover. Trends in ice duration in 65 water-bodies across the Great Lakes region (Minnesota, Wisconsin, Michigan, Ontario and New York) during a period of rapid climate warming (1975–2004) show that average ice duration decreased by 5.3 days per decade. Average temperatures from fall through spring in this region increased by 0.7 degrees Celsius. The average number of days with snow decreased by 5.0 days per decade, and the average snow depth on those days decreased by 1.7 centimeters per decade.[1]

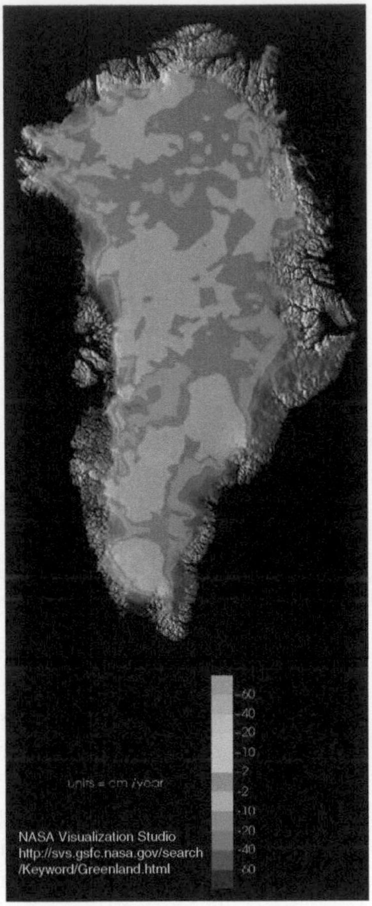

There is mounting evidence of rapidly shrinking glaciers. These processes are occurring faster in the Polar Regions. The Arctic is expected to become a new permanent sea route from the Atlantic to the Pacific. Ice in Greenland is disappearing.

A tropical virus has caused an epidemic in Italy, when several hundreds of cases of chikungunya, a form of dengue fever normally found in the Indian Ocean region, have been registered in Castiglione di Cervia in Northern Italy. In this case the disease was spread by insects: tiger mosquitoes, who can now thrive in a warming Europe. Tiger mosquitoes are now found across southern Europe and even in France and Switzerland.

The drought conditions in south-eastern Australia seem to be permanent now. For eleven years in a row temperatures have been above normal. Sydney's nights are its warmest since records were first kept 149 years ago. Sydney had its wettest year since 1998, receiving 1499 millimeters, well above the long-term average of 1215. Much of it was coastal, rain that fell at the wrong time for farmers, soaked into drought-parched soils or evaporated during scorching days. Sydney had its stormiest year since 1963, with 33 thunderstorms, historic average – 28.

Figure C.1   The Shrinking Ice Cover in Greenland.

NASA Visualization Studio
http://svs.gsfc.nasa.gov/search
/Keyword/Greenland.html

units = cm/year

---

[1] http://www.nsf.gov/discoveries/disc_summ.jsp?cntn_id=110967&govDel=USNSF_1

The list of these changes can be continued. Coral reefs are bleached and are degrading. Hurricanes have become more powerful and frequent. Floods and droughts are becoming more severe. Most disturbing are the numerous positive feedback effects involved in the above, and that drive the climatic machine of this planet.

According to the International Panel on Climate Change (IPCC) it is "very unlikely" that we will avoid the coming era of "dangerous climate change". Most likely we should expect water shortages, crop failures, disease, damages from extreme weather events, collapsing infrastructures, and breakdowns in the democratic process. Our first experience of re-engineering the planet seems to be producing quite ugly results. Unintentionally we may have triggered too many positive feedbacks that tend to get out of control. If we can't stop it – we will need to adapt to it. Any adaptation will require additional resources.

Unfortunately the **resource base** also does not look very promising. As we have already seen there is mounting evidence that oil reserves are approaching the threshold when extraction will consume almost as much energy as energy produced. It becomes meaningless to produce oil as an energy source after that. At the same time there is growing demand, especially in South-East Asia.

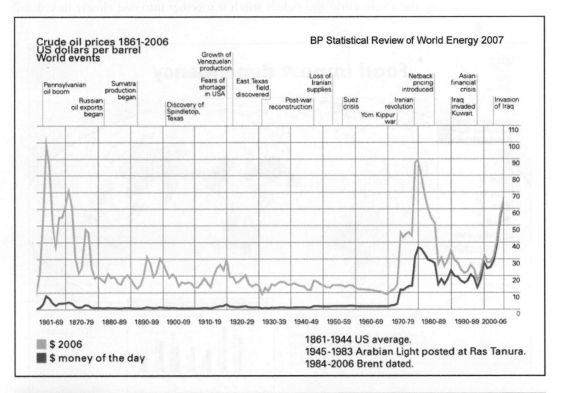

Crude oil prices 1861-2006
US dollars per barrel
World events

BP Statistical Review of World Energy 2007

$ 2006
$ money of the day

1861-1944 US average.
1945-1983 Arabian Light posted at Ras Tanura.
1984-2006 Brent dated.

Figure C.2   The growing price of oil. This time there seems to be no other reason except that supply cannot catch up with demand.

In the twenty-first century oil prices have gone up over 800%. There was a previous price spike in the 1970s, but at that time it was a deliberate decision of OPEC to decrease oil exports to get a price hike. There is no such policy pursued today, yet prices are steadily growing. Why is that? We have entered the era where supply can no longer keep pace with demand. Supply is stagnating, while demand continues to grow.

Similarly there are already visible horizons for many other non-renewable resources. If we continue to consume at present global average rates we are estimated to have 61 more years of copper, 45 years for gold, 13 – for indium, 40 – for tin, 29 – for silver, 59 – for uranium, etc. If the global consumption rate increases to ½ of the US consumption rate, these numbers go down to 38, 36, 4, 17, 9 and 19, respectively.

Most disturbingly, we are running out of our most important resource, that is the pollution assimilation capacity of the natural ecosystems. We already have to pay billions to clean up what our fathers and grandfathers left us.[2] At this same time we are hastily consuming whatever is left to leave our own children in even worse shape.

In the most optimistic of cases we have 1–2 generations to address these problems, possibly much less. That is not that long a timeframe to figure out the alternatives and, what is no less important, – to develop the appropriate infrastructure to produce these alternatives. These will require significant investment of energy and other resources, which will make them only scarcer. This is another positive feedback to consider. Once again, as we have seen above, it may actually be too late to invest in these alternatives.

Most disturbing is that this time we are dealing with a **global** phenomenon. We are living in an increasingly globalized system, when flows of material and information grip the whole world and tightly stitch it together into one closely linked and

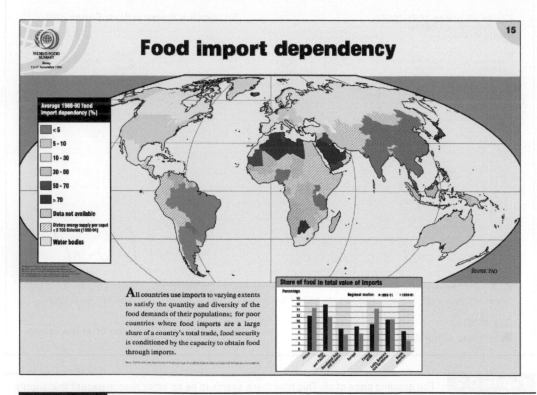

Figure C.3   Most of these countries are no longer self-sufficient in food products. They expect food imports and rely on exports.

---

[2] The Superfund was created in the USA to pay for toxic waste cleanup at sites where no other parties could be found responsible. The Congressional Budget Office base-case estimate is that Superfund cleanups will cost the public and private sectors about $75 billion from fiscal year 1993 onward (http://www.cbo.gov/ftp-doc.cfm?index=4845&type=0). There are currently 1,240 sites listed on the Superfund National Priority List.

interdependent system. The top 15 World oil producers deliver over 63 million barrels of oil per day. At the same time the top 15 oil exporters ship more than 39 million barrels of oil per day, meaning that almost 2/3 of all oil produced is destined to some other location, in many cases traveling many miles across the oceans. Most of the developed countries are dependent on foreign energy supplies.

Almost all countries depend on food imports. Sometimes as much as 70% of food supply has to be delivered. While in developed countries foreign imports are largely for exotic and luxury items, in some of the Middle East and African countries they are a necessity.

Even for many conventional items we see that trade flows circle the Earth in many cases going in both directions, as is the case with, say, oranges.

Financial flows further connect the World. An estimated 150 million migrants worldwide have sent some US$300 billion to their families in developing countries during 2006 through more than 1.5 billion separate financial transactions.

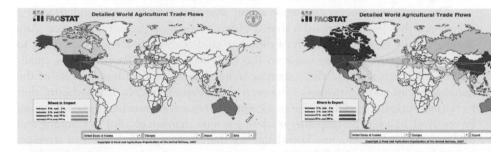

Figure C.4   Flows of imports and exports of many food items go in both directions.

At this point we are not looking at positive and negative impacts of globalization. What is important is to realize that this system is in place, and that as a result, the world is completely interconnected. Local crises will spread around swiftly; overconsumption in the developed countries will not be contained only to the areas of those countries. Just like depletion of oil reserves in, say, the 48 states of the USA will not stop oil consumption in the country, climate change triggered by greenhouse gas emissions is not going to be limited only to the locations where these gases are emitted.

The environments that we have created are facing considerable risks, and the safety net once provided by the favorable natural environment on planet Earth seems to be eroding. Since humans have taken control, to shape the environment to our own use rather than adapt to what was offered, we now have a fiduciary responsibility for the results of our efforts. In many cases the natural environments that were there to provide humans with resources needed and to absorb the waste and pollution that humans created, are no longer in place. Furthermore, they could never provide the carrying capacity needed to maintain the current size of the human population at the comfort levels that it has become accustomed to.

The paradigm shift, if it comes, needs to be based on an understanding of how systems work, of how we got here, and what the indirect and delayed responses of the system can be. The one resource that does not seem to have any limits is information. Moreover, by sharing information, we do not subtract from it. If I have a bucket of popcorn and want to share it with my neighbors, I will have to give them some of the popcorn from the bucket. As a result, there will be less left for me. This is not the case with information. If I share with you what I know, I do not then know less, probably more, because while communicating I might understand my information better. If it is in our genetic heritage to grow, to consume more, to expand, then probably the only area where we can do it safely – is with information.

The planet is limited: there is only that much of land, oil, water, tin, copper and gold. No matter how efficiently we use it, if there are more and more users, we will eventually run out of the goods. Information is limitless. We can explore, research, study, learn as much as we wish. Vernadskii dreamt of a system he called "noosphere" – a biosphere driven by human intellect, spirituality, knowledge, and understanding.

Models are an important part of this understanding. They are building blocks of our worldview. The models can be simple or complex, conceptual or numerical, formal or verbal, but for models to be good they need to be based on a culture of modeling – on good modeling practice. That is what we tried to learn in this book. If we have common standards for our models, it will be easier for us to communicate our understanding, to find common ground, to avoid conflict and make the right decisions.

The modeling process can work as our shared fact-finding and understanding experience that leads us toward a shared vision of the past, present and future. Any dispute can be treated as a clash of different models. Stakeholders contributing to a dispute resolution exercise come to the table with their different models, qualitative and quantitative, of the system at stake. The dispute evolves because of the inconsistencies and controversies between the different models. I hypothesize that by harmonizing the models for use in a common framework, much of the conflict can be resolved. In a way participatory modeling is a mechanism of joint fact finding and understanding when data and knowledge are shared among stakeholders in attempts to build a common model. When the participants mutually educate each other about the models they use, and arrive at a shared model of a system there remains less reason for conflict and dispute.

As the book goes to print, we are witnessing a burst of the housing bubble in the USA and a slide of the US economy towards recession. For a systems scientist this actually may be a positive trend. The economy is well overdue to slow down, giving people pause to reconsider some of our priorities. However, instead, another stimulus package is going to be passed by the US government, simply putting more money in the hands of people to ensure that they spend more to fuel further growth. The system is further forced into overdrive towards a collapse. Instead of investing in education, in retraining, in research, in the future, again we are choosing to invest in consumption, for the present. If we could only share our models and reach a common understanding…

Index

Printed and bound by CPI Group (UK) Ltd, Croydon, CR0 4YY

03/10/2024

01040312-0008